Plants of Central Asia

Volume 4

Plants of Central Asia

Plant collections from China and Mongolia

(*Editor-in-Chief*: V. I. Grubov)

Volume 4

Gramineae
(*Grasses*)

N.N. Tzvelev

CRC Press
Taylor & Francis Group
Boca Raton London New York

CRC Press is an imprint of the
Taylor & Francis Group, an **informa** business

A SCIENCE PUBLISHERS BOOK

ACADEMIA SCIENTIARUM URSS
INSTITUTUM BOTANICUM nomine V.L. KOMAROVII
PLANTAE ASIAE CENTRALIS
(secus materies Instituti botanici nomine V.L. Komarovii)
Fasciculus 4
GRAMINEAE
N.N. Tzvelev confecit

First published 2001 by Science Publishers Inc.

Published 2019 by CRC Press
Taylor & Francis Group
6000 Broken Sound Parkway NW, Suite 300
Boca Raton, FL 33487-2742

ISBN 13: 978-0-367-44713-7 (pbk)
ISBN 13: 978-1-57808-115-8 (hbk)
ISBN 13: 978-1-57808-062-5 (Set)

Visit the Taylor & Francis Web site at
http://www.taylorandfrancis.com

and the CRC Press Web site at
http://www.crcpress.com

Library of Congress Cataloging-in-Publication Data
Rasteniia TSentral'noi Azii. English
  Plants of Central Asia: plant collections from China
  and Mongolia
  /[editor-in-chief. V.I. Grubov].
    p. m.
  Research based on the collections of the V.L.
  Komarov Botanical Institute.
  Includes bibliographical references.
  Contents: V.4. Gramineae
  ISBN 1-57808-115-7 (v.4)
  1. Botany-Asia, Central. I. Grubov, V.I. II.
Botanicheskii institut im. V.L. Komarova. IV. Title.
QK374, R23613 2000
581.958-dc21
                                     99-36729
                                        CIP

Translation of: Rasteniya Central'nov Asii, vol. 4, 1968;
                    Nauka Publishers, Leningrad.
                    (updated for the English edition in 2000)

ANNOTATION

This is the fourth volume of the illustrated list of Central Asian plants (within the People's Republics of China and Mongolia) published by the Botanical Institute of the Academy of Sciences of the USSR based on the Central Asian collections of leading Russian travellers and naturalists (N.M. Przewalsky, G.N. Potanin and others) as well as of Soviet expeditions, and preserved in the Herbarium of the Institute.

This volume describes the family Gramineae—one of the largest families in Central Asian flora, whose members play a leading role in the formation of its plant cover and represent the characteristic species of vast arid, hill and desert steppes. In the Central Asian countries, they enjoy extremely great economic importance (pasture and fodder plants).

Editorial Board

I.T. Vassilczenko, V.I. Grubov (Editor-in-Chief), I.A. Linczevsky and S.Yu. Lipschitz

Volume Editor
V.I. Grubov

CONTENTS

INTRODUCTION

This volume, fourth in the series 'Plants of Central Asia', is wholly devoted to the family Gramineae, which represents one of the largest and economically most important families of flowering plants. A study of the vast herbarial collection preserved in the herbaria of the V.L. Komarov Botanical Institute, Academy of Sciences, USSR, and the literature available showed that Central Asia outside the USSR, covered in this publication, is represented by 74 genera and 354 species of this family, being clearly second only to the family Compositae in this respect. This number would greatly swell were the genera and species of grasses from the part of the USSR which, according to V.I. Grubov, falls in the Central Asian subregion of the Mediterranean (see map 1, vol. 1, p. 12*) included. The part of Kazakhstan falling in this subregion alone would raise the above numbers by 95 species and 11 genera (*Imperata* Cyr., *Eriochloa* H.B.K., *Leersia* Sw., *Apera* Adans., *Arundo* L., *Molinia* Schrank, *Sclerochloa* Beauv., *Scolochloa* Link, *Pholiurus* Trin., *Henrardia* C.E. Hubb. and *Aegilops* L.).

Apart from the species indicated above, this publication covers 18 more species of grasses, without serial numbers, half of which are only cultivated plants and the rest found in the border regions of Northern Mongolia. It is interesting that the inclusion in this publication of all species of grasses from Northern Mongolian, in spite of the predominantly Siberian character of its flora, would raise the above numbers by only two genera (*Arctagrostis* Griseb. and *Scolochloa* Link) and 13 species.

Regarding the distribution of genera of Central Asia, it may be said that six genera comprise over 20 species each (*Stipa* L. 34, *Poa* L. 33, *Elymus* L. 27, *Calamagrostis* Adans. and *Agropyron* Gaertn. 22 each and *Festuca* L. 21). These together constitute almost half of the Central Asian grass flora while 34 genera contain only one species each.

The endemism of the grass flora of the Central Asian subregion of the Mediterranean covered in this publication can be recognised as very high and undoubtedly suggests its prolonged and essentially autochthonous development. As justifiably pointed out by V.I. Grubov (vol. 1, pp. 15 to 23), this is particularly true of the Mongolian province of this subregion,

---

*Page numbers those of the English edition published by Science Publishers, Inc., Enfield, NH, 1999—General Editor.

while in the other provinces distinguished by V.I. Grubov, the migrations of species which occurred frequently from the adjoining territories had a far greater impact. It must further be pointed out that narrow endemic species of progressive character (excluding species of hybrid origin) are almost altogether absent in the family Gramineae as also in the family Cyperaceae, unlike in several other groups of angiosperms. This, in our view, only suggests some antiquity of the entire family and enhances the specific importance of the endemism of Gramineae.

Generic endemism is exhibited by four monotypic genera—*Psammochloa* Hitchc., *Timouria* Roshev., *Pappagrostis* Roshev. and *Sinochasea* Keng—which are highly specialised and at the same time very old members of the Central Asian flora. Genus *Psammochloa* of the tribe Stipeae Dum. is particularly archaic. The only species of this genus—*P. villosa* (Trin.) Bor—is a characteristic plant of drifting Mongolian and partly Qaidam sand (see map 2). This species exhibits several primitive morphological features, taking it closer to some genera of tribe Arundineae Dum. which, on the whole, is more primitive than tribe Stipeae and probably even ancestral. The stray species belonging to the more advanced (from the evolutionary point of view) genera *Timouria* and *Pappagrostis* possess remarkable similarity in general plant appearance and have nearly similar distribution ranges (see maps 2 and 6) but are not closely related; the former is closer to genus *Achnatherum* Beauv. of tribe Stipeae while the latter belongs to tribe Aveneae Dum. and is considerably more isolated. The development of these genera evidently proceeded convergently over a prolonged period and in the same territory. At some point of time, their contiguous distribution range was separated by Beishan desert upland and Alashan desert into three segments: Eastern and partly even Central Tien Shan with Kashgar hilly regions adjoining it, northern foothills of Nanshan and the Alashan hills. The fourth genus, *Sinochasea*, closer to *Pappagrostis*, is also very isolated but less interesting since it originates from adjoining Qinghai (a transitory province with respect to flora) which is abundant in East Asian species. The probability of occurrence in even much less studied adjoining hilly regions of China is quite high.

Among the Central Asian endemic species of other genera, which number about 50, over half are narrowly endemic. Most of those latter species have been described from floristically extremely rich Qinghai and Weitzan regions and partly also from Southern Tibet, however, and their narrow endemism perhaps merely reflects that the adjoining hilly regions of China have not been well studied. As for the rest of the few narrowly endemic species, most appear to be of hybrid origin but their origin is far from clear [for example, *Calamagrostis alexeenkoana* Litv., *Puccinellia roborovskyi* Tzvel., *Agropyron kokonoricum* (Keng) Tzvel. and others].

Furthermore, they hold little interest from the phytogeographical point of view.

Most of the more widely distributed endemic species are relatively poorly isolated while the vicarious species with close genetic affinity emerge beyond the Central Asian boundaries. Thus, the endemic species *Aeluropus micrantherus* Tzvel. distributed in the southern part of the Mongolian subprovince, Kashgar and Altay section of Junggar is replaced in the east by the very close species *A. sinensis* (Debeaux) Tzvel. and in the west by *A. intermedius* Regel, which is very widely distributed in Kazakhstan. Among the isolated endemic species of Central Asia, *Ptilagrostis pelliotii* (Danguy) Grub. is most interesting; its extensive range (map 3) does not run beyond its boundaries outside the USSR. This species is not only confined to the older forms of topography such as low hills and hummocks of Mongolia, Junggar and Northern Kashgar, but is also more primitive compared to all the other, far more mesophyllous species of the genus which form two groups of vicarious, genetically very close species. Three species of one of these groups—*P. mongholica* (Trin.) Griseb., *P. dichotoma* Keng and *P. tibetica* (Mez) Tzvel.—represent high-altitude or barren peak species. Their distribution ranges surround that of *P. pelliotii* from almost all sides and run far beyond the boundaries of Central Asia (map 3). The lone American species of the genus, *P. porteri* (Rydb.) Tzvel. (Nor. Cordillera), also has its origin here. It is almost indistinguishable from *P. mongholica* and migrated to America undoubtedly from Asia through the Bering Strait. Another group is made up of the far more advanced *P. concinna* (Hook. f.) Roshev. and *P. junatovii* Grub. The former species is distributed in the high mountains of Tibet, Pamir and Tien Shan and the latter on the barren peaks of Altay, Sayan and Hangay. It is highly probable that genus *Ptilagrostis* as a whole is of Central Asian origin.

Apart from the species that are endemic in the Central Asian subregion of the Mediterranean, there are still a large number of species (about 60) highly characteristic of this subregion and of Central Asian origin but more or less extending beyond its boundaries into China, Southern Siberia (sometimes including central Yakutia), middle Asia or the Himalayas. They contain quite a few morphologically distinct desert, desert-steppe and solonchak species, e.g., *Achnatherum splendens* (Trin.) Nevski, *A. inebrians* (Hance) Keng, *Stipa breviflora* Griseb., *Calamagrostis salina* Tzvel., *Enneapogon borealis* (Griseb.) Honda, *Cleistogenes squarrosa* (Trin.) Keng, *C. songorica* (Roshev.) Ohwi, *Poa tibetica* Munro ex Stapf, group of *P. ochotensis* Trin. s.l., *Puccinellia tenuiflora* (Griseb.) Scribn. et Merr. and *Leymus secalinus* (Georgi) Tzvel. Many of them (especially *Achnatherum splendens* and *Leymus secalinus*) occur equally abundantly on the Mongolian plains as well as in the high mountains of Pamir and Tibet (see map 10).

Species of the genus *Cleistogenes* Keng and *Enneapogon borealis* (Griseb.) Honda possess the remarkable biological characteristic of forming additional cleistogamic spikelets concealed within the sheath of cauline leaves (*Cleistogenes*) or within the highly contracted (reniform) shoots in the mat base (*Enneapogon borealis*). This guarantees possible propagation even after intense grazing and under very severe climatic conditions. Although the genus *Cleistogenes* is represented by relatively primitive species in the Mediterranean [*C. Serotina* (L.) Keng] and Japan [*C. hackelii* (Honda) Honda], it is highly probable that it is of Central Asian origin. This assumption is confirmed by the presence of the very close but more primitive genus *Orinus* Hitchc. in Central Asia as well as the existence there of *Cleistogenes songorica* (Roshev.) Ohwi, which is almost endemic in Central Asia; the latter is evidently the most primitive in the genus and has been separated by us into a special section *Pseudorinus* Tzvel., transitory to the genus *Orinus*. It is interesting that *C. songorica*, like *Ptilagrostis pelliotii* mentioned above, is confined to the much older forms of topography (hummocks) and its range largely coincides with that of the latter species (see map 7). Unlike this species, another very widely distributed species in Central Asia, *Cleistogenes Squarrosa* (Trin.) Keng (see map 8), is the most specialised in the genus with undoubted remarkable antiquity.

It is possible that the wholly isolated and fairly primitive group of species of genus *Stipa* L., the so-called "chee grasses" (*S. glareosa* P. Smirn., *S. caucasica* Schmalh., *S. gobica* Roshev., *S. tianschanica* Roshev., *S. klemenzii* Roshev. and also the more isolated species *S. orientalis* Trin.), are also of Central Asian origin. Of them, species *S. glareosa*, which is almost endemic in Central Asia, is particularly widely distributed (see map 4) and only one species, *S. caucasica*, extends far beyond its boundaries (into Middle Asia and the Near East). Three other species of chee grasses which are genetically close are vicarious: the more mesophyllous of them, *S. klemenzii*, is distributed-Transbaikal and Mongolia (in the south up to Gobi Altay); the second species, *S. gobica*, occurs in Mongolia including Khesi but also enters the eastern part of Junggar; the third species, *S. tianschanica*, endemic in Central Asia, is distributed in the hills of Tien Shan and Kashgar but also enters Pamir and Qinghai (Altyntag mountain range).

It is interesting that many steppe and desert-steppe species of grasses found in Central Asia have extremely close vicarious species in the steppe regions of North America, thus confirming the existence of fairly intimate floristic relations between the steppes of Asia and America, a view repeatedly expressed in the literature (especially by B.A. Yurtsev)[1].

---

[1]B.A. Yurtsev (1962). Floristic relations between the steppes of Siberia and prairies of North America. Bot. Zhurn., 47 (3): 317-336.

Examples are the species *Helictotrichon schellianum* (Hack.) Kitag., *Enneapogon borealis* (Griseb.) Honda and *Agropyron aegilopoides* Drob., replaced respectively in North America by the nearly indistinguishable species *Helictotrichon hookeri* (Scribn.) Henr., *Enneapogon desvauxii* Beauv. And *Agropyron spicatum* (Pursh) Scribn. et J.G. Smith.

There are a few Arctic and Arctic-Alpine species of grasses in Central Asia: *Phleum alpinum* L., *Alopecurus alpinus* Smith, *Trisetum spicatum* (L.) Richt., *Poa alpina* L., *Festuca brachyphylla* Schult. et Schult. f. and *Elymus kronokense* (Kom.) Tzvel., they mainly enter the high mountains of Mongolian Altay and Tien Shan. Only *Trisetum spicatum* and *Festuca brachyphylla* reach up to Pamir and Tibet in the south.

The widely distributed boreal species of grasses, which number about 45 in Central Asia, include forest (*Milium effusum* L., *Melica nutans* L., *Poa nemoralis* L. and others) as well as meadow (*Agrostis gigantea* Roth, *Poa pratensis* L., *Festuca rubra* L. and others) and water-marsh species (*Phragmites* Adans., *Catabrosa* Beauv. and *Glyceria* R. Br.). They enter predominantly the northern and western hilly regions of Central Asia and partly also the steppe plains of north-eastern Mongolian province (in particular Siberian taiga species such as *Agrostis clavata* Trin., *Poa sibirica* Roshev and *Elymus sibiricus* L.).

The fairly abundant steppe and desert-steppe species in Central Asian flora can be differentiated as Eurasian (about 20) and north Asian (about 15). The north Asian species, for example *Stipa baicalensis* Roshev., *Festuca dahurica* (St.-Yves) Krecz. et Bobr., *Agropyron cristatum* (L.) Beauv. and others, enter predominantly the north-eastern steppe regions of the Mongolian province and often represent vicarious ecogeographical races relative to the species of Eurasian steppes. The latter, on the contrary, are widely represented in Junggar-Turan province of the Central Asian subregion of the Mediterranean and their intrusion into non-Soviet Central Asia is usually restricted to Junggar, as shown in particular by E.M. Lavrenko and N.I. Nikol'skaya (Bot. Zhurn., 1965, 50 (10): 1419-1429) on the example of some species of *Stipa* L. The desert-steppe ephemeroids *Poa bulbosa* L. and *Catabrosella humilis* (M.B.) Tzvel. also behave in the same manner in non-Soviet Central Asia. Their extensive range here includes only the Chinese Junggar. However, such extensively distributed species as *Koeleria cristata* (L.) Pers., *Poa angustifolia* L. and *Festuca valesiaca* Gaud. are found in the Mongolian province too.

The well-known paucity of ephemerals in the Central Asian flora is very strikingly manifest among grasses. Only a few species of genera *Bromus* (L.), *Eremopyrum* (Griseb.) Jaub. et Spach, *Anisantha tectorum* (L.) Nevski and *Eremopoa songorica* (Schrenk) Roshev. are typical palaeo-Mediterranean ephemerals in the non-Soviet part of Central Asia and their distribution here is usually restricted to Chinese Junggar. If annuals such as *Aristida heymannii* Regel (tribe Aristideae C.E. Hubb.), species of genera *Crypsis* Ait. (tribe Sporoboleae Stapf) and *Eragrostis* Wolf (tribe Eragrosteae Stapf), *Chloris virgata* Sw. (tribe Chlorideae Agardh), *Enneapogon borealis* (Griseb.) Honda (tribe Pappophoreae Kunth), *Schismus arabicus* Nees (tribe Danthonieae Nevski) belonging to tropical and sub-tropical tribes can be acknowledged as "ephemerals", they are ephemerals of an altogether special type whose origin is associated not with the palaeo-Mediterranean, but with the desert and steppe regions of the Southern Hemisphere. Unlike the palaeo-Mediterranean ephemerals, they represent not winter but spring plants and, in a favourable ecological environment persist deep into autumn.

Among the rest of the species of grasses, a good proportion consists of hill species, barely entering in general the non-Soviet Central Asia from the adjoining regions of Middle Asia, East and South-East Asia or Indo-Malay region. Apart from them, there are also a few weeds of northern (for example, *Avena fatua* L.) as well as southern [for example, *Cynodon dactylon* (L.) Rich.] origin.

Insofar as the role of grasses in the vegetation cover of Central Asia is concerned, it is indeed immense, this family coming second only to Chenopodiaceae and Compositae. While the members of Chenopodiaceae and wormwoods (*Artemisia* L.) are usually assigned the predominant role in desert groups of different types, grasses represent the most common characteristic species in steppe and desert-steppe groups which are also extensively distributed in Central Asia. Here particularly distinguished species of *Stipa* L. are found, representing in particular the characteristic species of chee grass desert steppes (*S. glareosa* P. Smirn., *S. gobica* Roshev. and others), grasses of arid and desert steppes (*S. capillata* L., *S. sareptana* Beck., *S. krylovii* Roshev. and others), false chee grass of high-altitude steppes [*S. purpurea* Griseb., *S. subsessiliflora* (Rupr.) Roshev. and others], as well as narrow-leaved species of *Festuca* L. (in particular, *F. valesiaca* Gaud., characteristic of sheep's fescue steppes), species of *Cleistogenes* Keng (characteristic of desert steppes) and species of *Koeleria* Pers. Special groups such as chee grass thickets, not infrequently covering large expanses, form the major chee grass *Achnatherum splendens* (Trin.) Nevski, which is widely distributed in Central Asia.

Moreover, the characteristic plants of Central Asian sand deserts are *Psammochloa villosa* (Trin.) Bor, *Aristida pennata* Trin. and *A. grandiglumis* Roshev., and some species of the genus *Agropyron* Gaertn.; of these, the two genetically close species of spider grass inhabiting sand dunes—*Aristida pennata* and *A. grandiglumis*—represent vicarious species. The first of these is extensively distributed in Junggar-Turan province of the Central Asian subregion, being replaced by a special variant (*A. pennata* var. *minor* Litv.) in the Central Asian deserts, while the second is found in the deserts of Kashgar and partly Mongolia (Khesi).

Many species of genera *Leymus* Hochst., *Puccinellia* Parl. and *Aeluropus* Trin. as well as *Psathyrostachys junceus* (Fisch.) Nevski represent characteristic or subcharacteristic species of solonchak and saline meadows. The role of some species of grasses is also very significant in the composition of meadow, pebble, rocky and talus groups of vegetation.

New species found in Central Asia after the publication of Russian edition (1968) are incorporated in the main text and marked with the special sign (O). For such species in distribution only regions are indicated (without specimens citation) and general distribution is omitted.

The scheme of phytogeographical division of Central Asia adopted by V.I. Grubov has been maintained in this volume. The abbreviations of names of regions remain the same as in the first volume (1999, pp. 78-80) but the sequence of their enumeration has been slightly modified commencing from the second volume (1966, p. 7): regions of the [former] USSR territory falling in the Central Asian subregion of the Mediterranean have been cited first, followed by the other regions in the already established order.

Artist T.N. Shishlova prepared the drawings in the plates presented in this volume.

With utmost sincerity, the author acknowledges the very significant assistance rendered by the senior laboratory assistant O.I. Starikova in studying the material of Chinese collectors (translation of labels, identifying the geographic locations, etc.).

This study of the vast material covering family Gramineae from Central Asia was conducted under the constant guidance of the initiator and editor-in-chief of the series, V.I. Grubov, to whom the author expresses his sincere gratitude.

Note of the series editor-in-chief

In the current series the subspecies category is not adopted and is not used because of its uncertainty, which is specially stated in the introduction to the vol. 1. However, the authors of addendum, both to this volume and volume 3, did not follow this instruction. It is necessary to take this fact into account when using these volumes of the series.

V.I. Grubov

TAXONOMY

SPECIAL ABBREVIATIONS

Abbreviations of the Names of Collectors

A. Reg. — A. Regel
Bar. — V.I. Baranov
Chaff. — J. Chaffanjon
Ching — R.C. Ching
Chu — C.N. Chu
Czet. — S.S. Czetyrkin
Divn. — D.A. Divnogorskaya
Fet. — A.M. Fetisov
Glag. — S.A. Glagolev
Gr.-Grzh. — G.E. Grum-Grzhimailo
Grombch. — B.L. Grombchevski
Grub. — V.I. Grubov
Gus. — V.A. Gusev
Ik-Gal. — N.P. Ikonnikov-Galitzkij
Ivan — A.F. Ivanov
Kal. — A.V. Kalinina
Keng — Y.L. Keng
Klem. — E.N. Klements
Krasch — I.M. Krascheninnikov
Kryl. — P.N. Krylov
Kuan — K.C. Kuan
Lad. — V.F. Ladygin
Lavr. — E.M. Lavrenko
Lee — A.R. Lee (1959)
Li — S.H. Li (1951)
Lis. — V.I. Lisovsky
Litw. — D.I. Litwinow
Mois. — V.S. Moiseenko
Nov. — V.F. Novitski
Pal. — J.W. Palibin
Pavl. — N.V. Pavlov

Petr.	— M.P. Petrov
Pias.	— P.Ya. Piassezki
Pob.	— E. Pobedimova
Pop.	— M. Popov
Pot.	— G.N. Potanin
Przew.	— N.M. Przewalsky
Rhins	— J.L. Dutreuil de Rhins
Rob.	— V.I. Roborowsky
Sap.	— V.V. Sapozhnikov
Schisch.	— B.K. Schischkin
Serp.	— V.M. Serpukhov
Shukh.	— V.N. Shukhardin
Shum.	— E.M. Shumakov
Sold.	— V.V. Soldatov
Tug.	— A.Ya. Tugarinov
Wang.	— K.C. Wang
Yun.	— A.A. Yunatov
Zab.	— D.K. Zabolotnyi
Zam.	— B.M. Zamatkinov

Family 19. GRAMINEAE Juss.

1. Spikelets unisexual: male spikelets forming terminal panicle and female axillary spadix. Cultivated annuals .... **Zea** L.
+ Spikelets usually bisexual, less often unisexual; in latter event, female spikelets not forming spadix .... 2.
2. General inflorescence consists of several spikes disposed palmately or subpalmately; its rachis, if present, 1.5 times shorter than spikes .... 3.
+ Branches of general inflorescence not disposed palmately; rachis of inflorescence usually longer than lateral branches, less often, almost as long .... 7.
3. Leaf blades broadly lanceolate or narrowly ovate; annual plant with prostrate surface shoots rooted at nodes .... 5. **Arthraxon Beauv.**
+ Leaf blades linear .... 4.
4. Spikelets prominently awned .... 5.
+ Spikelets not awned .... 6.
5. Spikelets paired: one sessile, the other stalked. Perennials .... 6. **Botriochloa** Kuntze.
+ Spikelets subsessile, occurring singly in two rows. Annuals .... 40. **Chloris** Sw.

6.   Spikelets in pairs and threes: one sessile, others stalked. Annuals without creeping subsurface shoots .... 12. **Digitaria** Heist. ex Fabr.

+   Spikelets subsessile, occurring singly in two rows. Perennials with creeping subsurface shoots .... 39. **Cynodon** Rich.

7(2)   Highly shortened branches of spicate panicles, besides spikelets, also bearing bristles usually longer than spikelets or less often almost as long; bristles coarse due to presence of spinules .... 8.

+   Branches of general inflorescenes without coarse bristles .... 9.

8.   Annuals. Fruiting spikelets shedding without surrounding bristles .... 13. **Setaria** Beauv.

+   Perennials. Fruiting spikelets shedding together with involucre comprising surrounding bristles .... 14. **Pennisetum** Rich.

9.   Upper glumes as long as spikelets with 5 highly prominent ribs bearing large hooked spines and much smaller spinules; general inflorescence spicate with spikelets in groups of two (three) on up to 2.5 mm long stalks. .... 8. **Tragus** Hall.

+   Glumes without large hooked spines .... 10.

10.   Branches of general inflorescence in axils of terminal leaves mostly with reduced blades .... 11.

+   Branches of general inflorescence not axillary (very rarely, besides terminal panicles, axillary branches also with cleistogamous spikelets) .... 12.

11.   Spikelets few on branches, in pairs: one sessile, the other stalked .... 7. **Cymbopogon** Spreng.

+   Spikelets many on branches, sessile and in four rows on highly thickened rachis of branches .... 4. **Hemarthria** R. Br.

12.   General inflorescence—spike: spikelets in fairly regular longitudinal rows on undeveloped rachis, or sessile, or on very short (up to 1.5 mm long) thick stalks .... 13.

+   General inflorescence paniculate or racemose, usually very dense, spicate or capitate, but then at least some of its branches with 2 or more spikelets .... 24.

13.   All spikelets, except terminal, with only one lanceolate glume .... 66. Lolium L.

+   All spikelets with 2 glumes; glumes sometimes subulate and secund .... 14.

14.   Ovary glabrous with stigmas on long styles; spikes 5–15 cm long, very narrow. Perennial, densely caespitose plants with up to 1.5 mm broad leaf blades .... 41. **Tripogon** Roem. et Schult.

+   Ovary short but profusely pilose at top with subsessile stigmas . ... 15.

15.   Annuals.. 16.

+ Perennials ....19.

16. Spikelets in groups of three each on rachis but lateral spikelets in each group may be underdeveloped ....71. **Hordeum** L. (sect. Hordeum).

+ Spikelets singly on rachis ....17.

17. Spikes (including very short awns) up to 4 (4.5) cm long, with two rows of pectinately arranged spikelets. Plants up to 30 (40) cm tall .... 68. **Eremopyrum** (Ledeb.) Jaub. et Spach.

+ Spikes (ignoring awns which are usually very long) more than 5 cm long; spikelets not pectinately arranged. Plants usually more than 40 cm tall ....18.

18. Glumes linear-subulate, without lateral teeth .... 70. **Secale** L.

+ Glumes oblong or elliptical, truncated obliquely at top, with 1 or 2 lateral teeth .... 69. **Triticum** L.

19(15). Rachis near fruits disintegrating at joints even in dry state; spikelets in groups of 3 each; 1–2(3)–flowered ....20.

+ Rachis without joints, not disintegrating into segments; spikelets single or in groups of 2 to 6; (2) 3–7 (12)–flowered ....21.

20. All spikelets sessile, uniformly developed; 1–2 (3)–flowered .... 72. **Psathyrostachys** Nevski.

+ Lateral spikelets in groups of 3 on short (0.8–1.5 mm long) stalks, usually underdeveloped or with single male floret; central ones sessile, with single bisexual floret .... 71. **Hordeum** L. (sect. Stenostachys Nevski).

21. Glumes linear- or lanceolate-subulate, without distinct ribs; spikelets usually in groups of 2–6 each, less often single. Plants usually with creeping subsurface shoots, less often without them. ....73. **Leymus** Hochst.

+ Glumes lanceolate or elliptical, with 3–7 distinct ribs (only lower glume sometimes with single rib) ....22.

22. Spikes usually with less than 10 spikelets; lemma with 7 ribs, very distinct in upper part; palea 0.6–0.9 mm broad at tip; spikelets single, on very short (0.7–1.5 mm long) stalks .... 65. **Brachypodium** Beauv.

+ Spikes usually with more than 10 spikelets; lemma with 5 ribs, distinct only in upper one-fourth part; palea 0.2–0.4 mm broad at tip ....23.

23. Transverse depressions present between minute stalk of spikelet and base of glumes; one rib of glume better developed than rest, forming poorly or well-developed keel, usually with spinules; callus of lemma rounded; spikelets single. Plants with or without creeping subsurface shoots .... 67. **Agropyron** Gaertn.

+   Transverse depressions absent between very short but well-developed stalk of spikelet and base of glumes; glumes without keel and usually with scabrous ribs; callus of lemma obtusely conical; spikelets single or in groups of 2–4 each. Plants without creeping subsurface shoots .... 74. **Elymus** L.

24(12).   Branches of fairly dense panicles, at least in upper part, disintegrating into segments around fruits even in dry state; spikelets 2–3 each on branches; one spikelet sessile or subsessile but rest stalked and enveloped in long (more than half length of spikelet) hairs, diverging from stalks of spikelets and lower part of glumes. Tall perennials .... 25.

+   Branches of general inflorscence not disintegrating into segments; spikelets not enveloped in long hairs .... 27.

25.   Spikelets without awns, enveloped in hairs 2–5 times longer than spikelets .... 1. **Saccharum** L.

+   Spikelets awned, enveloped in hairs not longer, or not more than 1.5 times longer than spikelet .... 26.

26.   Hairs longer than spikelets. Leaf blades linear, usually 3–8 mm broad .... 2. **Erianthus** Michx.

+   Hairs shorter than spikelets. Leaf blades lanceolate-linear, usually 10–15 mm broad .... 3. **Spodiopogon** Trin.

27.   Glumes greatly swollen on back; spikelets subsessile, in 2 adjacent rows on one side of spicate branches, forming secund general inflorscence .... 42. **Beckmannia** Host.

+   Glumes not greatly swollen on back; if slightly so, general inflorescence not secund .... 28.

28.   Spikelets with single normally developed flower; rachilla above base of this solitary flower either altogether undeveloped or extends in form of glabrous or pilose shaft without lemma and palea at tip .... 29.

+   Spikelets usually with 2 or more florets, less often with single normally developed floret but then rachis not only extends above base of this floret in form of a shaft, but carries at tip 1 or more lemma and palea of reduced upper florets of spikelet .... 52.

29.   Florets with 6 stamens; 2 sheath-like outgrowths near base of spikelets represent rudiments of glumes. Cultivated annuals ....
.... **Oryza** L.

+   Florets with 2 or 3 stamens .... 30.

30.   Lemma at tip with 3 subequal awns diverging from single points
.... 18. **Aristida** L.

+   Lemma not awned or with single awn, less often with 3, of which lateral ones few times shorter than central and not diverging from single point .... 31.

31. Leaf without ligules: in their place, not even a transverse row of hairs. Annuals. .... 11. **Echinochloa** Beauv.

+. Ligules invariably present in form of hairy transverse shaft or very narrow membranous margin .... 32.

32. Spikelets in pairs on branches of dense or largely diffuse panicles: one sessile with bisexual floret, the other on short stalk, underdeveloped or with male floret. Tall, cultivated annuals .... **Sorghum** Moench.

+ All spikelets of inflorescence with single bisexual floret and not arranged in pairs .... 33.

33. Besides 2 glumes and lemma and palea, additional pair of scales present above glumes .... 34.

+ Only 2 glumes and (less often, 1) lemma and palea on spikelet .. .. 35.

34. Plants 40–150 cm tall, with creeping subsurface shoots. Panicles very dense with scabrous branches .... 15. **Typhoides** Moench.

+ Plants 15–40 cm tall, without creeping subsurface shoots. Panicles spicate, with glabrous or subglabrous branches .... 16. **Anthoxanthum** L.

35. Panicles very dense, capitate or spicate, with regular cylindrical or ovate form; spikelets 2–6 mm long, strongly laterally flattened; callus of lemma glabrous. .... 36.

+ General inflorescence different in shape; very rarely with regular shortly cylindrical form (some species of *Calamagrostis* Adans.) but then callus of lemma with tufts of long hairs; spikelets not flattened or only very slightly laterally flattened .... 38.

36. Panicles capitate or ovate, with sheaths of upper cauline leaves at base. Ligules up to 0.2 mm long, transforming into row of hairs almost from base .... 26. **Crypsis** Alt.

+ Panicles cylindrical or ovate, emerging from sheath of upper cauline leaf. Ligules more than 1 mm long, membranous, without hairs along margin .... 37.

37. Lemma not awned; palea present .... 27. **Phleum** L.

+ Lemma usually with developed (often not emerging from spikelet) awn; palea absent .... 28. **Alopecurus** L.

38. Lemma glabrous but often with generally pilose callus .... 39.

+ Lemma usually pilose throughout entire surface or only in lower part (ignoring callus) .... 43.

39. Lemma not awned, coriaceous, lustrous and largely different in consistency from that of glumes .... 40.

+ Lemma awned or not, differing little in consistency from glumes .... 41.

40. Perennials. Sheath and leaf blade glabrous; ligule 2.5–8 mm long, glabrous .... 25. **Milium** L.

+ Annuals. Sheath and often leaf blade pilose; ligule up to 1 mm long, pilose .... 10. **Panicum** L.

41. Lemma awned; callus with tufts of hairs longer than lemmas or not less than 1/5 their length .... 32. **Calamagrostis** Adans.

+ Lemma awned or not, with or without hairs on callus; when present, hairs more than five times shorter than lemma .... **42.**

42. Palea as long as lemma or slightly shorter; stalks of spikelets jointed at base or in central part .... 29. **Polypogon** Desf.

+ Palea 1.5 times or more shorter than lemma, sometimes absent; stalks of spikelets without joints .... 30. **Agrostis** L.

43(38). Palea with 5–7 ribs; lemma 9–13 mm long, not awned or with short, early shedding awn. Fairly large plant on drifting or weakly fixed sand .... 19. **Psammochloa** Hitch.

+ Palea with 2 ribs ....**44.**

44. Lemma not awned. High-altitude plants with creeping subsurface shoots ....
.... 57. **Paracolpodium** (Tzvel.) Tzvel.

+ Lemma awned, sometimes readily shedding. Plants without creeping subsurface shoots ....**45.**

45. Awn diverging from back of lemma below centre; spikelets 3–4 mm long .... 31. **Pentatherum** Nabel.

+ Awn diverging from tip of lemma ....**46.**

46. Awn persistent, diverging between two lobes equal to 1/4 or 1/3 entire length of lemma ....**47**.

+ Awn shedding or persistent, diverging directly from tip of lemma, less often between small (1/6 or more shorter than lemma) teeth **48.**

47. Stigmas 3; hairs in upper part of lemma only slightly longer than in lower part .... 34. **Sinochasea** Keng.

+ Stigmas 2; hairs in upper part of lemma considerably longer, forming transverse crown of hairs at base of lobes of lemma ....
.... 33. **Pappagrostis** Roshev.

48. Callus of lemma more than 0.7 mm long, cylindrical at base, later incised obliquely, long and acuminate; lemmas usually envelop caryopsis, completely overlapping one margin by the other; awn scabrous or pilose .... 23. **Stipa** L.

+ Callus of lemma up to 0.6 mm long, conical from base, obtuse or short and acuminate; lemmas not totally enveloping caryopsis and their margins not overlapping ....**49.**

49. Lemma slightly coriaceous, coriaceous in fruit, dark brown, lustrous, with very short (up to 0.3 mm long) obtuse callus,

somewhat isolated from glumes; awn short, scabrous, readily shedding .... 24. **Piptatherum** Beauv.

+ Lemma coriaceous-membranous, not hardening in fruit, light brown, not lustrous, with acute or rather pointed callus, well isolated from glumes .... 50.

50. Awn pilose throughout length, not shedding.................................
.... 22. **Ptilagrostis** Griseb.

+ Awn scabrous .... 51.

51. Glumes with 3 very prominent ribs, keeled; panicles spicate, 2.5–4.5 cm long; awn short, erect, readily shedding ....
.... 20. **Timouria** Roshev.

+ Glumes with 1–5 faint ribs, without or with faint keel; panicles usually diffuse or dense, but then very long; awn persistent or readily shedding, often genuflexed ....
.... 21. **Achnatherum** Beauv.

52(28). Lemma with 9 awns, 1.5–2 times longer than lemma; panicles very dense, spicate .... 43. **Enneapogon** Desv. ex Beauv.

+ Lemma not awned or with solitary awn .... 53.

53. Lemma with highly elongated slender callus covered with very long (as long or longer than lemma) hairs. Stems 0.5–3 m high ..
.... 44. **Phragmites** Adans.

+ Callus of lemma very short, glabrous or covered with very short (less than half length of lemma) hairs .... 54.

54. Spikelets with 2–3 florets, of which 1 or 2 lower florets staminate and upper bisexual; lemmas of staminate and bisexual florets differ in consistency, colour and disposition of trichome (spinules and hairs); rachis very short, indistinct. Perennials with creeping subsurface shoots .... 55.

+ All developed florets of spikelet bisexual, less often (in unisexual species of *Festuca* L.) either staminate or pistillate; lemmas of all florets of spikelet similar, differing only in size; rachis normally developed, distinctly visible .... 56.

55. Spikelets with 3 florets; creeping subsurface shoots with rapidly disintegrating membranous scale-like leaves.................................
.... 17. **Hierochloe** R. Br.

+ Spikelets with 2 florets; creeping subsurface shoots with persistent coriaceous scale-like leaves ....
.... 9. **Arundinella** Raddi.

56. Leaf ligules shorter than 1 mm, with row of hairs longer than ligule .... 57.

+ Leaf ligules longer or shorter than 1 mm but, along margin, glabrous or with very short (visible only on powerful magnification) cilia, shorter than ligule .... 61.

57. Lemma with 7–11 distinct ribs .... 61.
+ Lemma with 3–5 distinct ribs .... 58.
58. Perennials, with creeping subsurface shoots forming large beds. Panicles very dense, somewhat secund; tip of lemma entire ....
.... 51. **Aeluropus** Trin.
+ Annuals, forming small mats. Panicles small, dense, not secund; lemma bilobed at tip .... 53. **Schismus** Beauv.
59. Annuals, less often (some cultivated species) perennials. Lemma 1.3–3 mm long, not awned .... 47. **Eragrostis** Wolf.
+ Perennials. Lemma 3.2–7 mm long, awned or not .... 60.
60. Very short branches with cleistogamous spikelets in axils of middle and upper cauline leaves .... 46. **Cleistogenes** Keng.
+ Cleistogamous spikelets absent in axils .... 45. **Orinus** Hitchc.
61(56). Lemma keeled .... 62.
+ Lemma without keel .... 67.
62. Panicles very dense, secund; lemma lanceolate, often with cusp at tip. Perennials without creeping subsurface shoots ....
.... 52. **Dactylis** L.
+ Panicles lax, usually diffuse or dense (up to spicate), but then not secund .... 63.
63. Lemma with awn (sometimes very short!), diverging slightly below bicuspid tip .... **36. Trisetum** Pers.
+ Tip of lemma without awn .... 64.
64. Branches of very dense, often spicate panicles profusely covered with very short hairs. Perennial, without creeping subsurface shoots. .... 48. **Koeleria** Pers.
+ Branches of dense or generally diffuse panicles glabrous, smooth, or scabrous due to spinules .... 65.
65. Lemma with 3–5 relatively faint ribs; palea glabrous or pilose along keel but invariably with spinules, quite often visible only on powerful magnification; branches of panicles scabrous or smooth .... 54. **Poa** L.
+ Lemma with three highly prominent ribs; palea glabrous or pilose along keel but invariably without spinules; branches of panicles smooth .... 66.
66. Plants forming small but dense mats with vegetative shoots bulbous, thickened at base. Lemma pilose along ribs in lower part .... 56. **Catabrosella** (Tzvel.) Tzvel.
+ Plants with creeping subsurface or underwater shoots. Lemma glabrous .... 49. **Catabrosa** Beauv.
67(61). Annuals .... 68.
+ Perennials .... 71.

68. Spikelets 3–6 mm long. Lemma 1.8–3.5 mm long, without awn ... . 55.**Eremopoa** Roshev.

\+ Spikelets more than 12 mm long. Lemma more than 6 mm long, usually awned, less often not .... 69.

69. Sheath laciniate almost up to base. Awn, if present, diverging from back of lemma .... 38. **Avena L.**

\+ Sheath closed, or almost so throughout its length. Awn, if present, diverging from tip (or slightly below it) of lemma .... 70.

70. Lower glume with 1, upper with 3 ribs; spikelets broadening towards top .... 63. **Anisantha** C. Koch.

\+ Lower glume with 3–5, upper with 5–9 ribs; spikelets narrowing towards top .... 64. **Bromus L.**

71(67). Lemma with awn diverging from back but quite often close to tip .... 72.

\+ Lemma not awned or with awn diverging from tip .... 73.

72. Spikelets 2.3–6 mm long; lemma 1.8–5 mm long; ovary glabrous .... 35. **Deschampsia** Beauv.

\+ Spikelets 7–20 mm long; lemma 6.5–14 mm long; ovary puberulent at tip .... 37. **Helictotrichon** Bess.

73. Sheath of cauline leaves closed for not more than 1/4 its length from base .... 74.

\+ Sheath of cauline leaves closed almost throughout its length .... 76.

74. Lemma 11–30 mm long with broadly membranous lateral margins not involuted: ovary puberulent at tip .... 61. **Littledalea** Hemsl.

\+ Lemma up to 9.5 mm long with narrowly membranous, generally involuted lateral margins .... 75.

75. Lemma 1.5–4 mm long, obtuse or somewhat pointed at tip; ovary glabrous; caryopsis with orbicular or oval hilum .... 59. **Puccinellia** Parl.

\+ Lemma 2.3–9.5 mm long, with cusp or awn at tip, less often with only cusp; in latter case, more than 4.2 mm long; ovary glabrous or pilose at tip; caryopsis with long linear hilum .... 60. **Festuca L.**

76(73). Lemma 10–15 mm long, lanceolate, acuminate at tip, usually with cusp or awn; ovary puberulent at tip .... 62. **Zerna** Panz.

\+ Lemma 2–10 mm long, from ovate to oblong-lanceolate, usually obtuse at tip, less often rather pointed or sharp, without cusp or awn; ovary glabrous .... 77.

77. Spikelets with 1–3(4) developed florets with usually isolated ellipsoidal or pyriform appendage of lemma of underdeveloped

florets on top; stalk of spikelets usually pilose in upper part, less frequently scabrous .... 50. Melica L.

+ Spikelets with 3–10 (15) developed florets, without isolated appendage of lemma of underdeveloped florets at tip; stalks of spikelets in upper part scabrous or smooth ....
.... 58. Glyceria R. Br.

Zea L.
Sp. pl. (1753) 971.

Z. mays L. Sp. pl. (1753) 971; Forbes and Hemsley, Index Fl. Sin. 3 (1904) 346; 1318; Roshevitz in Fl. SSSR, 2 (1934) 4; Hand.-Mazz. Symb. Sin. 7, 5 (1936) 1318; Fl. Kazakhst. 1 (1956) 119; Fl. Tadzh. 1 (1957) 489; Keng, Fl. ill. sin., Gram. (1959) 851; Bor, Grasses Burma, Ceyl., Ind. and Pakist. (1960) 270.—I.c.: Fl. Kazakhst. 1, Plate 8; Fl. Tadzh. 1, Plate 59.

Described from America. Type in London (Linn.) .

Cultivated in oases.

IB. Kashgar: *West.* (near Egiz-Ir settlement, July 31; near Yangishar, Aug. 3—1913, (Knorring).

IIA. Junggar: *Dzhark.* (in Kul'dzha town, 1875—Larionov).

General distribution: widely cultivated in many countries of both the hemispheres; origin Cent. and South America.

1. Saccharum L.
Sp. pl. (1753) 54.

1. S. spontaneum L. Mant. pl. alt. (1771) 183; Forbes and Hemsley, Index Fl. Sin. 3 (1904) 349; Roshevitz in Fl. SSSR, 2 (1934) 9; Hand.-Mazz. Symb. Sin. 7, 5 (1936) 1307; Fl. Kirgiz. 2 (1950) 19; Fl. Kazakhst. 1 (1950) 120; Fl. Tadzh. 1 (1957) 480; Keng, Fl. ill. sin., Gram. (1959) 762; Bor, Grasses Burma, Ceyl., Ind. and Pakist. (1960) 214.—Ic.: Fl. SSSR, 2, Plate 1, fig. 3.

Described from India. Type in London (Linn.).

Along banks of irrigation ditches, in various types of plantations and in tugais; up to lower mountain belt.

IB. Kashgar: *South.* (25 km east of Guma town, Sept. 21, 1959—Petr.).

General distribution: Near East, Middle Asia, China (Cent., East., South-West., South., Hainan, Taiwan), Indo-Malay., Afr. (nor.-east.).

2. Erianthus Michx.
Fl. Bor. Amer. 1 (1803) 54.

1. E. ravennae (L.) Beauv. Ess. Agrost. (1812) 14; Fl. Tadzh. 1 (1957) 481; Bor, Grasses Burma, Ceyl., Ind. and Pakist. (1960) 151.—*Andropogon*

ravennae L. Sp. pl., ed. 2, 2 (1763) 1481.—*Erianthus purpurascens* auct. Non Anderss.: Roshevitz in Fl. SSSR, 2 (1934) 9; Fl. Kirgiz. 2 (1950) 20; Fl. Kazakhst. 1 (1956) 122.—Ic.: Fl. SSSR, 2, Plate 1, fig. 4.

Described from Italy. Type in London (Linn.).

In sand and pebble beds in river valleys and along banks of water reservoirs; up to lower mountain belt.

IB. **Kashgar:** *West.* (48 km east of Yangigissar, Sept. 18, 1959—Petr.), *South.* (nor. Kunlun foothills, 1600 m, Sept. 13, 1889—Rob.).

General distribution: Balkh. region (south); Mediterr., Balk.-Asia Minor, Near East, Caucasus, Middle Asia, Himalayas, Indo-Malay. (nor.-west.).

3. Spodiopogon Trin.
Fund. Agrost. (1820) 192.

1. S. sibiricus Trin. Fund. Agrost. (1820) 192; Franch. Pl. David. (1884) 327; Forbes and Hemsley, Index Fl. Sin. 3 (1904) 359; Pavlov in Byull. Mosc. obshch. ispyt. prir. 38 (1929) 11; Roshevitz in Fl. SSSR, 2 (1934) 11; Hand.-Mazz. Symb. Sin. 7, 5 (1936); Kitag. Lin. Fl. Mansh. (1939) 94; Grubov, Consp. fl. MNR (1955) 60; Keng, Fl. ill. sin., Gram. (1959) 766.—Ic.: Fl. SSSR, 2, Plate 1, fig. 6, Keng, l.c. fig. 710.

Described from East. Siberia. Type in Leningrad.

In dry meadows, steppes and on rocky slopes; up to lower mountain belt.

IA. **Mongolia:** *Cis-Hing.* (Khalkhin-Gol somon, Dege-Gol—Nomuryg-Gol interfluve region, Aug. 27, 1963—Dashnyam); *East. Mong.* (near Khailar town, 1959—Ivan.; 49 km east of Dashibalbar somon, Zagal hills, July 29, 1963—Dashnyam).

General distribution: East. Sib. (south), Far East (south), Nor. Mong. (Hent., Mong.-Daur.), China (Dunbei, Nor., Nor.-West., Cent.), Korea, Japan.

4. Hemarthria R. Br.
Prodr. Fl. Nov. Holl. (1810) 207.

1. H. sibirica (Gand.) Ohwi in Bull. Sci. Mus. Tokyo, 18 (1947) 1.—*H. japonica* (Hack.) Roshev. in Fl. SSSR, 2 (1934) 13; Kitag. Lin. Fl. Mansh. (1939) 78.—*Rottboellia compressa* var. *japonica* Hack. Monogr. Andropog. (1889) 288; Forbes and Hemsley, Index Fl. Sin. 3 (1904) 361.—*R. sibirica* Gand. in Bull. Soc. bot. France, 66 (1920) 302.—*R. japonica* (Hack.) Honda in Bot. Mag. Tokyo, 41 (1927) 8.—*Hemarthria compressa* auct. non R. Br.: Keng, Fl. ill. sin., Gram. (1959) 800, p.p.—Ic.: Fl. SSSR, 2, Plate 1, fig. 8.

Described from Far East (around Khabarovsk town). Type in Paris. Isotype in Leningrad.

In meadows, along banks of water reservoirs, sometimes as weed around irrigation ditches and along farm boundaries.

IA. **Mongolia:** *East. Mong.* (near Khailar town, 1959—Ivan.).

General distribution: Far East (south), China (Dunbei, Nor., East.), Korea, Japan.

5. Arthraxon Beauv.

Ess. Agrost. (1812) 111.

1. **A. hispidus** (Thunb.) Makino in Bot. Mag. Tokyo, 26 (1912) 214; Kitag. Lin. Fl. Mansh. (1939) 61; Keng, Fl. ill. sin., Gram. (1959) 813.—*A. ciliaris* subsp. *langsdorffii* var. *genuinus* Hack. Monogr. Andropog. (1889) 355; Forbes and Hemsley, Index Fl. Sin. 3 (1904) 360.—*A. hispidus* var. *centrasiaticus* (Griseb.) Honda in Bot. Mag. Tokyo, 39 (1925) 278; Keng, l.c. 814.—*A. langsdorffii* (Trin.) Roshev. in Fl. SSSR, 2 (1934) 13; Fl. Kirgiz. 2 (1950) 23; Fl. Kazakhst. 1 (1956) 123; Fl. Tadzh. 1 (1957) 487.—*A. centrasiaticus* (Griseb.) Gamajun. in Opred. zlakov Kazakhst. [Key to Grasses of Kazakhstan] (1948) 10 and in Fl. Kazakhst. 1 (1956) 123. —*Phalaris hispida* Thunb. Fl. Jap. (1784) 44.—*Pleuroplitis langsdorffii* Trin. Fund. Agrost. (1820) 175.—*P. centrasiatica* Griseb. in Ledeb. Fl. Ross. 4 (1853) 477.—*Arthraxon ciliaris* auct. non Beauv.: Franch. Pl. David. 1 (1884) 328.—Ic.: Fl. SSSR, 2, Plate 1, fig. 9; Keng. l.c. fig. 765.

Described from Japan. Type in Uppsala.

In lawns and pebble beds, along banks of water reservoirs, often as weed around irrigation ditches and among field crops; up to lower mountain belt.

IA. **Mongolia:** *Ordos* (Ulan-Morin river, Aug. 11, 1884—Pot.; 75 km south of Dzhasak town, bank of Taitykhaitsy lake, Aug. 17, 1957—Petr.).

IB. **Kashgar:** *Nor.* (near Bai town, Aug. 14, 1929—Pop.; 20 km east of Tsarga village, No. 8176, Sept. 3, 1958—Lee and Chu), *East.* (south. fringe of Khami oasis, Aug. 19, 1895—Rob.).

IIA. **Junggar:** *Jung. Gobi* (south.: 3 to 4 km nor. of Staryi Kuitun settlement on Shikho-Manas road, June 30, 1957—Yun.; near Kuitun settlement, No. 379, July 6, 1957—Huang), *Balkh.-Alak.* (near Chuguchak town, Aug. 10, 1957—Huang).

General distribution: Balkh. region (south); Asia Minor, Near East, Caucasus, Middle Asia, Far East (south), China, Korea, Japan; introduced in many other countries.

Note: Almost all specimens cited above belong to the relatively more xerophilous variety *A. hispidus* var. *centrasiaticus* (Griseb.) Honda (l.c.). This variety, widely distributed in the Near East and Middle Asia, probably merits the rank of subspecies. Unlike the type distributed mainly in East. Asia (only V.I. Roborowsky's specimens from Khami oasis from among those listed belong to the type, this variety has relatively much smaller (usually 1-3 cm long) leaf blades generally covered with sparse hairs (and not glabrous on both sides).

6. Botriochloa Kuntze

Rev. Gen. 2 (1891) 762.

1. **B. ischaemum** (L.) Keng in Contribs Biol. Lab. Sci. Soc. China, Bot. 10, 2 (1936) 201; Fl. Tadzh. 1 (1957) 487; Keng, Fl. ill. sin., Gram. (1959) 825; Bor, Grasses Burma, Ceyl., Ind. and Pakist. (1960) 108.—*Andropogon ischaemum* L. Sp. pl. (1759) 1047; Henderson and Hume, Lahore to

Yarkand (1873) 342; Forbes and Hemsley, Index Fl. Sin. 3 (1904) 374; Simpson in J. Linn. Soc. London (Bot.) 41 (1913) 451; Krylov, Fl. Zap. Sib. 2(1928) 148; Pampanini, Fl. Carac. (1930) 70; Roshevitz in Fl. SSSR, 2 (1934) 14; Kitag. Lin. Fl. Mansh. (1939) 60; Fl. Kirgiz. 2 (1950) 24; Fl. Kazakhst. 1 (1956) 124.—Ic.: Fl. SSSR, 2, Plate 1, fig. 10; Keng, l.c. fig. 773.

Described from South. Europe. Type in London (Linn.).

In steppes, riverine pebble beds, on rocky and rubble slopes, rocks and talus; sometimes as weed along roadsides and around irrigation ditches; up to midmountain belt.

IB. **Kashgar:** *Nor.* (Sairek village near Aksu, July 8, 1929—Pop.; near Aksu 3 km south of Sazkhari village, No. 8956, Sept. 29, 1958—Lee and Chu); *West.* ("Yarkand Plains"—Henderson and Hume, l.c.); *South.* (nor. Kunlun foothills, June 13, 1889—Rob.); *East.* (Bagrash-Kul' lake region, Tsinshuikhe village, No. 7637, July 28; same site, Bortu village, No. 8938, Aug. 2—1958—Lee and Chu).

IIA. **Junggar:** *Alt. region* ("Sharasume, leg. Price"—Simson, l.c.); *Jung. Alt., Tien Shan, Jung. Gobi* (Manas river 2–3 km nor.-west of Manas town, July 8, 1957—Yun.; 50 km east of Fukan, Oct. 3, 1959—Petr.; on road from Shara-Sume to Karamai, No. 10801, July 30, 1959—Lee et al.).

IIIA. **Qinghai:** *Nanshan* (south. slope of Yuzhno-Tetungsk mountain range, July 28, 1908—Czet.).

General distribution: Aral.-Casp. (Mangyshlak), Balkh. region, Jung.-Tarb., Nor. and Cent. Tien Shan; Europe (south), Mediterr., Balk.-Asia Minor, Near East, Caucasus, Middle Asia, West. Sib. (south-west. Altay), China (Dunbei, Nor., Nor.-West., Cent., South-West.), Himalayas, Korea (south); introduced in other countries.

7. **Cymbopogon** Spreng.
Pl. Pugill. 2 (1815) 14.

1. C. tibeticus Bor in Kew Bull. (1953) 275; ej. Grasses Burma, Ceyl., Ind. and Pakist. (1960) 132.

Described from South. Tibet. Type in London (K).

In sand, less frequently on talus and rocky slopes; in upper mountain belt.

IIIB. **Tibet:** *South.* ("Kyichu valley, 15 miles east of Lhasa, 3000 m, Aug. 1904, leg. H. Walton, type!; Tsangpo valley, No. 12022, July 19, 1935, leg. F. Kingdon-Ward; vicinity of Lhasa, 3000 m, No. 273, July 1939, leg. H. Richardson; hills south of Lhasa, 4000 m, No. 9745, July 11, 1943, leg. Ludlow and Sheriff"—Bor, l.c.).

General Distribution: China (South-west).

Sorghum Moench
Meth. pl. (1794) 207, nom. conserv.

1. Plant 0.5–1.5 m tall with linear 0.5–1.8 cm broad leaf blades. Panicles spreading at anthesis; internodes of their branches between sessile spikelets 2.5–4 mm long, profusely pilose;

spikelets usually lanceolate-ovate, 5–7 mm long ....
.... S. sudanense (Piper) Stapf.

+ Plants 1–3 m tall with lanceolate-linear 2–8 cm broad leaf blades. Panicles mostly dense; internodes of their branches between sessile spikelets 0.5–2.5 mm long, usually less pilose .... 2.

2. Sessile spikelets broadly lanceolate or lanceolate-ovate; twice or more longer than broad at anthesis; panicles invariably erect .....
.... 3.

+ Sessile spikelets ovate or obovate; less than twice longer than broad at anthesis .... 4.

3. Glumes coriaceous, generally herbaceous in upper half, with prominent ribs; panicles usually ellipsoidal; spikelets 3.8–5 mm long; caryopsis usually emerging from glumes ....
.... S. nervosum Bess.

+ Glumes coriaceous almost throughout length; ribs relatively faint only near tip; panicles from ellipsoidal to subflabellate; spikelets 4.5–6.5 mm long; caryopsis not emerging from glumes ....
.... S. saccharatum (L.) Moench.

4. Panicles relatively more lax, erect; glumes coriaceous almost up to tip, lustrous, wrinkle-free; ribs relatively faint at tip ....
.... S. bicolor (L.) Moench.

+ Panicles very dense, drooping or erect; glumes coriaceous in lower half, lustrous, usually with distinct wrinkles; herbaceous and largely concave (with depression) in upper part with distinct ribs .... S. cernuum (Ard.) Host.

S. bicolor (L.) Moench, Meth. Pl. (1794) 207; Roshevitz in Fl. SSSR, 2 (1934) 20; Bor, Grasses Burma, Ceyl., Ind. and Pakist. (1960) 227.—*Holcus bicolor* L. Mant. Pl. alt. (1771) 301.—*Andropogon sorghum* (L.) Brot. Fl. lusit. 1 (1804) 88, quoad Pl. Henderson and Hume, Lahore to Yarkand (1873) 342.—*A. sorghum* subsp. *sativus* var. *bicolor* (L.) Koern. ex Hack. Monogr. Andropog. (1889) 519; Forbes and Hemsley, Index Fl. Sin. 3 (1904) 368.

Described from Iran. Type in London (Linn.).

Cultivated in oases.

IB. Kashgar: *West.* ("Jarkand"—Henderson and Hume, l.c.); *East.* (Pichan river valley, Sept. 27, 1898—Klem.).

IIA. Junggar: *Dzhark.* (on Dumandzhi river, 1874—Larionov).

General distribution: only as cultivated plant in many countries of the world; Balkh. region (south); Mediterr., Balk.-Asia Minor; Near East, Caucasus, Middle Asia, Far East (south), China, Korea, Japan, Indo-Malay., Afr. (north); origin evidently Nor. Africa and Mediterranean.

S. cernuum (Ard.) Host, Gram. Austr. 4 (1809) 2; Roshevitz in Fl. SSSR, 2 (1934) 20; Fl. Kirgiz. 2 (1950) 31; Fl. Kazakhst. 1 (1956) 126; Fl. Tadzh. 1 (1957) 484.—*S. vulgare* Pers. Synops. Pl. 1 (1804) 101, p.p. (incl. Typo)

—*Holcus sorghum* L. Sp. pl. (1753) 1047, p.p. (incl. typo).—*H. cernuus* Ard. in Saggi, Sci. Lett. Ac. Padova, 1 (1786) 128.

Described from Italy. Type probably in Padua.

Cultivated in oases.

IB. **Kashgar:** *East.* (on Algoi river near Turfan, Sept. 14, 1879—A. Reg.).

IIA. **Junggar:** *Dzhark.* (in Kul'dzha town, 1875—Larionov).

General distribution: only as cultivated plant in many countries of Eurasia and Africa (south of Aral-Casp.; south of Balkh. region, Jung.-Tarb., Nor. Tien Shan; Mediterr., Balk.-Asia Minor, Near East, Middle Asia, Indo-Malay., Nor. Afr.); origin evidently Nor.-East. Africa and Near East.

S. nervosum Bess. in Schult. and Schult. f. Mant. 3 (1827) 669.—*S. japonicum* (Hack.) Roshev. in Fl. SSSR, 2 (1934) 20, quoad pl.

Described from cultivated specimens. Origin China. Type probably in Kiev.

Cultivated in oases.

IA. **Mongolia:** *Ordos* (near Linchzhou town, Oct. 6, 1884—Pot.).

IB. **Kashgar:** *West* (near Ak-Tash village, Aug. 5, 1913—Knorring).

IIA. **Junggar:** *Jung Gobi* (Sandzhi settlement, Nov. 6, 1879—A. Reg.; between Santakhu and Dzhimuchi, Aug. 19, 1898—Klem.).

General distribution: only as cultivated plant in East. Asia (south. Far East, China, Korea, Japan), evidently its place of origin.

S. saccharatum (L.) Moench, Meth. Pl. (1794) 207; Roshevitz in Fl. SSSR, 2 (1934) 21; Fl. Kirgiz. 2 (1950) 31; Fl. Tadzh. 1 (1957) 484. —*S. technicum* (Koern.) Battand. et Trabut, Fl. Algér (1895) 128; Roshevitz, l.c. 20; Fl. Kirgiz. 2 (1950) 31; Fl. Kazakhst. 1 (1956) 126; Fl. Tadzh. 1(1957) 484. —*S. dochna* (Forsk.) Snowden in Kew Bull. (1935) 234; Bor, Grasses Burma, Ceyl., Ind. and Pakist. (1960) 234. —*S. dochna* var. *technicum* (Koern.) Snowden, l.c. 235; Bor, l.c. 234.—*Holcus saccharatus* L. Sp. pl. (1753) 1047, p.p. (incl. typo).—*H. dochna* Forsk. Fl. aegypt.-arab. (1775) 174.— *Andropogon saccharatus* (L.) Roxb. Hort. Beng. (1814) 7; Henderson and Hume, Lahore to Yarkand (1873) 342. —*A. sorghum* var. *technicus* Koern. Syst. Uebers. Cereal. u. Legum. (1873) 20.

Described from India. Type in London (Linn.).

Cultivated in oases.

IA. **Mongolia:** *Khesi* (Sachzhou oasis [An'si], 1200 m, Aug. 1895—Rob.).

IB. **Kashgar:** *West.* ("near Jarkand City"—Henderson and Hume, l.c.).

IIA. **Junggar:** *Dzhark.* (on Dumandzhi river, 1874; in Kul'dzha town, 1875—Larionov).

General distribution: only as cultivated plant in many countries of the world : Aral-Casp., Balkh. region, Jung.-Tarb., Nor. Tien Shan; Europe (south), Mediterr., Balk.-Asia Minor, Near East, Caucasus, Middle Asia, Far East (south), China, Korea, Japan, Indo-Malay., Afr. (nor.); origin evidently Nor. Africa and Mediterranean.

Note: Technical varieties (cultivars) of this species are cultivated predominantly in Cent. Asia. These have highly elongated lower branches of panicles and belong to the group *S. saccharatum* convar. *technicum* (Koern.) Tzvel. comb. nova (= *Andropogon sorghum* var. *technicus* Koern. l.c.).

S. sudanense (Piper) Stapf in Prain, Fl. Trop. Afr. 9 (1917) 113; Roshevitz in Fl. SSSR, 2 (1934) 22; Fl. Kirgiz. 2 (1950) 28; Fl. Kazakhst. 1 (1956) 125; Fl. Tadzh. 1(1957) 485.—*Andropogon sorghum sudanense* Piper in Proc. Biol. Soc. Wash. 28 (1915) 33.

Described from cultivated specimens. Origin Sudan. Type in Washington. Introduced in cultivation as drought-resistant fodder plant.

IA. **Mongolia:** *East. Gobi* (1 km south-east of Dalan-Dzadagad in experimental plantations, July 28; same site, Sept. 12—1951—Kal.).

General distribution: Afr., as cultivated plant in many other countries.

8. Tragus Hall.
Hist. Stirp. Helv. 2 (1768) 203, nom. conserv.

1. T. mongolorum Ohwi in Acta Phytotax. et Geobot. 10, 4 (1941) 268.— *T. racemosus* auct. non All.: Franch. Pl. David. 1 (1884) 327; Forbes and Hemsley, Index Fl. Sin. 3 (1904) 343, p.p.; Wang in Acta Phytotax. Sin. 1 (1951) 32; Grubov, Consp. fl. MNR (1955) 60; Keng, Fl. ill. sin., Gram. (1959) 739.—*T. berteronianus* auct. non Schult.: Kitag. Lin. Fl. Mansh. (1939) 95.—Ic.: Keng, l.c. fig. 686 (sub *T. racemosa*).

Described from Inner Mongolia. Type in Tokyo.

In coastal sand, pebble beds, on rubble and rocky slopes and along roadsides; up to lower mountain belt.

IA. **Mongolia:** *East. Mong.* (near Shilin-Khoto, 1959—Ivan.); *Gobi-Alt.* (Bain-Tukhum area, Aug. 7; foothills of Bain-Tsagan mountain range, Aug. 29—1931—Ik.-Gal.); *East. Gobi* ("urbe Raoto, No. 10019"—Ohwi, l.c. type !; Udinsk lowland before Ubugun-Tsagan-Obo, Aug. 24; Argali mountain range and around Khodotyin-Khuduk well, Sept. 5—1928—Shastin; Kalgan road, Dzamyn-Ude, Aug. 26, 1931—Pob.; 16 km nor.-east of Sain-Shanda on road to Baishintu, Sept. 1, 1940—Yun.); *Alash. Gobi* (Tengeri sand around Chzhunvei, July 24, 1957; Tengeri sand near Bayan-Khoto, Aug. 15, 1958—Petr.); *Ordos* (Ulan-Morin river valley, Aug. 9, 1884—Pot.; 40 km south of Dalat town, Khantaichuan river valley, Aug. 9, 1957—Petr.).

General distribution: China (Dunbei, Nor.).

Note: This species differs from the Mediterranean species *T. racemosus* (L.) All. in very short branches of spicate panicles bearing usually only two spikelets and upper glumes with five, very prominent, subidentical longitudinal ribs covered with large number of spines. Species *T. berteronianus* Schult. (= *T. tcheliensis* Debeaux) with spikes nearly half smaller and widely distributed in East. and South. China, may be found in Cent. Asia.

9. Arundinella Raddi
Agrost. Brasil. (1823) 36.

1. A. hirta (Thunb.) Tanaka in Bull. Sci. Fak. Terkult. Kjusu Univ. 1, 4 (1925) 195; Keng, Fl. ill. sin., Gram. (1959) 724.—*A. anomala* Steud. Synops.

Pl. Glum. 1 (1854) 116; Franch. Pl. David. 1 (1884) 325; Forbes and Hemsley, Index Fl. Sin. 3 (1904) 341; Roshevitz in Fl. SSSR, 2 (1934) 24; Bor, Grasses Burma, Ceyl., Ind. and Pakist. (1960) 426.—*A. hirta* var. *ciliata* (Thunb.) Koidz. in Bot. Mag. Tokyo, 39 (1925) 303; Hand.-Mazz. Symb. Sin. 7, 5 (1936) 1317; Kitag. Lin. Fl. Mansh. (1939) 62.—*Poa hirta* Thunb. Fl. Jap. (1784) 49.—*Agrostis ciliata* Thunb. l.c., non *Arundinella ciliata* Nees (1851).—Ic.: Fl. SSSR 2, Plate 1, fig. 14; Keng, l.c. fig. 673.

Described from Japan. Type in Uppsala.

In meadows, on rocky slopes, in forest glades, sometimes as weed along roadsides, around irrigation ditches and along farm boundaries; up to lower mountain belt.

IA. **Mongolia:** *Cis-Hing.* (Dege-Gol river on road from Bain-Buridu to Numurgin-Gol, May 5, 1944—Yun.); *Cent. Khalkha* (Dashi-balbar somon, 5 km west of Zagal hills, July 30, 1963—Dashnyam); *East. Mong.* (near Hailar town, 1960—Ivan.); *Ordos* (in Huang He river valley, July 28, 1871—Przew.).

General distribution: East. Sib. (Daur.), Far East (south), China, Korea, Japan, Indo-Malay. (?).

Note: Only the relatively more continental variety *A. hirta* var. *ciliata* (Thunb.) Koidz. (= *A. anomala* Steud.) with all leaf sheaths glabrous or pilose only along margin is found in Cent. Asia. The type, however, has sheaths in which all or only lower leaves are covered with dense and fairly long hairs.

10. Panicum L.

Sp. pl. (1753) 55.

1. **P. miliaceum** L. Sp. pl. (1753) 58; Henderson and Hume, Lahore to Yarkand (1873) 339; Forbes and Hemsley, Index Fl. Sin. 3 (1904) 331; Krylov, Fl. Zap. Sib. 2 (1928) 151; Roshevitz in Fl. SSSR 2. (1934) 36; Kitag. Lin. Fl. Mansh. (1939) 84; Norlindh, Fl. Mong. Steppe, 1 (1949) 53; Fl. Kirgiz. 2 (1950) 35; Grubov, Consp. fl. MNR (1955) 60; Fl. Kazakhst. 1 (1956) 131; Fl. Tadzh. 1 (1957) 465; Keng, Fl. ill. sin., Gram. (1959) 656; Bor. Grasses Burma, Ceyl., Ind. and Pakist. (1960) 327.—*P. miliaceum* var. *ruderale* Kitag. in Bot. Mag. Tokyo, 51 (1937) 153; ej. Lin. Fl. Mansh. (1939) 84.—*P. spontaneum* Lyssov in Tr. po prikl. bot., gen. i sel. 29, 3 (1952) 112, diagn. ross.—Ic.: Fl. SSSR, 2, Plate 2, Fig. 9; Fl. Tadzh. 1, Plate 58, figs, 7–9; Keng, l.c. fig. 596.

Described from India. Type in London (Linn.).

Widely cultivated in oases as fodder and food plant but also found as weed or wild plant along roadsides and around irrigation ditches, in wastelands, in field crops and in sand and pebble beds in river valleys; up to midmountain belt.

IA. **Mongolia:** *East. Mong.* (plain south of Khukh—Khoto town, July 21, 1884—Pot.; Kulun—Buir—Nurskaya plain, Nomon-Khan area, Aug. 10, 1899—Pot. and Sold.; *Bas. lakes* (Ara-Tarkhalik river in Ubsu-Nur lake region, Oct. 1, 1879—Pot.); *Val.* lakes (in saxaul thickets along road from Khobdo somon to Dalan-Dzadagad near its crossing with the road to Baindel, July 27, 1951—Kal.); *Gobi—Alt.* (south. trail of Dundu-Saikhan hills, abandoned field, Aug. 25, 1950—A.T. Ivanov); *East. Gobi* (Baishintu khure, on sand hillocks, Sept. 4, 1931—Ik.-Gal.); *Alash. Gobi* (Dynyuan' in oasis [Bayan-Khoto], August 22, 1908—Czet.); *Ordos* (valley of Shukhen-Gol river, Aug. 20; south of Shine-Sume monastery, Aug. 29—1884—Pot.); *Khesi.*

IB. **Kashgar:** *West.* ("near Jarkand"—Henderson and Hume, l.c.); *South.* (nor. foothills of Kunlun, about 1600 m, Sept. 13, 1889—Rob.); *East.* (in Turfan town, Sept. 21, 1879—A. Reg.).

IIA. **Junggar:** *Tien Shan* (Talki river, July 18, 1877: Tsagan-Usu, 1000 m, June 18, 1879; Dzhisumtal on Kash river, about 1000 m, July 2, 1879—A. Reg.); *Jung Gobi* (nor.-West. : Bulun-Tokhoi settlement, Aug. 5 and 8; near crossing of Dyurbeldzhin and Ch. Irtysh river, Aug. 12—1876—Pot.; Chingil' river valley, July 29, 1906—Sap.); *Zaisan* (right bank of Ch. Irtysh river near Burchum river estuary, tugai, June 14, 1914—Schisch.); *Dzhark.* (in Kul'dzha town, July 12, 1878—Larionov).

General distribution: Aral-Casp., Balkh. region, Jung.-Tarb., Nor. Tien Shan; Europe, Mediterr., Balk.-Asia Minor, Near East, Caucasus, Middle Asia, West. Sib. (south), East. Sib. (south), Far East (south), Nor. Mong. (Mong.-Daur.), China, Himalayas, Korea, Japan, Indo-Malay.; also cultivated occasionally in many countries of both hemispheres; China and Mongolia evidently represent its native region.

Note: Millet growing wild as weed is found in Cent. Asia along with cultivated types of millet. The wild type is characterised by the presence of joints at base of at least some spikelets of spreading invariably panicles and, on average, with very narrow (near fruits 1.3–1.6 mm broad) lemma. This millet, probably representing the immediate ancestor of cultivated types, has been described as a distinct variety *P. miliaceum* var. *ruderale* Kitag. (l.c.) and an independent species *P. spontaneum* Lyssov (l.c.). From among the specimens listed above, only the collections of Potanin (of Aug. 12, 1876), Schischkin and Kalinina belong here.

11. Echinochloa Beauv.
Ess. Agrost. (1812) 53.

1. Spikelets (without awn) 3.7–4.8 mm long, only a few with joint near base; lemma in fruit 3.5–4.5 mm long ....
.... 2. **E. oryzoides** (Ard.) Fritsch.

+ Spikelets (without awn) 2.3–3.6 mm long; usually more than 1/4 with joint in each panicle near base shedding easily in fruits; lemma in fruit 2.3–3.5 mm long .... 1. **E. crusgalli** (L.) Beauv.

1. **E. crusgalli** (L.) Beauv. Ess. Agrost. (1812) 53; Forbes and Hemsley, Index Fl. Sin. 3 (1904) 328; Roshevitz in Fl. SSSR, 2 (1934) 32; Kitag. Lin. Fl. Mansh. (1939) 72, p.p.; Norlindh, Fl. mong. steppe, 1 (1949) 52; Fl. Kirgiz. 2 (1950) 35; Grubov. Consp. fl. MNR (1955) 60; Fl. Kazakhst. 1 (1956) 130; Fl. Tadzh. 1 (1957) 466; Keng. Fl. ill. sin., Gram. (1959) 673, p.p.; Bor, Grasses Burma, Ceyl., Ind. and Pakist. (1960) 310.—*E. spiralis* Vasing. in Fl. SSSR, 2 (1934) 34 and 739.—*E. caudata* Roshev. in Fl. SSSR, 2 (1934) 35 and in Tr. Bot. inst. AN SSSR, ser. 1, 2 (1936) 91.—*E. crusgalli* var. *caudata* (Roshev.) Kitag. Lin. Fl. Mansh. (1939) 73—*Panicum crusgalli*

L. Sp. pl. (1753) 56; Franch. Pl. David. (1884) 322; Krylov, Fl. Zap. Sib. 2 (1928) 150.—Ic.: Fl. SSSR, 2, Plate 2, fig. 8; Fl. Tadzh. 1, Plate 58, figs. 4–6; keng, fig. 616.

Described from Europe and Nor. America. Type in London (Linn.). In wet meadows, riverine pebble beds, on banks of water reservoirs, often as weed along roadsides and around irrigation ditches, in farms and field crops and in wastelands; up to midmountain belt.

IA. **Mongolia:** *Khobd.* (between border and Kobdo town, 1870—Kalning); *Mong. Alt.* (east. bank of Tonkhil'-Nur lake, July 16, 1947—Yun.); *East. Mong., Bas, lakes, East. Gobi, Alash.* Gobi (west of Bayan-Khoto near west. Border of Tengeri sand, July 13, 1958—Petr.); *Ordos, Khesi.*

IB. **Kashgar:** *Nor., West., South., East., Takla Makan* (Cherchen oasis, June 3, 1890—Rob.).

IIA. **Junggar:** *Alt. Region* (near Koktogoi settlement, No. 1695, Aug. 11, 1956—Ching); *Jung. Alt., Tien Shan, Jung. Gobi, Dzhark., Balkh.-Alak.*

General distribution: Aral-Casp., Balkh. region, Jung.-Tarb., Nor. and Cent. Tien Shan; Europe (except nor.-east), Mediterr., Balk.-Asia Minor, Near East, Caucasus, Middle Asia, West. Sib. (south), East. Sib. (south), Far East (south), Nor. Mong. (Mong.-Daur.), China, Himalayas, Korea, Japan, Indo-Malay., Nor. Amer. (except north), South Amer., Afr., Austral., New Zealand.

Note: This species is characterised by high polymorphism (variability) especially in length of awns, which may be altogether absent (in *E. crusgalli* var. *crusgalli*) or measure 2.5–4 cm (in *E. crusgalli* var. *aristata* S.F. Gray). Specimens with relatively very small (2.3–3.2 mm long) spikelets with highly branched panicles are found in oases in the more southern regions of Cent. Asia. Species *E. spiralis* Vasing (l.c.) has been described from such specimens distributed mainly in East. Asia and America but they merit distinction only as a variety. Establishing its correct nomenclature is very difficult due to extremely confused synonyms within species *E. crusgalli* (L.) Beauv. s.l. The herbarium of the Botanical Institute, Academy of Sciences, USSR, holds a specimen from East. Mongolia (Kerulen river valley near Bo-Tszangin-Sume, 1899—Pal.), constituting yet another variety, *E. crusgalli* var. *caudata* (Roshev.) Kitag. (l.c.), distributed predominantly in Amur basin. This specimen with very long (3–5 cm) awns differs from the aforesaid variety by long awns in almost all spikelets having a joint at the base and nearly wholly reduced palea of the lower neuter floret of the spikelet.

2. **E. oryzoides** (Ard.) Fritsch in Verh. Zool.-Bot. Gesellsch. Wien, 41 (1891) 742.—*E. macrocarpa* Vasing. in Fl. SSSR, 2 (1934) 34 and 739; Fl. Kirgiz. 2 (1950) 32; Fl. Kazakhst. 1 (1956) 131.—*E. coarctata* Kossenko in Bot. mater. Gerb. Bot. inst. AN SSSR, 9, 1 (1941) 28; Fl. Tadzh. 1 (1957) 467.—*E. crusgalli* var. *macrocarpa* (Vasing.) Ohwi in J. Jap. Bot. 18 (1942) 541.—*E. hostii* (M.B.) Stev. in Bull. Soc. nature. Moscou, 30, 3 (1857) 120.—*Panicum oryzoides* Ard. Animadv. Bot. 2 (1764) 16.—*P. hostii* M.B. Fl. taur.-cauc. 3 (1819) 56.—Ic.: Sorn. rast. SSSR [Weed Plants of the USSR], 1 (1934) fig. 28.

Described from specimens grown from caryopsis contaminating rice of unknown origin. Type probably in Florence.

Weed in rice fields, less often in other field crops, along roadsides and around irrigation ditches; up to lower mountain belt.

IB. **Kashgar:** *Nor.* (Bugur village, Aug. 20, 1929—Pop.; 14 km west of Kel'pin, No. 7443, Sept. 8, 1958—Lee and Chu); *West.* (near Kashgar, July 26, 1929—Pop.); *South.* (nor. foothills of Kunlun, Sept. 13, 1889—Rob.).

IIA. **Junggar:** *Tien Shan* (Nizhn. Borotala, 1000 m, Aug. 27, 1878—A. Reg.).

General distribution: Aral-Casp., Balkh. region, Nor. Tien Shan; Cent. and South. Europe, Mediterr., Balk.-Asia Minor, Near East, Caucasus, Middle Asia, Far East (south), China, Himalayas, Korea, Japan, Indo-Malay.; introduced in many other countries.

12. **Digitaria** Heist. ex Fabr.
Enum. Meth. pl. (1759) 207.

1. Spikelets 2.5–4 mm long, broadly lanceolate, in pairs; lower glume of spikelet 0.3–0.6 mm long but distinctly visible, deltoid-ovate, herbaceous; upper glume (ignoring lemma of totally reduced lower floret externally similar to scale of spikelet!) 1.5–2 times shorter than spikelet. Sheaths and leaf blades usually pilose for most part .... 2.

+ Spikelets 1.7–2.4 mm long, lanceolate-ovate, in groups of 3 each, less often in pairs; lower glume of spikelet in form of faint sheath-like membranous scale; upper glume as long as spikelet or not more than 1/4 shorter. Sheaths and leaf blades invariably glabrous .... 3.

2. Lemma of totally reduced lower floret smooth along ribs; usually profusely pilose between ribs; upper glume of spikelet almost 1.5 times shorter than spikelet .... 1. D. adscendens (H.B.K.) Henr.

+ Lemma of totally reduced lower floret with spinules along ribs; relatively faintly pilose between ribs; upper glume of spikelet almost twice shorter than spikelet ..... 2. D. sanguinalis (L.) Scop.

3. Upper glume of spikelet and also lemma of totally reduced lower floret glabrous, less often subglabrous (with slightly shorter hairs along sides of lemma) .... 5. D. stewartiana Bor.

+ Upper glume of spikelet and also lemma of lower floret pilose almost throughout surface .... 4.

4. Pubescence of glume consists of relatively long simple hairs; spikelets 1.6–2.2 mm long .... 3. D. asiatica Tzvel.

+ Pubescence of glume consists of very short hairs, many of which (sometimes almost all) thickened, clavate at tip; spikelets 1.8–2.5 mm long .... 4.. D. ischaemum (Schreb.) Muhl.

Section 1. Digitaria

1. **D. adscendens** (H.B.K.) Henr. in Blumea, 1 (1934) 92; Keng, Fl. ill. sin., Gram (1959) 704; Bor, Grasses Burma, Ceyl., Ind. and Pakist. (1960) 298.—*D. chinensis* Hornem. Suppl. Hort. Bot. Hafn. (1819) 8; Keng. l.c.—*D. sanguinalis* auct. non Scop.: Forbes and Hemsley, Index Fl. Sin. 3 (1904) 325; Kitag. Lin. Fl. Mansh. (1939) 72.—Ic.: Keng, l.c., figs., 650 and 651.

Described from South America. Type in Berlin.

As weed in oases; up to lower mountain belt.

IB. **Kashgar:** *East.* (Lyukchunsk basin, Sept. 29, 1895—Rob.; Burlyuk village 12 km nor. of Turfan, Sept. 29, 1957—Yun.).

General distribution: Near East, China, Korea, Japan, Indo-Malay., Nor. America (south), South America, Afr., Austral.; introduced in other countries.

Note: A widely distributed tropical weed, this species is very closely related to the next species but differs in its considerable polymorphism. It is classified into several units whose taxonomic rank is still not wholly clear. The variety with dense and long hairs between lateral ribs of lemma of reduced lower florets in spikelets, *D. adscendens* var. *criniformis* Henr., most common in China, is found in Cent. Asia.

2. **D. sanguinalis** (L.) Scop. Fl. Carniol., ed. 2, 1 (1772) 52; Roshevitz in Fl. SSSR, 2 (1934) 29; Fl. Kirgiz. 2 (1950) 32; Fl. Kazakhst. 1 (1956) 129; Fl. Tadzh. 1 (1957) 476; Keng, Fl. ill. sin., Gram. (1959) 702; Bor, Grasses Burma, Ceyl., Ind. and Pakist. (1960) 304.—*Panicum sanguinale* L. Sp. pl. (1753) 57.—Ic.: Fl. SSSR , 2, Plate 2, fig. 5.

Described from South. Europe. Type in London (Linn.).

As weed in oases.

IB. **Kashgar:** *West.* (45 km north of Yarkand on road to Maralbashi, Sept. 30, 1958—Yun.).

General distribution: Aral-Casp., Balkh. region (south); Europe, Mediterr., Balk.-Asia Minor, Near East, Caucasus, Middle Asia, Nor. America; introduced in many other countries of both the hemispheres.

Section 2. Ischaemum Ohwi

3. **D. asiatica** Tzvel. in Bot. mater. Gerb. Bot. inst. AN SSSR, 22 (1963) 64.—*D. ischaemum* subsp. *asiatica* (Tzvel.). Tzvel. in Spiske rast. Gerb. fl. SSSR, 17 (1966) 29.—*D. linearis* auct. non Crép. : Roshevitz in Fl. SSSR, 2 (1934) 28, p.p.; Fl. Kirgiz. 2 (1950) 31; Fl. Kazakhst. 1 (1956) 129, p.p.—*D. ischaemum* auct. non Muhl.: Kitag. Lin. Fl. Mansh. (1939) 71, p. max. p.; Fl. Tadzh. 1 (1957) 475, p. max. p.; Keng. Fl. ill. sin., Gram. (1959) 699, p.p.; Bor, Grasses Burma, Ceyl., Ind. and Pakist. (1960) 302, p.p.

Described from Japan. Type in Leningrad.

As weed in oases.

IB: **Kashgar**: *Nor.* (near Ishme village, Aug. 21, 1929—Pop.; 20 km east of Tsarga village, No. 8176, Sept. 3; near Kucha settlement, No. 8731, Sept. 11—1958—Lee and Chu).

General distribution: Aral.-Casp. (south), Balkh. region (south); Mediterr., Balk.-Asia Minor, Near East, Caucasus, Middle Asia, Far East (south), China, Korea, Japan, Indo-Malay.; introduced in many countries of both hemispheres.

Note: This and the next two species represent closely related ecogeographical races and hence can be regarded as subspecies of a single polytypic species, *D. ischaemum* (Schreb.) Muhl. s.l. Among them, *D. asiatica* is more thermophilic and evidently the source for the remaining two. The other speices—*D. ischaemum* (Schreb.) Muhl. s.s.—is mainly distributed in Europe and southern Siberia, which underwent Pleistocene glaciation, but is later seen in hilly regions of Middle and Central Asia. The third speices, *D. stewartiana* Bor, is the highest altitude species and its distribution is extremely restricted.

4. **D. ischaemum** (Schreb.) Muhl. Descr. Gram. (1817) 131; Persson in Bot. notiser (1938) 273; Kitag. Lin. Fl. Mansh. (1939) 71, p.p.; Fl. Tadzh. 1 (1957) 475, p.p.; Keng, Fl. ill. sin., Gram. (1959) 699, p.p.; Bor. Grasses Burma, Ceyl., Ind. and Pakist. (1960) 302, p.p.—*D. humifusa* Rich. in Pers. Synops. Pl. 1(1805) 85; Forbes and Hemsley, Index Fl. Sin. 3 (1904) 324, p.p.—*D. linearis* (L.) Crép. Manuel Fl. Belg. ed. 2 (1866) 335, p.p. (excl. typo); Roshevitz in Fl. SSSR, 2 (1934) 28, p.p.; Fl. Kazakhst. 1 (1956) 129, p.p.—*Panicum ischaemum* Schreb. in Schweigg. Specim. Fl. Erlang. (1804) 16.—**Ic.**: Fl. SSSR, 2, Plate 2, fig. 4 (sub *D. lineari*); Fl. Tadzh. 1, Plate 58, fig. 3; Keng, l.c. fig. 644.

Described from West Germany. Type location not known.

As weed in oases.

IA. **Mongolia**: *East. Gobi* (near Bailinmyao, 1959—Ivan.).

IB. **Kashgar**: *Nor.* (in Bai town, Aug. 14, 1929—Pop.); *West* ("Jarkend, 1350 m, Sept. 5, 1931; Kaschgar, 1330 m, Aug. 31, 1934"—Persson, l.c.; 45 km nor. of Yarkand along road to Maralbashi, Sept. 30, 1958—Yun); *South* (Sampula oasis in the vicinity of Khotan town, about 1400 m, Aug. 28, 1885—Przew.).

General distribution: Aral-Casp., Balkh. region; Europe, Mediterr., Balk.-Asia Minor, Near East, Caucasus (less often and predominantly in Fore-Caucasus), Middle Asia (less often), West. Sib. (south), East. Sib. (south), Far East (south, as introduced plant), China (Dunbei, Nor.), Himalayas (west. Kashmir), Nor. Amer.; as introduced plant in many other countries of both hemispheres.

5. **D. stewartiana** Bor in Kew Bull. (1951) 166; ej. Grasses Burma, Ceyl., Ind. and Pakist. (1960) 305.

Described from Kashmir. Type in London (K).

In pebble beds, river and lake alluvia and along roadsides; in midmountain belt.

13. Setaria Beauv.
Ess. Agrost. (1812) 51, nom. conserv.

1. Branches of panicles usually with only solitary spikelet, less often frequently with additional 1–2 underdeveloped spikelets; bristles surrounding spikelets yellowish or orange; spikelets 2.8–3.4 m long, with distinct joint at base; upper glume 1/4 to 1/3 shorter than spikelet; lemma in fruit transversely distinctly rugose .... 1. **S. glauca** (L.) Beauv.

+ Branches of panicles with 2–15 (or more) fully developed spikelets; bristles surrounding spikelets greenish or violet; spikelets 1.8–2.8 mm long; upper glume almost as long as spikelet or not more than 1/6 shorter; lemma in fruit finely tuberculate, only very weakly transversely rugose or not .... 2.

2. All spikelets without joints at base and not shedding in fruit; distinct joint seen only under lemma in fruit; panicles usually large (8–30 cm long), often highly branched. Plants 50–200 cm tall with 6–30 mm broad leaf blades .... 2. S. italica (L.) Beauv.

+ All spikelets with distinct joints at base and easily shedding entirely in fruit (in dry state as well) ....3.

3. Plants 60–150 cm tall with (6) 8 to 20 (25) mm broad leaf blades; stems usually solitary or few, with 5–10 nodes .... 3. **S. pycnocoma** (Steud.) Henr. ex Nakai.

+ Plants (5) 10–60 (80) cm tall with 2–10 (15) mm broad leaf blades; stems, usually many, forming mats, with 2–6 nodes ....**4. S. viridis** (L.) Beauv.

1. **S. glauca** (L.) Beauv. Ess. Agrost. (1812) 51; Henderson and Hume, Lahore to Yarkand (1879) 330; Franch. Pl. David. 1 (1884) 323; Forbes and Hemsley, Index Fl. Sin. 3 (1904) 335; Danguy in Bull. Mus. nat. hist. natur. 17 (1911) 6; Simpson in J. Linn. Soc. London (Bot.) 41 (1913) 450; Krylov, Fl. Zap. Sib. 2 (1928) 153; Roshevitz in Fl. SSSR, 2 (1934) 39; Fl. Kirgiz. 2 (1950) 36; Fl. Kazakhst. 1 (1956) 132; Fl. Tadzh. 1 (1957) 469; Bor, Grasses Burma, Ceyl., Ind. and Pakist. (1960) 360.—*S. lutescens* (Weig.) F.T. Hubb. in Rhodora, 18 (1916) 232; Hand.-Mazz. Symb. Sin. 7 (1936) 1305; Kitag. Lin. Fl. Mansh (1939) 93; Keng, Fl. ill. sin., Gram. (1959) 712.— *Panicum glaucum* L. Sp. pl. (1755) 56, p.p. and Syst. nat., ed. 10, 2 (1759) 870.—*P. lutescens* Weig. Obs. Bot. (1772) 20.—**Ic.**: Fl. SSSR, 2, Plate 2, fig. 12; Keng, l.c. fig. 660.

Described from Europe. Type in London (Linn.).

In river sand and pebble beds, often also as weed along roadsides and around irrigation ditches and in field crops; up to midmountain belt.

IA. Mongolia: *East. Mong.* (Huang He bank below Khekou town, 1884—Pot.; near Shilin-Khoto, 1959—Ivan.).

IB. Kashgar: *Nor., West.* ("Jarkand City"—Henderson and Hume, l.c.); *South* (nor. foothills of Kunlun, 2000 m, Sept. 5, 1889—Rob.); *East.*

IIA. Junggar: *Alt. region* ("Shara-Sume, No. 149, leg. Price"—Simpson l.c.); *Tien Shan, Jung. Gobi.*

General distribution: Aral-Casp., Balkh. region, Jung.-Tarb., Nor. and Cent. Tien Shan; Europe (excluding nor.-east.), Mediterr., Balk.-Asia Minor, Near East, Caucasus, Middle Asia, West. Sib. (south), East. Sib. (south), Far East (south), China, Himalayas, Korea, Japan, Indo-Malay., Nor. and South. Amer., Afr., Austral.

2. S. italica (L.) Beauv. Ess. Agrost. (1812) 51; Forbes and Hemsley, Index Fl. Sin. 3 (1904) 335; Simpson in J. Linn. Soc. London (Bot.) 41 (1913) 450; Krylov, Fl. Zap. Sib. 2 (1928) 154; Roshevitz in Fl. SSSR, 2 (1934) 43; Norlindh, Fl. mong. steppe, 1 (1949) 472; Keng, Fl. ill. sin., Gram. (1959) 711; Bor, Grasses Burma, Ceyl., Ind. and Pakist. (1960) 362.—*Panicum italicum* L. Sp. pl. (1753) 56.—Ic.: Fl. SSSR, 2, Plate 2, figs. 14 and 15; Keng, l.c. fig. 658.

Described from India. Type in London (Linn.).

Widely cultivated in oases as fodder and food plant but also found sometimes as introduced or wild plant in field crops, along roadsides, around irrigation ditches and in wastelands; up to midmountain belt.

IA. Mongolia: *East. Mong.* (plain near Khukh-Khoto town) Sartu village, July 21, 1884—Pot.); *Alash. Gobi* (Dynyuan' in oasis [Bayan-Khoto], Aug. 22, 1908—Czet.); *Ordos* (near Chunir village in Dzhasak town region, Aug. 16, 1957—Petr.); *Khesi* (near Koutai town, July 28, 1875—Pias.; Sachzhou oasis [An'si], 1200 m, Aug. 2, 1895— Rob.; between Baiduntszy and Sayan'tszyn villages, 1500–1700 m, Sept. 20, 1900— Lad.).

IB. Kashgar: *Nor.* (between Bai and Aksu near Taz-Lyangar village, Aug. 11, 1929—Pop.).

IIA. Junggar: *Alt. region.* ("Sharasume, Kran River, leg. Price"—Simpson, l.c.; Kran river valley near Shara—Sume, Aug. 10, 1906—Sap.); *Tien Shan* (near Kash river, 1000 m, Sept. 6, 1878—A. Reg.); *Jung. Gobi* (nor. of Guchen, Oct. 1890—Gr.-Grzh.; near Bulun-Tokhoi town, Aug. 7, 1876—Pot.; around Sandzhi settlement, Aug. 13; between Sandzhi and Dzhimisar, Aug. 17—1898—Klem.); Dzhark. (near Kul'dzha, 1874 and 1875—Larionov).

IIIA. Qinghai: *Nanshan* (south. slope of Yuzhno-Tetungsk mountain range, July 28, 1908—Czet.).

General distribution: Aral-Casp., Balkh. region, Europe (south), Mediterr., Balk.-Asia Minor, Near East, Caucasus, Middle Asia, West. Sib. (south), Far East (south), China, Himalayas, Korea, Japan, Indo-Malay.; also cultivated in many other countries of both hemispheres.

3. S. pycnocoma (Steud.) Henr. ex Nakai in J. Jap. Bot. 15 (1939) 399.— S. viridis var. major (Gaud.) Pospich. Fl. Oesterr. Küstenl. 1 (1897) 51.—S. gigantea (Franch. et Savat.) Makino in Bot. Mag. Tokyo, 25, (1911) 227; Kitag. Lin. Fl. Mansh. (1939) 93.—S. ketzchovellii Menabde and Eritsian in Tr. Tbil. bot. inst. An GruzSSR, ser., 2, 11 (1947) 196.—*Panicum viride majus* Gaud. Agrost. Helv. 1 (1811) 18.—*P. pycnocomum* Steud. Synops. Pl. Glum.

1 (1854) 462.—*P. viride* var. *gigantea* Franch et Savat. Enum. Pl. Jap. 2 (1879) 162.

Described from Japan. Type in Leiden.

As weed in oases; predominantly in fields of preceding species; up to midmountain belt.

IIA. **Junggar**: *Tien Shan* (between Ebi-Nur and Sairam-Nur lakes, No. 4551, Aug. 19, 1957—Huang).

General distribution: Balkh. region (south); Europe (south), Mediterr., Balk.-Asia Minor, Near East, Caucasus, Middle Asia, Far East (south), China, Korea, Japan, Indo-Malay.; introduced in many other countries.

Note: Evidently, this species is very widely distributed in Cent. Asia, being found in many oases where the preceding species is cultivated. Occupies a somewhat intermediate status between *S. italica* (L.) Beauv. and *S. viridis* (L.) Beauv. and, according to some authors, represents their hybrid.

4. S. viridis (L.) Beauv. Ess. Agrost. (1812) 51; Franch. Pl. David. 1 (1884) 323; Forbes and Hemsley, Index Fl. Sin. 3 (1904) 336; Danguy in Bull. Mus. nat. hist. natur. 6 (1911) 6 and 20 (1914) 145; Simpson in J. Linn. Soc. London (Bot.) 41 (1913) 451; Krylov, Fl. Zap. Sib. 2 (1928) 152; Pavlov in Byull. Mosk. obshch. ispyt. prir. 38 (1929) 11; Pampanini, Fl. Carac. (1930) 71; Roshevitz in Fl. SSSR, 2 (1934) 40; Persson in Bot. notiser (1938) 273 and 274; Kitag. Lin. Fl. Mansh. (1939) 93; Ching in Contribs U.S. Nat. Herb. 28 (1941) 598; Norlindh, Fl. mong. steppe, 1 (1949) 54; Fl. Kirgiz. 2 (1950) 36; Grubov. Consp. fl. MNR (1955) 60; Fl. Kazakhst. 1 (1956) 133; Fl. Tadzh. 1 (1957) 470; Keng, Fl. ill. sin., Gram. (1959) 710; Bor, Grasses Burma, Ceyl., Ind. and Pakist. (1960) 365; Ikonnikov, Opred. rast. Pamira (Key to Plants of Pamir) (1963) 46.—*S. weinmannii* Roem. et Schult. Syst. Veg. 2 (1817) 490.—? *S. arenaria* Kitag. in Rep. Inst. Sci. Res. Manchoukuo, 4, 7 (1940) 77.—*Panicum viride* L. Syst. nat., ed. 10, 2 (1759) 870.—Ic.: Fl. SSSR, 2, Plate 2, fig. 11; Fl. Tadzh. 1, Plate 58, figs. 10–13.

Described from Europe. Type in London (Linn.).

In river sand, pebble beds, along banks of water reservoirs, often as weed in field crops, along roadsides and in wastelands; up to midmountain belt.

IA. **Mongolia**: all regions.
IB. **Kashgar**: all regions.
IC. **Qaidam**: *plains*.
IIA. **Junggar**: all regions.

General distribution: Aral-Casp., Balkh. region, Jung.-Tarb., Nor. and Cent. Tien Shan, East. Pamir (very rare); Europe, Mediterr., Balk.-Asia Minor, Near East, Caucasus, Middle Asia, West. Sib. (south), East. Sib. (south), Far East, Nor. Mong., China, Himalayas, Korea, Japan, Indo-Malay., Nor. and South Amer., Afr., Austral.

Note: *S. virdis* var. *breviseta* (Doell) Hitchc. in Rhodora, 8 (1906) 210 = *Panicum viride* var. *brevisetum* Doell. Rhein. Fl. (1843) 128, with bristles barely longer than spikelets and *S. viridis* var. *weinmannii* (Roem. et Schult.) Borb. in Math. Termesz. Közlem. 15 (1878) 310 = *S. weinmannii* Roem. et Schult. l.c. with violet-coloured bristles are reported along with the type in Cent. Asia. Species *S. arenaria* Kitag. (l.c.) described from around Hailar ("in arenosis circa Hailar, 4 VIII 1939, M. Kitagawa"), according to the original diagnosis, has relatively large (up to 3 mm long) spikelets but does not differ significantly from *S. viridis* in other respects.

14. Pennisetum Rich.
in Pers. Syn. pl. 1 (1805) 72.

1. **P. centrasiaticum** Tzvel. sp. nova.—? *P. flaccidum* var. *interruptum* Griseb. in Nachr. Gesellsch. Wissensch. u. Univ. Goett. 3 (1868) 86. —*P. flaccidum* auct. non Griseb.: Franch. Pl. David. 1 (1884) 323; Hemsley in J. Linn. Soc. London (Bot.) 30 (1894) 120; Deasy, in Tibet and Chin. Turk. (1901) 399; Hemsley in J. Linn. Soc. London (Bot.) 35 (1902) 202; Forbes and Hemsley, Index Fl. Sin. 3 (1904) 339; Roshevitz in Fl. SSSR, 2 (1934) 44, p.p.; Hao in Engler's Bot. Jahrb. 68 (1938) 584; Persson in Bot. notiser (1938) 274; Kitag. Lin. Fl. Mansh. (1939) 85; Ching in Contribs, U.S. Nat. Herb. 28 (1941) 597; Norlindh, Fl. mong. steppe, 1 (1949) 56; Fl. Kazakhst. 1 (1956) 134; Fl. Tadzh. 1 (1957) 472, p.p.; Keng, Fl. ill. sin., Gram. (1959) 714; Bor, Grasses Burma, Ceyl., Ind. and Pakist. (1960) 344, p.p.—*P. mongolicum* Franch. ex Roshev. in Fl. Aziatsk. Rossii, 6 (1914) 77, in syn.—Ic.: Fl. SSSR, 2, Plate 2, fig. 16 (sub. *P. flaccido*); Keng, l.c. fig. 662 (sub *P. flaccido*). —Planta perennis 15–80 cm alta, laxe caespitosa; culmi erecti vel adscendentes, glabri; ligulae 0.7–1.2 mm lg., margine dense ciliatae; laminae 2–5 mm lt., planae vel laxe convolutae, plus minusve scabrae. Paniculae 5–18 cm lg. et 0.5–1.3 mm lt., spiciformes, sed sat laxa, rachide scabra; spiculae 3.5–6 cm lg., in ramis vulgo solitariae, involucro e setis scabris cinctae; articulationes supra basin ramorum positae; gluma inferior 1/5–1/7 spiculae aequans; gluma superior circa 2/3 spiculae aequans; lemmata spiculam aequantes, late lanceolata, apice acuminata; antherae 2.8–3.8 mm lg., atropurpureae.

Typus: Tadzhikistania, Pamir occidentalis, distr. Vachan-Ischkaschim prope pag. Schitcharj ad fl. Pjandzh, No. 1653, 6 VIII 1935, C. Afanasjev et P. Ovczinnikov. In Herb. Inst. Bot. Acad. Sci. URSS (Leningrad) conservatur.

Affinitas. A specie proxima—*P. orientale* Rich. haec species involucri setis solum solitariis pennatis et paniculae rachide scabris differt. A specis alia—*P. flaccidum* Griseb. articulationibus supra basin ramorum positis et involucri setis solitariis pennatis bene differt.

Described from West. Pamir. Type in Leningrad.

On rocky and rubble slopes, in coastal sand and pebble beds, often as weed along roadsides and around irrigation ditches and in farms and field crops; up to midmountain belt.

IA. **Mongolia:** *East. Mong.* (right bank of Huang He below Hekou town, July 26; Termin-Bashin area, July 29; near Dzhungor-Beili, Aug. 3; south of Tuin-Gol river,

Aug. 5; Baga-Edzhin-Khoro area, Aug. 6; Ulan-Morin river, Aug. 10—1884—Pot.;
"Khongkhor-Obo, No. 221, Aug. 14, 1926, leg. Erikkson"—Norlindh, l.c.; near Shilin-
Khoto town, 1959—Ivan.); *East. Gobi* (Jichi-Ola, 1200 m, No. 431, 1925—Chaney;
"inter Khongkorin-Gol and Khashiatu-Hutuk, Aug. 10–11, 1927, leg. Hummel; Lao-
Hu-Ku, dist. Dunga-Gung, No. 6950, Aug. 29, 1927, leg. "Söderbom"—Norlindh, l.c.;
Tumur-Hada, Roehrich Exped., No. 541, July 28, 1935—Keng; Khan-Bogdo somon,
nor. fringe of Galba massif, Sept. 28, 1940—Yun.); *Alash. Gobi* ("Tukhumin-Gol, No.
1513, Aug. 31, 1927, leg. Hummel"—Norlindh, l.c.; Tengeri sand near Bayan-Khoto,
Aug. 4, 1958—Petr.); *Ordos, Khesi* (near Shan'dan' town, July 22, 1875—Pias.).

IB. Kashgar: *Nor.* (near Kucha town, No. 8736, 1959—Lee et al.); *West.* (Yarkand
oasis, 1000 m, June 4, 1889—Rob.; "Jarkend-Kardong, ca. 1800 m, July 25; Jarkend,
1350 m, Aug. 14, 1931"—Persson, l.c.); *South.* (nor. slope of Russky mountain range,
June 6; Keriya oasis near Agai village, June 10; near Lyushi river estuary, 2600 m, June
18—1885—Przew.; Budiya settlement south-west. of Khotan, Sept. 27, 1958—Yun.).

IC. Qaidam: *plains* ("Tsaidam, leg. Thorold"—Hemsley, l.c. [1894]; Khatu gorge
on nor. slope of Burkhan-Budda mountain range, 3500 m, June 25, 1901—Lad.).

IIIA. Qinghai: *Nanshan* (on Tetung river, 2500 m, July 15, 1880—Przew.;
"Kokonor, Kia-Po-Kia, 3400 m; am Ufer des Flüsses Ara-Gol und des Sees Da-Lian-
Nor, 1930"—Hao, l.c.).

General distribution: Near East (Afghanistan), Middle Asia (along Syr-Darya
river and West. Pamir), China (Dunbei, Nor., Nor.-West., Cent.), Himalayas.

Note: A review of the Himalayan material on *P. flaccidum* Griseb. (including
Thomson's specimens—kindly furnished by Kew at our request—from which this
specimen was described) showed that not one, but two outwardly similar but actually
altogether different species have been cited from the Himalayas as well as
Tadzhikistan under the same name. One of them, for which the name *P. flaccidum*
Griseb. should be retained, has branches of spicate panicle with joints at its base, 1–3
spikelets on branches and some distinctly pinnate; inner bristles of the involucre; hence
it should be placed in the type section of the genus—sect. Pennisetum (= sect.
Setipenna Griseb.; = sect. Cenchropsis Leeke). The distribution range of this species is
evidently restricted to the hilly regions of Tadzhikistan, Afghanistan and nor. India.
Another species widely distributed in Cent. Asia—*P. centrasiaticum* Tzvel. to which the
name *P. flaccidum* var. *interruptum* Griseb. (l.c.) probably belongs—has joints close to
the middle of the panicle branch, one spikelet each on branches and all involucre,
bristles scabrous; hence it should be placed in the other section of the genus—sect.
Gymnothrix (Beauv.) Benth. and Hook. f. The two species differ even with respect to
ecology; *P. flaccidum* inhabits only rocky slopes and talus while *P. centrasiaticum* is also
found very often as a weed.

Oryza L.
Sp. pl. (1753) 333.

O. sativa L. Sp. pl. (1753) 333; Henderson and Hume, Lahore to Yarkand
(1873) 334; Franch. Pl. David. 1 (1884) 324; Forbes and Hemsley, Index Fl.
Sin. 3 (1904) 344; Danguy in Bull. Mus. nat. hist. natur. 17 (1911) 6;
Roshevitz in Fl. SSSR, 2 (1934) 47; Persson in Bot. notiser (1938) 274; Fl.
Kirgiz. 2 (1950) 39; Fl. Kazakhst. 1 (1956) 134; Fl. Tadzh. 1 (1957) 461;
Keng, Fl. ill, sin., Gram. (1959) 629; Bor, Grasses Burma, Ceyl., Ind. and

Pakist. (1960) 605.—Ic.: Fl. SSSR, 2, Plate 2, fig. 2; Fl. Tadzh. 1, Plate 57, figs., 3–4; Keng, l.c. fig. 566.

Described from India. Type in London (Linn.).

Cultivated in oases; up to midmountain belt.

IA. **Mongolia**: *Ordos* (near Linchzhou town, Oct. 6, 1884—Pot.).

IB. **Kashgar**: *Nor.* ("Kourla, 1000 m, Sept. 10, 1907"—Danguy, l.c.; near Aksu town, Aug. 10, 1929—Pop.; Uchturfan—unknown coll.); *West.* ("about Jarkand"—Henderson and Hume, l.c.; near Yangishar settlement, July 3, 1913—Knorring; "Kashgar, 1330 m, Aug. 10, 1925"—Persson, l.c.); *South.* (Khotan oasis, 1600 m, Sept. 23, 1889—Rob.).

IIA. **Junggar**: *Tien Shan* (lower Borotaly, Aug. 24, 1878—A. Reg.); *Dzhark.* (around Kul'dzha, July 18, 1875—Larionov).

General distribution: widely cultivated in many countries of the world; origin tropical and subtropical regions of Eurasia and Africa.

15. **Typhoides** Moench
Méth. pl. (1794) 201.

1. **T. arundinacea** (L.) Moench, Méth. pl. (1794) 202; Hand.-Mazz. Symb. Sin. 7, 5 (1936) 1299.—*Phalaris arundinacea* L. Sp. pl. (1753) 55; Franch. Pl. David. 1 (1884) 328; Forbes and Hemsley, Index Fl. Sin. 3 (1904) 379; Danguy in Bull. Mus. nat. hist. natur. 20 (1914) 145; Pavlov in Byull. Mosk. obshch. ispyt. prir. 38 (1929) 11; Kitag. Lin. Fl. Mansh. (1939) 85; Keng, Fl. ill. sin., Gram. (1959) 627; Bor, Grasses Burma, Ceyl., Ind. and Pakist. (1960) 615.—*Ph. japonica* Steud. Synops. pl. glum. 1 (1854) 11.—*Ph. arundinacea* var. *japonica* (Steud.) Hack. in Bull. Herb. Boiss. 7 (1899) 646.—*Digraphis arundineaea* (L.) Trin. Fund. Agrost. (1820) 127; Krylov, Fl. Zap. Sib. 2 (1928) 155; Roshevitz in Fl. SSSR, 2 (1934) 55; Fl. Kirgiz. 2 (1950) 40; Grubov, Consp. fl. MNR (1955) 61; Fl. Kazakhst. 1 (1956) 135; Fl. Tadzh. 1 (1957) 392.—Ic.: Fl. SSSR, 2, Plate 2, fig. 7 (sub *Digraphis arundinacea*); Keng, l.c. fig. 565.

Described from Europe. Type in London (Linn.).

Along banks of water reservoirs and in marshy meadows; up to midmountain belt.

IA. **Mongolia**: *Mong. Alt.* ("entre la lac Oulioun-Gour et Kobdo, 2930 m, 1896, leg. Chaffanjon"—Danguy, l.c.); *Cis-Hing.* (left bank of Khalkhin-Gol tributary near upper Numuryg, Aug. 6-7, 1949—Yun.).

IIA. **Junggar**: *Alt. region, Tien Shan, Jung. Gobi* (south.: Savan region, Shamyn'tszy village, June 17, Dauyuan'gou, July 4—1957, Huang; Chugoi settlement on left bank of Manas river, June 18, 1957—Yun); *Dzhark.* (left bank of Ili south-west of Kul'dzha town, June 29, 1877—A. Reg.).

General distribution: Aral-Casp., Balkh, region, Jung.-Tarb., Nor. and Cent. Tien Shan; Europe, Mediterr., Balk.-Asia Minor, Near East, Caucasus, Middle Asia, West. and East. Sib., Far East, Nor. Mong., China, Himalayas, Korea, Japan, Nor. Amer., Nor. and South Afr.

Note: The above-cited specimen from Cis-Hingan belongs to a distinct variety *Typhoides arundinacea* var. *japonica* (Steud.) Tzvel. Comb. nova (=*Phalaris japonica* Steud. l.c.) distributed in East. Asia (in the north up to Amur basin, Sakhalin and Kuril islands), with glumes having distinct winged keel, unlike the type.

16. Anthoxanthum L.
Sp. pl. (1753) 28.

1. **A. odoratum** L. Sp. pl. (1753) 28; Krylov, Fl. Zap. Sib. 2 (1928) 156; Pavlov in Byull. Mosk. obshch. ispyt. prir. 38 (1929) 11; Roshevitz in Fl. SSSR, 2 (1934) 56; Kitag. Lin. Fl. Mansh. (1939) 61; Fl. Kirgiz. 2 (1950) 40; Grubov, consp. fl. MNR (1955) 61; Fl. Kazakhst. 1 (1956) 136; Keng, Fl. ill., sin., Gram. (1959) 624; Bor, Grasses Burma, Ceyl., Ind. and Pakist. (1960) 431.—*A. alpinum* A. et D. Löve in Reports Depart. Agric. Univ. Reykjavik, ser. B, 3 (1948) 105.—*A. odoratum* subsp. *alpinum* (A. et D. Love) B. Jones et Meld. in Proc. Bot. Soc. Brit. 5, 4 (1964) 376; Böcher et al. Grønl. Fl. (1957) 299, nom. nud.—I.c. Fl. SSSR, 2, Plate 4, fig. 8; Fl. Kazakhst. 1, Plate 9, fig. 12; Keng, l.c. fig. 563.

Described from Europe. Type in London (Linn.).

In meadows, on rocky slopes and in pebble beds; in upper mountain belt.

IA. Mongolia: *Mong. Alt.*

IIA. Junggar: *Alt. region, Jung. Alt., Tien Shan* (Arystyn, 3000–3300 m, July 16, 1879—A. Reg.; nor. slope of Narat mountain range between Dagit and Tsanma valley crossing, Aug. 7, 1958—Yun.).

General distribution: Balkh. region (Chu-Ili hills), Jung.-Tarb., Nor. and Cent. Tien Shan; Arct. (Eur.), Europe, Mediterr., Balk.-Asia Minor, Near East, Caucasus, West. and East. Sib., Far East (probably introduced), Nor. Mong. (Hent., Mong.-Daur.), China (Nor., East.), Himalayas, Korea, Japan; introduced in many other countries of both hemispheres.

Note: Only subspecies *A. odoratum* subsp. *alpinum* (A. et D. Löve) B. Jones et Meld. (l.c.) is found within Cent. Asia. It is widely distributed in high-altitude Eurasia and the European Arctic and differs from the predominantly European subspecies *A. odoratum* subsp. *odoratum* in glabrous, smooth (or nearly so) stalks of spikelets and also invariably in glabrous and usually smooth glumes.

17. Hierochloë R. Br.
Prodr. Fl. Nov. Holl. (1810) 208, nom. conserv.

1. Plant usually laxly caespitose with creeping subsurface shoots; leaf blades 0.5–3 mm broad, often longitudinally folded. Spikelets 4.5–7 mm long; lemma of second male floret with geniculate awn arising below middle and surpassing its tip by 2-4 mm. .... **H. alpina** (Sw.) Roem. et Schult.

+ Plants with long creeping subsurface shoots; leaf blades 2–8 mm broad, flat. Spikelets 3–5.5 mm long; lemma of both male florets at emarginate tip or very shortly bilobed, not awned or with minute (up to 1 mm long) awn arising near tip .... 2.

2. Sheath of cauline leaves usually pilose. Lemma of two lower male florets marginally ciliate and glabrous and smooth elsewhere all along (or nearly so) surface; callus of lemma also glabrous..........

.... 1. **H. glabra** Trin.

+ Sheath of cauline leaves glabrous but usually scabrous due to spinules. Lemma marginally similarly ciliate, covered with spinules in upper 1/4 to 2/3 of surface, mixed with short hairs. Callus of lemma glabrous or covered with short stiff hairs ....

....**2. H. odorata** (L.) Beauv.

H. alpina (Sw.) Roem. et Schult. Syst. Veg. 2 (1817) 515; Krylov, Fl. Zap. Sib. 2 (1928) 160; Pavlov in Byull. Mosk. obshch. ispyt. prir. 38 (1929) 11; Roshevitz in Fl. SSSR, 2 (1934) 60; Grubov, Consp. fl. MNR (1955) 61; Fl. Kazakhst. 1 (1956) 137; Keng, Fl. ill. sin., Gram. (1959) 623.—*Holcus alpinus* Sw. in Willd. Sp. pl. (1806) 937.—**Ic.:** Fl. SSSR, 2, Plate 4, fig. 10; Keng, l.c. fig. 560.

Described from Scandinavia (Lapland). Type in Berlin.

In meadows, on rocky slopes and rocks; in upper mountain belt.

Found in border regions of Nor. Mongolia.

General distribution: Arct. (Circumpolar), Europe (Scandinavian peninsula, Khibiny hills, Nor. Urals), West. Sib. (Altay), East. Sib., Far East, Nor. Mong. (Fore Hubs., Hent., Hang.), China (Dunbei), Korea, Japan, (north), Nor. Amer. (in south up to northern USA).

1. **H. glabra** Trin. in Spreng. Neue Entdeck. 2 (1821) 66; Forbes and Hemsley, Index Fl. Sin. 3 (1904) 380; Pavlov in Byull. Mosk. obshch. ispyt. prir. 38 (1929) 11; Roshevitz in Fl. SSSR, 2 (1934) 62; Kitag. Lin. Fl. Mansh. (1939) 79; Grubov, Consp. fl. MNR (1955) 61; Keng, Fl. ill. sin., Gram. (1959) 621.—*H. dahurica* Trin. Mém. Ac. Sci. St.-Pétersb., sér, 6, 5, 2 (1840) 80, nom. illeg., Franch. Pl. David. 1 (1884) 329; Danguy in Bull. Mus. nat. hist. natur. 20 (1914) 145.—*H. bungeana* Trin. l.c. (1840) 82; Roshevitz, l.c. 62; Kitag. l.c. 78.—*H. odorata* f. *pubescens* Kryl. Fl. Alt. (1914) 1553; Krylov, Fl. Zap. Sib. 2(1928) 159.—*H. odorata* subsp. *dahurica* (Trin.) Printz, Veg. Sib.-Mong. Front. (1921) 119; Norlindh, Fl. mong. steppe, 1 (1949) 56.—**Ic.:** Norlindh, l.c. fig. 3; Keng, l.c. fig. 559.

Described from Transbaikal. Type in Leningrad.

In relatively arid, often solonetz meadows, solonetz, in steppes and on rubble and rocky slopes, sand and pebble beds; up to midmountain belt.

IA. Mongolia: *Khobd.* (nor.-east. slope of Bairimen-Daban pass, June 8; hills nor.-east of Ureg-Nur lake, June 10; Tszusylan river, June 29—1879—Pot.); *Mong. Alt. Cis-*

Hinggan, Cent. Khalkha ("Bordsdu Kéroulen, 25 V 1896, leg. Chaffanjon"—Danguy, l.c.); *East. Mong.*

IIA. Junggar: *Tien Shan* (nor. foot of Tien Shan, May 30, 1877—Przew.; Yuldus plateau, 2400-3000 m, June 5, 1877; Zagastai river, 3000 m, Sept. 5, 1879—A. Reg.; on Urumchi-Karashar road, 2800 m, No. 6121, July 22, 1958—Lee and Chu).

IIIA. Qinghai: *Nanshan* (near Lauvachen town, April 7; San'-'chuan', May 27—1885—Pot.).

General distribution: West. Sib. (Salair range, Chui steppe of Altay), East. Sib., Far East (south), Nor. Mong., China (Dunbei, Nor., Nor.-West.), Korea, Japan.

Note. This species is closely related to the next one and may be merged as a subspecies. The type specimen of *H. bungeana* Trin. ("prope Tan-Schan ad margines agrorum, no. 2, leg. A. Bunge") preserved in Leningrad (LE) has spikelets 3–3.5 mm long and in this respect, as also in other features, does not differ at all from type specimens of *H. glabra*.

2. H. odorata (L.) Beauv. Ess. Agrost. (1812) 62 and 164; Krylov, Fl. Zap. Sib. 2 (1928) 158, p.p.; Pavlov in Byull. Mosk. obshch. ispyt. prir. 38 (1929) 11; Roshevitz in Fl. SSSR, 2 (1934) 61; Kitag. Lin. Fl. Mansh. (1939) 79; Fl. Kirgiz. 2 (1950) 41; Fl. Kazakhst. 1 (1956) 137; Keng, Fl. ill. sin., Gram. (1959) 620.—*H. borealis* (Schrad.) Roem. et Schult. Syst. Veg. 2 (1817) 513; Forbes and Hemsley, Index Fl. Sin. 3 (1904) 380.—*Holcus odoratus* L. Sp. pl. (1753) 1048.—*H. borealis* Schrad. Fl. Germ. 1 (1806) 252, nom. illeg.—Ic.: Fl. SSSR, 2, Plate 4, fig. 11; Keng, l.c. fig. 558.

Described from Europe. Type in London (Linn.).

In meadows, on rocky slopes, in pebble beds, sand and meadow steppes; up to upper mountain belt.

IA. Mongolia: *Khobd.* (left bank of Kharkiry river, June 27; valley of Khar-Tarbagatai river-Burgassutai river tributary, July 13—1879—Pot.); *Mong. Alt.* (Indertiin-Gol valley, July 24; nor. slope of Adzhi-Bogdo mountain range, Mainigtu-Ama creek valley, Aug. 7—1947—Yun.); *Cent. Khalkha* (Ulan Bator-Tsetserleg road, along south. margin of Tsagan-Nur lake lowland, June 25, 1948—Grub.); *East. Mong.* (Nunlin'tun' village near Hailar, No. 681, June 1, 1951—Wang; near Hailar town, No. 571, June 8, 1951—Lee et al.; same site, 1960—Ivan.).

IIA. Junggar: *Alt. region, Tarbag.* (south. slope of Saur mountain range, valley of Karagaitu river, Bain-Tsagan creek valley, June 23, 1957—Yun.); *Jung. Alt., Tien Shan.*

IIIA. Qinghai: *Nanshan* (Humboldt mountain range, Kuku-Usu region, June 6, 1894—Rob.); *Amdo* (upper course of Huang He, Churmyn river estuary, 2800–3000 m, April 26, 1880—Przew.).

IIIB. Tibet: *Weitzan* (nor. slope of Burkhan-Budda mountain range, Nomokhun region, 4000–4300 m, May 20; bank of Russky lake, 4500 m, June 20—1900—Lad.).

General distribution: Aral-Casp. (nor.), Balkh. region (nor.), Jung.-Tarb., Nor. and Cent. Tien Shan; Arct., Europe (excluding south.), Asia Minor (nor.-east), Caucasus (hilly), Middle Asia (Alay), West. and East, Sib., Far East (excluding south.), Nor. Mong., China (Dunbei, Nor., Nor.-West), Himalayas, Nor. Amer. (excluding south).

Note: Typical specimens of this species, widely distributed within the forest zone of Eurasia and Nor. America, in Cent. Asia are known only from Tien Shan mountains (region of lake Sairamnur) and Junggar-Alatau and, in Nor. Mongolia, from barren peaks of Hentey. Their characteristic feature is that the lemma of two lower male florets is almost smooth only in lower 1/4-1/3 ignoring the short stiff hairs of callus which are invariably present. Nevertheless, in the rest of the Cent. Asian specimens the lemma of two lower florets of spikelet is smooth (or almost so) for 1/2-3/4 its length from base and without stiff hairs on callus. In the steppe belt from Hungary to the western foothills of Altay (including Aral-Casp. and Balkh. region), a distinct subspecies, *H. odorata* subsp. *pannonica* Chrtek et Jiras. (=*H. stepporum* P. Smirn.), is encountered.

The herbarium of V.L. Komarov Botanical Institute, Academy of Sciences, USSR, has two herbarium sheets on *Hierochloë* R. Br. with specimens from Gansu province (near Li-Dzha-Pu village, June 20, 1885—Pot.) belonging to a yet undescribed species of this genus: *H. potaninii* Tzvel. sp. nova—Planta perennis, 50–120 cm alta, laxe caespitosa, surculos subterraneos breviter repentes emittens; culmi erecti, leaves; vaginae glabrae, leaves vel scabriusculae; ligulae 4–8 mm lg., glabrae; laminae 6–14 mm lt., planae, glabrae, supra scabriusculae vel sublaeves, subtus scabrae, folii caulini supremi 8–15 cm lg. Paniculae 12–22 cm lg. et 2–6 cm lt., sat laxae, sed non effusae, ramis laevibus vel sublaevibus; spiculae 4–6 mm lg., laete virides triflorae; glumae membranaceae, glabrae et laeves, ovato-lanceolatae, apice acutiusculae vel obtusatae, inferior uninervis spicula sesqui vel duplo brevior, superior trinervis, spicula paulo vel subsequi brevior, lemmata florum masculorum lanceolata, subcoriacea, 5-nervia, extus scabra, margine ciliata, apice vulgo emarginata, inferius exaristata vel brevissime aristata, superius infra apicem breviter aristata, arista ad 4 mm lg.; lemma floris bisexualis lemmatis inferioribus simile, sed prope apicem tantum scabrum, apice obtusa, exaristata; paleae submembranaceae, carinis scabris; antherae 2–3.5 mm lg. Typus et isotypus: "China borealis, Kansu occidentali, ad pg. Li-Dsha-Pu in declivitatibus inter frutices, 20 VI 1885, G. Potanin", in Herb. Inst. Bot. Acad. Sci. URSS—LE conservatur.

This species, evidently one of the more primitive in the genus, is easily differentiated from all other Eurasian species in the very large plant size on the whole, long blades of upper cauline leaves, highly uneven glumes (of which lower significantly shorter than spikelet) and uniformly scabrous lemma of two lower florets all along surface. Among the other species of the genus described from China, *H. pallida* Hand.-Mazz. [Anz. Akad. Wiss. Wien, 57 (1920) 273] belongs to genus *Anthoxanthum* L. and should be named *Anthoxanthum pallidum* (Hand.-Mazz.) Tzvel. comb. nova; another species, *H. elongata* Hand.-Mazz., also described by this author [Symb. Sin. 7 (1936) 1299], evidently belongs to some other genus of grasses. Concomitantly, specimens cited by Handel-Mazzetti as "*H. odorata* (L.) Beauv." [Symb. Sin. 7 (1936) 1299] clearly belong to *H. potaninii*.

18. Aristida L.
Sp. pl. (1753) 82.

1. Annual. Panicles compressed, with much shortened branches and approximate spikelets; lemma without joint in upper part separating base of three awns from rest of lemmas; all awns scabrous .... 1. **A. heymannii** Regel.
+ Perennial. Panicles during anthesis and thereafter spreading, with interrupted spikelets; lemma with joint in upper part separating base of three awns from rest of lemmas; all awns plumose-pilose ....2.
2. Lower glume of spikelet 23–30 mm long, upper 17–22 mm long; lemmas (excluding awn), 6–8 mm long; their awns 18–24 mm long .... 2. **A. grandiglumis** Roshev.
+ Lower glume of spikelet 10–22 mm long, upper 8–17 mm long; lemmas 5–7 mm long; their awns 10–15 mm long .... 3. **A. pennata** Trin.

Section 1. Aristida

1. **A. heymannii** Regel in Acta Horti Petrop. 7, 2 (1881) 649.—*A. vulgaris* Trin. et Rupr. in Mém. Ac. Sci. St.-Pétersb., ser. VI, 7, 2, Sci. nat. 5 (1842) 131, p.p., nom. superfl.—*A. vulgaris* var. *mongholica* Trin. et Rupr. l.c. 133; Danguy in Bull. Mus. nat. hist. natur. 20 (1914) 145.—*A. adsensionis* auct. non L.: Forbes and Hemsley, Index Fl. Sin. 3 (1904) 381; Pavlov in Byull. Mosk. obshch. ispyt. prir. 38 (1929) 11; Roshevitz in Fl. SSSR, 2 (1934) 66; Kitag. Lin. Fl. Mansh. (1939) 61; Norlindh, Fl. mong. steppe, 1 (1949) 58; Fl. Kirgiz. 2 (1950) 41; Grubov, Consp. fl. MNR (1955) 61; Fl. Kazakhst. 1 (1956) 138; Fl. Tadzh. 1(1957) 456; Keng, Fl. ill. sin., Gram. (1959) 619; Bor, Grasses Burma, Ceyl., Ind. and Pakist. (1960) 407. —Ic.: Fl. SSSR, 2, Plate 5, fig. 1; Fl. Tadzh. 1, Plate 56, Fig. 1; Keng, l.c. fig. 557; Bor, l.c. fig. 43.

Described from East. Kazakhstan (lli river valley). Type in Leningrad. Map 1.

In arid sandy, rubble and rocky sites, pebble beds, on rocks, sometimes as weed at parking sites and in oases; up to midmountain belt.

IA. **Mongolia:** *Cent. Khalkha* (70 km south of Choiren, Aug. 21, 1926—Lis.); *East. Mong.* (in locis subarenosis Mongoliae Chinensis, 1831—Bunge); *Bas. lakes* (Chon-Kharikha river, Aug. 1, 1879—Pot.; Tuguryuk village, No. 208, Aug. 15, 1930—Bar.); *Val. lakes* (in Tuin-Gol river valley, Sept. 4; in Orok-Nur lake basin, Sept. 11—1924;—Pavl.; Ongiin-Gol river, July 21, 1926—Lis.; left bank terrace of Ongiin-Gol, Aug. 2, 1931—Ik. Gal.); *Gobi-Alt., East. Gobi, West. Gobi, Alash. Gobi.*

IB. Kashgar: *Nor., West.* (Yarkand and Kargalyk oases on Yarkand-Darya, 1300–2000 m, June 8, 1889—Rob.; 32 km west of Yarkand, Sept. 18 1959—Petr.); *East., South.* (nor. foot of Kunlun, about 1600 m, Sept. 13, 1889—Rob.).

IIA. Junggar: *Tien Shan, Jung. Gobi.*

General distribution: Aral-Casp. (Mangyshlak peninsula), Balkh. region (south-east), Jung.-Tarb., Nor. and Cent. Tien Shan; Mediterr. (east.), Asia Minor, Near East, Caucasus (south of Transcaucasus), Middle Asia (Ferghana valley and adjoining hilly regions, Pyandzh river basin), China (Dunbei, Nor.), Himalayas (west. Kashmir), Afr. (nor.-east).

Note: This species, commonly merged along with several other closely related species into a single, extremely large polytypic species *A. adsensionis* L. s.l., although not widely distributed everywhere, represents a characteristic plant of many desert and savanna-like groups of vegetation in Eurasia, Africa and America. Among these species, *A. adsensionis* L. s.s., in addition to Voznesensk island in the Atlantic Ocean from where it was described, is widely distributed in the tropical and subtropical regions of Africa as well as Nor. and South. Amer. This species differs from *A. heymannii* Regel in very short (not surpassing upper glume of spikelet) lemmas which are scabrous not only along the keel. Another, evidently more primitive species of the group, *A. coerulescens* Desf., found in the Mediterranean, represents an obligate perennial. Yet one more relatively primitive species, *A. vulpioides* Hance, found in eastern provinces of China outside Central Asia differs from *A. heymannii* in the very large overall plant size, very lax panicles and very long awns. *A. heymannii* found in Cent. Asia and occurring in more severe climatic conditions and may thus be regarded as more advanced (in the evolutionary sense) among species of this group.

Section 2. Arthratherum
(Beauv.) Reichb.

2. **A. grandiglumis** Roshev. in Bot. mater. Gerb. Bot. inst. AN SSSR, 11 (1949) 18.

Described from Kashgar. Type in Leningrad. Map 1.

In hummocky sand in sandy deserts.

IA. Mongolia: *Khesi* (nor. part of Sachzhou [An'si] oasis, 1100 m, Aug. 5, 1895—Rob.

IB. Kashgar: *Nor.* (on Pichan-Luksun road, No. 6716, June 21, 1958—Lee and Chu); *South.* (20 km south-east of Keria, Aug. 20, 1959—Chzhou); *Takla-Makan* (5 km south of Cherchen in Cherchen river valley, 1400 m, June 9, 1959—Yun.); *Lobnor* (nor. foothills of Altyntag mountain range, about 1600 m, Aug. 4, 1895—Rob., lectotype !; Kumtag sand 65 km west of Dun'khuan, Aug. 4, 1958—Yun.).

General distribution: endemic.

Note: This species, endemic in Takla-Makan desert and adjoining desert regions, is very close to *A. pennata* Trin. but concomitantly shows a close relationship with *A. karelinii* (Trin. et Rupr.) Roshev., widely distributed in Middle Asian deserts.

3. **A. pennata** Trin. in Mém. Ac. Sci. St. - Pétersb. 6 (1815) 488; Krylov, Fl. Zap. Sib. 2 (1928) 161; Roshevitz in Fl. SSSR, 2 (1934) 67; Fl. Kazakhst. 1 (1956) 138; Fl. Tadzh. 1 (1957) 458; Keng, Fl. ill. sin., Gram. (1959) 614.

—Ic.: Trin. l.c. Table 10; Fl. SSSR, 2, Plate 5, figs. 6–8; Fl. Kazakhst. 1, Plate 11, fig. 1; Fl. Tadzh. 1, Plate 56, fig. 3; Keng, l.c. fig. 553.

Described from West. Kazakhstan (between Volga and Ural). Type in Leningrad. Map 1.

In hummocky sand in sandy deserts.

IIA. Junggar: *Jung. Gobi, Zaisan, Dzhark.* (right bank of Ili river near Suidun settlement, June 1, 1877—A. Reg.).

General distribution: Aral.-Casp., Balkh. region; Near East (nor.-east), Middle Asia.

Note: Typical specimens of this species with spikelets 15–22 mm long are found in Junggar and desert regions of Kazakhstan. In Middle Asian deserts, however (mainly in Kyzylkums and Karakums), a distinct variety, *A. pennata* var. *minor* Litv. (in Spisok rast. Gerb. russk. fl. [List of Plants in the Herbarium of Russian Flora] 3 (1901) 82), with much smaller (10–15 mm long) spikelets is encountered.

19. Psammochloa Hitchc.
in J. Wash. Ac. Sci. 17 (1927) 140.

1. P. villosa (Trin.) Bor in Kew Bull. (1951) 191; Grubov, consp. fl. MNR (1955) 66.—*P. mongolica* Hitchc. in J. Wash. Ac. Sci. 17 (1927) 140; Keng, Fl. ill. sin., Gram. (1959) 584.—*Arundo villosa* Trin. Sp. Gram. Icon. et Descr. 3 (1836) Tab. 352.—*Timouria mongolica* (Hitchc.) Roshev. in J. Wash. Ac. Sci. 18 (1928) 500 and in Izv. Glavn. bot. sada SSSR, 27 (1928) 353.— *T. villosa* (Trin.) Hand.-Mazz. in Oesterr. bot. Z. 86 (1937) 302. —Ic.: Trin. l.c.; Keng, l.c., fig. 519.

Described from Inner Mongolia. Type in Leningrad. Plate 1, fig. 1; map 2.

In sand; up to midmountain belt.

IA. Mongolia: *East. Mong.* (in arena mobili prop Durm et Kobur, VII 1831—Bunge, type!; around Kalgan, May 8, 1831—Kuznetsov; right bank of Huang He above Hekou town, July 26, 1884-Pot.; "Ordos Hailiutuhwa towards Sikitan by Borobalgassun, No. 6831, July 28, 1922; Gobi, Kaongsteingol, Yendo Sume, No. 7490, June 23, 1924; East. Mongolia, No. 7510, June 27,1924, leg. Licent"—Bor, l.c.; Gatun Bologai, 1200 m, No. 443, 1925—Chaney; Ongon-Elisu sand around Baishintin Sume, Aug. 20, 1927—Zam.; same site, Sept. 14, 1931—Pob.; Moltsok-Elisu sand 3–4 km nor. of Moltsok-Khid, May 16, 1944; east. bank of Buir-Nur lake, Aug. 13, 1949—Yun.); *Bas. lakes* (near Ulangom, June 21, 1879—Pot.; from Ulyasutai to Kosh-Agach, June 15 to July 15, 1880—Pevtsov; left bank of Bogden-Gol near Gudzhirtei area, July 16, 1894—Klem.; valley of Dzapkhyn river near Dzur, Aug. 20, 1926—Smirnov); *Val. lakes* (Shar-Elisu sand near Baidarik river estuary, July 29, 1922—Pisarev; "Tsagan Nor, Outer Mongolia, 1300 m, No. 502, 1925; leg. Chaney"—Hitchcock, l.c.; 10–15 km east of Orok-Nur lake, Sept. 17, 1943—Yun.; Batsagad somon, Tavan-Elisu area, Aug. 17, 1949—Kal.); *Gobi-Alt.* (between Dzolen and Bain-Tsagan hills south of Bain-Tukhum lake, Aug. 6, 1931—Ik.-Gal.; nor. slope of Dzolen mountain range, July–Aug., 1933— Simukova); *East. Gobi* (around Muni-Ula hills, 1873—Przew.; Sain-Usu basin 20–25 km nor.-nor. east of Sain-Shanda town, Aug. 28; Khan-Bogdo somon, nor.-east. margin of Galba massif, Sept. 26, 1940; Ail'-Bain somon, 75 km south-east of Ulegei-Khid on road to Solong-Khere, July 2, 1941—Yun.); *Alash. Gobi* (on west. margin of Alashan hills, June 6, 1873; near south. margin of Alashan hills, Aug. 13, 1880—Przew.; Tengeri sand, Khoir-Khuduk area, July 11, 1909—Czet.; Khumein somon, 6 km nor.-nor.-east of Boso-Khuduk col., June 14, 1949—Yun.; 120 km nor. of Bayan-Khoto town, Tszilan'tai village, July 8, 1957; Yaburaiyan'chi village 37 km nor.-west of Bayan-Khoto, south. margin of Boran-Chilin sand, June 30, 1958; Tengeri sand in Bayan-Khoto town area, Bagetobukhu, Aug. 12, 1958—Petr.); *Ordos* (Huang He river valley east of Urgun-Nor lake, July 12, 1871—Przew.; 65 km nor. of Yuilin' town, July 14, 1957—Kabanov); *Khesi* (south-east of Suchzhou [Tszyutsyuan'] oasis, July 31, 1875—Pias.; between Yanchi and Shuan-Chzhin, June 19, 1886—Pot.).

IC. Qaidam: *hill.* (around Ergitsyul' marsh, 2900 m, Aug. 4, 1901—Lad.).

General distribution: endemic.

20. Timouria Roshev.

in Fl. Aziatsk. Rossii 12 (1916) 173.

1. **T. saposhnikovii** Roshev. in Fl. Aziatsk. Rossii, 12 (1916) 174 and in Fl. SSSR, 2 (1934) 118; Keng, Fl. ill. sin., Gram. (1959) 585.—*Achnatherum saposhnikovii* (Roshev.) Nevski in Tr. Bot. inst. AN SSSR, ser. 1, 4 (1937) 224.—Ic.: Fl. Aziatsk. Rossii, 12, Plate 12; Fl. SSSR, 2, Plate 5, fig. 20.

Described from Cent. Tien Shan (Sarydzhas river). Type in Leningrad. Plate 1, fig. 2; map 2.

On rocky slopes, rocks, in pebble beds and talus; in middle and upper mountain belts.

IA. Mongolia: *Alash. Gobi* (south. part of Alashan mountain range 50 km on Inchuan Bayan-Khoto road, 1800 m, June 10, 1958—Petr.); *Khesi* (15 km nor. of Yunchan town, slopes of Beidashan' hill, June 28, 1958—Petr.).

IB. Kashgar: *West.* (Jerzil, ca 2800 m, July 4, 1930—Persson; 10 to 12 km nor. of Baikurt settlement on Kashgar to Turugart road, June 20 and 21, 1959—Yun.; same site, 2300 m, June 2, 1959—Lee et al.).

IIA. Junggar: *Tien Shan* (Bogdoshan' hills near Turfan, 2000 m, No. 5729, June 18; near Lotogao village, 2550 m, No. 6278, Aug. 1; on road from Bartu village to timber works, 2160 m, No. 6978, Aug. 3—1958, Lee and Chu; Khanga river valley 25 km nor.-west of Balinte settlement on Karashar-Yuldus road, Aug. 1; Muzart river valley 10-12 km above its discharge into Bai basin, Chokarpa area, 2100 m, Sept. 7; same site, Sazlik area, Sept. 8—1958—Yun.).

IIIA. Qinghai: *Nanshan* (nor. foothills of Altyntag mountain range 5 km south of Aksai, 2330 m, Aug. 2, 1958—Petr.).

General distribution: Cent. Tien Shan.

21. Achnatherum Beauv.

Ess. Agrost. (1812) 19.

1. Lemma 6–7.5 mm long, with three awns at tip, of which middle one 7–10 mm long, weakly curved, lateral ones 1–1.5 mm long,

straight; spikelets 7.5–8.5 mm long; anthers glabrous at tip. Plants 50–120 cm tall; base of shoots covered with numerous coriaceous, scale-like leaves......................................1. **A. hookeri** (Stapf) Keng.

+ Lemma with solitary awn at tip, spikelets up to 7 mm long. Base of shoots without scale-like leaves ....2.

2. Lemma 3.4–5.7 mm long, with two distinct teeth at tip; fairly firm awn, usually curved, 5–14 mm long between teeth; callus acute, 0.3–0.6 mm long; tip of anthers with tufts of short hairs; spikelets 4–6 mm long..3.

+ Lemma 2–4.6 mm long, without distinct teeth at base of easily shedding awn; callus obtuse, 0.2–0.3 mm long ....4.

3. Panicles very dense and narrow, with closely packed spikelets; glumes of nearly uniform length. Ligules 0.5–1.2 mm long; leaf blades usually scabrous in lower part due to acute tubercles ......
.... 4. **A. inebrians** (Hance) Keng.

+ Panicles less dense, often broadly spreading, with spikelets less tightly packed; lower glume almost 1.5 times shorter than upper. Ligules 1–7 mm long; leaf blades glabrous below ....
.... 6. **A. splendens** (Trin.) Nevski.

4. Panicles 7–15 cm long, invariably compressed, with short branches; spikelets 6–7 mm long; lemma 3.8–4.6 mm long, with genuflexed awn 15–19 mm long, easily breaking at base; anthers glabrous at tip .... 5. **A. psilantherum** Keng.

+ Panicles 8–40 cm long, spreading widely during anthesis, with slender and long branches; spikelets 3.2–5 mm long; lemma 2–3.6 mm long, with 5.5–10 mm long awn, easily breaking at base, slightly curved; anthers with tufts of short hairs at tip ....5.

5. Ligules 0.1–0.2 mm long. Spikelets 3.2–4.2 mm long; lemma 2–2.5 mm long with 5.5–7 mm long awn ....
.... 3. **A. chinense** (Hitchc.) Tzvel.

+ Ligules 0.5–1 mm long, Spikelets 4–5 mm long; lemma 2.5–3.6 mm long with 6–10 mm long awn .... 2. **A caragana** (Trin.) Nevski.

Section 1. Trikeraia (Bor) Tzvel.[1]

1. **A. hookeri** (Stapf) Keng, Clav. Gram. Prim. Sin. (1957) 213 and Fl. ill. sin., Gram. (1959) 593.—*Stipa hookeri* Stapf in Hemsley in J. Linn. Soc. London (Bot.) 30 (1894) 120; Hemsley, Fl.Tibet (1902) 202; Pampanini, Fl. Carac. (1930) 72.—*Timouria aurita* Hitchc. in J. Wash. Ac. Sci. 23 (1933) 134.—*Trikeria hookeri* (Stapf) Bor in Kew Bull. (1955) 555 and Grasses

---

[1] *Achnatherum* sect. Trikeraia (Bor) Tzvel. comb. nova.—Genus *Trikeraia* Bor in Kew Bull. (1955) 555.

Burma, Ceyl., Ind. and Pakist. (1960) 647.—Ic.: Bor. l.c. (1960) fig. 78; Keng, l.c., fig. 530.

Described from Tibet. Type in London (K).

On rocky slopes and rocks; in upper mountain belt.

IIIB. Tibet: *South.* ("Tibet [probably in vicin. of Lhasa], ca 3900 m, No. 124, July to Sept., 1891, leg. Thorold"—Stapf, l.c. type!).

General distribution: China (South-West.: Kam), Himalayas.

Note. The most significant characteristic of section *Trikeraia* (Bor) Tzvel. to which this species belongs is, in our opinion, the singular presence of extravaginal shoots covered at the base with coriaceous scale-like leaves. The Middle Asian species *A. longearistatum* (Boiss. et Hausskn.) Nevski should also be placed in this section.

Section 2. Achnatherum

2. **A. caragana** (Trin.) Nevski in Tr. Bot. inst. AN SSSR, ser.1, 4 (1937) 224 and 337.—*Stipa caragana* Trin. in Mém. Ac. Sci. St.-Pétersb., sér. VI, 1 (1831) 74; Krylov, Fl. Zap. Sib. 2 (1928) 164; Bor, Grasses Burma, Ceyl., Ind. and Pakist. (1960) 644.—*Lasiagrostis caragana* (Trin.) Trin. et Rupr. in Mém. Ac. Sci. St.-Pétersb., sér. VI, 7, Sci. nat. 5 (1843) 90; Roshevitz in Fl. SSSR, 2 (1934) 72; Fl. Kirgiz. 2 (1950) 42; Fl. Kazakhst. 1 (1956) 140; Fl. Tadzh. 1 (1957) 412.—Ic.: Fl. SSSR, 2, Plate 6, figs. 10–12; Fl. Tadzh. 1, Plate 51, figs. 5–7.

Described from east coast of the Caspian Sea. Type in Leningrad.

On rocky and rubble slopes, in talus and pebble beds; up to midmountain belt.

IIA. Junggar: *Alt. region* (in the region of Koktogoi settlement, No. 2270, Aug. 20, 1956—Ching; 15 km south-south-east of Shara-Sume settlement along left bank of Kran river on road to Shipati, July 7, 1959—Yun.); *Jung. Alt.* (south-west. margin of Maili-Barlyk mountain range 15 km nor.-nor.-east of Kzyl-Tuz settlement on road to Karaganda-Daban pass, Aug. 14, 1957—Yun.); *Tien Shan* (Aktyube settlement near Kul'dzha, May 13, 1877—A. Reg.; Urumchinki river basin 7 km south-south-east of Urumchi town near Yanervo settlement, May 31; nor. slope of Ketmen' mountain range, Sarbushin-Gol river valley 1 km nor. of Sarbushin settlement, Aug. 21, 1957—Yun.); *Jung. Gobi, Balkh.-Alak.* (34 km east of Durbul'dzhin on road to Temirtam, Aug. 6; 30 km east of Chuguchak on road to Durbul'dzhin, Aug. 8—1957—Yun.).

General distribution: Aral-Casp., Balkh. region, Jung. Tarb., Nor. and Cent. Tien Shan; Near East (nor.), Caucasus, Middle Asia.

3. **A. chinense** (Hitchc.) Tzvel. comb. nova.—*Oryzopsis chinensis* Hitchc. in Proc. Biol. Soc. Washington, 43 (1930) 92; Keng, Fl. ill. sin., Gram. (1959) 582.—*Piptatherum parviflorum* Roshev. in Bot. mater. Gerb. Bot. inst. AN SSSR, 14 (1951) 126.—Ic.: Keng, l.c. fig. 516.

Described from China (Shaanxi province). Type in Washington.

On rocky and rubble slopes, in sand and pebble beds; up to midmountain belt.

IA. Mongolia: *Alash. Gobi* (Alashan mountain range, Kuku-Daba hills, on sand, May 8, 1908—Czet.).

General distribution: China (Nor., Nor.–West., Cent., East.).

Note: In all respects, this species is closer to *A. caragana* (Trin.) Nevski than to any other species of the genus *Piptatherum* Beauv.

4. **A. inebrians** (Hance) Keng, Clav. Gram. Prim. Sin. (1957) 213 and Fl. ill. sin., Gram. (1959) 593.—*Stipa inebrians* Hance in J. Bot. Brit. and For. 14 (1876) 212; Franch. Pl. David. 1 (1884) 330; Keng in Sunyatsenia, 6, 1 (1941) 72; Grubov, Consp. fl. MNR (1955) 63.—Ic.: Keng, l.c. (1959), fig. 529.

Described from Inner Mongolia (Alashan mountain range). Type in Leningrad. Map 2.

In upland steppes, on rocky slopes, in pebble beds and sand; up to upper mountain belt.

IA. Mongolia: *Gobi-Alt., Alash. Gobi* (Alashan: near Dyn'yuan'in [Bayan-Khoto] oasis, June 7, 1872; in Alashan hills, June 26, 1873—Przew.; in montibus Alashan, No. 19204, 1875, misit. Bretschneider—type!; Tszosto gorge, May 16, 1908—Czet.; Alashan promontory near Bayan-Khoto, July 6, 1957—Kabanov); *Khesi* (15 km nor. of Yunchan town, Beidashan' hill slopes, June 28, 1958—Petr.).

IIA. Junggar: *Tien Shan* (hills south of Urumchi, Nos. 354, 367 and 379, July 14, 1956—Ching; Turfan vicinity, on San'shan'kou-Shikhaotszy road, 2400 m, No. 5673, June 16; Yan'tsi village on Urumchi-Karashar road, 1900 m, No. 5966, July 21—1958— Lee and Chu).

IIIA. Qinghai: *Nanshan. Amdo* (along Mudzhik river, 3000–3100 m, June 4, 1880—Przew.; Dulan-Khit monastery and Tsagan-Nor lake, 3300 m, Aug. 12, 1901— Lad.).`

IIB. Tibet: *Weitzan* (vicinity of Kabchzha-Kamba village on Khichu river, 4000 m, July 21; Yangtze river basin, near Chzherku monastery, 3800 m, Aug. 8—1900—Lad.).

General distribution: China (nor.-West., Cent.).

5. **A. psilantherum** Keng, Fl. ill. sin., Gram. (1959) 595, diagn. sin. —Ic.: Keng, l.c., fig. 532 (type!). Planta perennis, 40–100 cm alta, dense caespitosa; culmi et vaginae glabrae et laeves; ligulae 0.1–0.3 mm lg., brevissime ciliolatae; laminae 1–2.5 mm lt., laxe convolutae, subtus laeves, supra scabra. Paniculae 7–15 cm lg., contractae, ramis conspique abbreviatis scabris; glumae subaequilongae, 6–7 mm lg., lanceolatae, 3-nerviae; lemmata 3.8–4.6 mm lg., breviter pilosa, apice edentata; aristae caducae, unigeniculatae, scabrae, 15–19 mm lg.; callus ca. 0.3 mm lg., acutiusculus; paleae lemmatis sesqui breviores; antherae 2–2.5 mm lg., apice glabrae.

Described from China. Type ("prov. Kansu, prope mon. Labrang, No. 5892") in Nanking.

In upland steppes, on rocky slopes and rocks; in midmountain belt.

IIIA. Qinghai: *Nanshan* (66 km west of Xining, rocky slopes of hillocks, 2800 m, Aug. 5, 1959—Petr.).

General distribution: China (Nor.-West., Cent., South-West.).

6. **A. splendens** (Trin.) Nevski in Tr. Bot. inst. AN SSSR, ser. 1, 4 (1937) 224; Ohwi in J. Jap. Bot. 17 (1941) 404; Keng, Fl. ill. sin., Gram. (1959) 589; Ikonnikov, Opred. rast. Pamira [Key to Plants of Pamir] (1963) 46.—*Stipa splendens* Trin. in Spreng. Neue Entdeck. 2 (1821) 54; Deasy, In Tibet and Chin. Turkestan (1901) 405; Keissler in Ann. Naturh. Hofmus. Wien, 22 (1907) 32; Danguy in Bull. Mus. nat. hist. natur. 20 (1914) 145; Krylov, Fl. Zap. Sib. 2 (1928) 164; Pavlov in Byull. Mosk. obshch. ispyt. prir. 38 (1929) 12; Pampanini, Fl. Carac. (1930) 72; Rehder in J. Arnold Arb. 14 (1933) 3; Persson in Bot. notiser (1938) 274; Ching in Contribs. U.S. Nat. Herb. 28 (1941) 599; Keng in Sunyatsenia, 6, 1 (1941) 70; Norlindh, Fl. mong. steppe, 1 (1949) 60; Bor, Grasses Burma, Ceyl. Ind. and Pakist. (1960) 647.—*S. altaica* Trin. in Ledeb. Fl. alt. 1 (1829) 80; Diels in Filchner, Wissensch. Ergebn. (1908) 248.—? *S. schlagintweitii* Mez in Feddes repert. 17 (1921) 208.—*S. kokonorica* Hao in Engler's Bot. Jahrb. 68 (1938) 583.— *Lasiagrostis splendens* (Trin.) Kunth, Rév. Gram. 1 (1829) 58; Franch. Pl. David. 1 (1884) 331; Kanitz in Széchenyi, Wissensch. Ergebn. 2 (1898) 736; Roshevitz in Fl. SSSR, 2 (1934) 72; Kitag. Lin. Fl. Mansh. (1939) 81; Fl. Kirgiz. 2 (1950) 42; Grubov, Consp. fl. MNR (1955) 62; Fl. Kazakhst. 1 (1956) 142; Fl. Tadzh. 1 (1957) 413.—*Aristella longiflora* Regel in Acta Horti Petrop. 7 (1880) 645.—Ic.: Fl. SSSR, 2, Plate 6, figs. 13–21; Fl. Kazakhst. 1, Plate 51, figs. 8 and 9; Keng, l.c. (1959), fig. 523.

Described from Transbaikal. Type in Leningrad.

In saline meadows, sasas and sand, along solonchak borders, floors of gorges, in river and lake valleys and on rocky slopes; forms pure chee grass thickets; up to upper mountain belt.

IA. **Mongolia:** *Mong. Alt., Cis-Hing.* (Khuna province, Sinbaerkhuchi district, No. 1075, June 30, 1951—Lee et al.); *Cent. Khalkha, East. Mong., Bas. Lakes* (Kholbo-Nur lake, July 15; Baga-nur lake, July 19; near Ulangom, Aug. 25—1879—Pot.); *Val. Lakes, Gobi-Alt., East. Gobi* (Khar-Sair, Roerich Exped., No. 464, July 22, 1935— Keng); *Alash. Gobi, Ordos* (in Huang He valley, July 19, 1871—Przew.); *Khesi.*

IB. **Kashgar:** *Nor.* (Keinsk basin nor.-west of Kucha town, Kyzyl river upper course, Sept. 4, 1958—Yun.); *West., South.* (nor. foothills of Kunlun, 1700–2000 m, June 13, 1889; nor. slope of Russky mountain range, Salganchi area, about 3300 m, June 23, 1890—Rob.; "Nordfuß des Kizil, Kurab-Su, 2950 m, June 18, 1906, Zugmayer"— Keissler, l.c.; in Cherchen district, No. 9064, June 7, 1959—Lee et al.); *East.* (near Lotogou village, No. 6277, Aug. 1, 1958—Lee and Chu).

IIA. **Junggar:** *Jung. Alt.* (Dzhair mountain range on Shikho-Chipeidzy road, July 18, 1951—Mois.); *Tien Shan, Jung. Gobi, Zaisan* (valley of Ch. Irtysh near Dyurbel'dzhin crossing, Aug. 14, 1876—Pot.); *Dzhark.* (near Kul'dzha, June 15; around Suidun town, July 16—1877—A. Reg.); *Balkh.-Alak.* (10 km south-east of Durbul'dzhin on road to Toli, Aug. 8, 1957—Yun.; Emel' river between Durbul'dzhin and Dachen, No. 2798, Aug. 10, 1957—Huang).

IIIA. **Qinghai:** *Nanshan* ("Sining-fu, Filchner"—Diels, l.c.; "Kokonor region, leg. Rock"—Rehder, l.c.; Nanshan, 1879—Przew. nor. slope of Humboldt mountain range, Chonsai-Yaodzy area, 3600–4000 m, June 23, 1895—Rob.); *Amdo* ("Kokonor, Sha-chu-yi, No. 1271, Sept. 12, 1930"—Hao, l.c.; "Shui-Chü, Jao-Chieh"—Ching, l.c.; 20 km west of Gunkho, 2980 m, Aug. 6, 1959—Petr.).

IIIB. **Tibet:** *Weitzan* (nor. slope of Burkhan-Budda mountain range, along Khatu river, 3400 m, Aug. 13, 1884—Przew.; Yangtze basin, along Ichu river, 4100 m, July 27, 1900; nor. slope of Burkhan-Budda mountain range, Khatu gorge, 2700–4000 m, July 29, 1901—Lad.).

IIIC. **Pamir** (Tagarma valley, Besh-Kurgan area, July 26, 1909—Divn.; Karasu river valley, July 22, 1913—Knorring).

General distribution: Aral-Casp., Balkh. region, Jung.-Tarb., Nor. and Cent. Tien Shan, East. Pam.; Europe (far south-east.), Near East (nor.-east.), Middle Asia, West. Sib. (south.), East. Sib. (south.) , Nor. Mong. (Hent., Mong.-Daur., Hang.), China (Nor., Nor.-West., Cent.), Himalayas.

Note: *Stipa kokonorica* Hao (l.c.) was described from very large specimens of *Achnatherum splendens* (Trin.) Nevski with up to 7.5 mm long spikelets.

22. Ptilagrostis Griseb.

in Ledeb. Fl. Ross. 4 (1852) 447.

1. Ligules 0.2–1.5 m long, densely enveloped in fairly long hairs. Glumes acute at tip, 5–6 mm long; light pink only near base. Lemma 3–4 mm long with minute teeth at tip and evenly pilose over entire outer surface; awns 18–30 mm long; panicles widely spread out .... 4. **P. pelliotii** (Danguy) Grub.

+ Ligules 0.3–2 mm long, glabrous, smooth or covered with spinules. Glumes obtuse or subobtuse at tip, usually pinkish-violet; lemma with two distinct teeth at tip .... 2.

2. Panicles 2.5–6 cm long, compressed or weakly spread out, with short, glabrous or subglabrous branches; stalk of spikelet (0.7) 1–10 (13) mm long, straight; spikelets 4.5–7 mm long; lemma 3–5 mm long, often glabrous upward. Leaf blades of vegetative shoots smooth or slightly scabrous at bottom .... 3.

+ Panicles 5–20 cm long, spreading, with long branches, stalk of spikelets (6) 10–25 (35) mm long, often sinuate. All leaf blades strongly scabrous at bottom due to short spinules .... 4.

3. Awn of lemma 7–13 mm long ....
.... 1. **P. concinna** (Hook. f. Roshev.).

+ Awn of lemma 14–20 mm long .... **P. junatovii** Grub.

4. Lemma uniformly puberulent over entire outer surface; awn 9–15 mm long, covered with 0.8–1.5 mm long hairs; spikelets 5–7.5 mm long .... 5. **P. tibetica** (Mez) Tzvel.

+ Lemma often glabrous in upper half .... 5.

5. Awn (15) 18–25 (28) mm long, weakly genuflexed; below flexure, straight or unevenly sinuate, covered with 0.8–1.5 mm long hairs; spikelets 4.5–6 mm long; anthers 2–3 mm long; branches of panicles usually covered with sparse long spinules, sometimes subglabrous .... 3. **P. mongholica** (Turcz. ex Trin.) Griseb.

+ Awn (7) 10–18 (20) mm long, distinctly genuflexed close to centre and below flexure, coiled several times and covered with 1.8–3 mm long; hairs hairs very short above flexure; spikelets 3.2–5.3 mm long; anthers, 1–2 m long .... 2. **P. dichotoma** Keng.

1. **P. concinna** (Hook. f.) Roshev. in Fl. SSSR, 2 (1934) 75; Fl. Kazakhst. 1 (1956) 157.—*Stipa concinna* Hook. f. Fl. Brit. Ind. 7 (1896) 230; Bor, Grasses Burma, Ceyl., Ind. and Pakist. (1960) 644.

Described from the Himalayas (Sikkim). Type in London (K). Isotype in Leningrad.

In grasslands and on rocky slopes; in upper mountain belt.

IIA. Junggar: *Tien Shan* (upper course of Danu-Gol river 7–8 km south of Danu crossing, 3450 m, No. 507, July 22, 1957—Huang).

IIIB. Tibet: *Chang Tang* (upper course of Tiznaf river 3–4 km east of Saryk crossing, June 4, 1959—Yun.); *Weitzan* (Yangtze river basin, Chamudug-La pass, 5200 m, July 26, 1900—Lad.).

IIIC. Pamir (interstream area of rivers Atrakyr and Tyuz-Utek, 4500–5000 m, July 20; in lower courses of Chon-Arek, left tributary of Kara-Dzhilga river, 4000–4500 m, July 22; Shor-Luk river gorge, 4000–5500 m, July 28—1942—Serp.).

General distribution: Jung.-Tarb. (?), Cent. Tien Shan; Himalayas (west., Kashmir).

Note. This species and *P. junatovii* Grub. represent ecogeographical races that are very closely related and belong to the more advanced species of the genus. The reference to *P. concinna* (Hook. f.) Roshev. for Junggar Alatau (Fl. Kazakhst. l.c.) probably pertains to *P. junatovii*.

2. **P. dichotoma** Keng, Fl. ill. sin., Gram. (1959) 598, diagn. sin.—*Stipa mongholica* auct. non Trin.: Forbes and Hemsley, Index Fl. Sin. 1 (1904) 382; Rehder in J. Arnold Arb. 14 (1933) 3; Keng in Sunyatsenia, 6, 1 (1941) 72.—Ic.: Keng, l.c. (1959) fig. 537 (type!).—Planta perennis, 15–45 cm alta, caespitosa; ligulae 1–2.5 mm lg., scabriusculae; laminae complicatae, 0.3–0.6 mm in diam., scabrae. Paniculae 4–10 cm lg., effusae, ramis sat longis laevibus vel sublaevibus saepe flexuosis; glumae 4–5.3 mm lg., apice acutiusculae, roseo-violaceo tinctae; lemmata 2.5–4.5 mm lg., inferne pilosa, superne scabra, apice bidenticulata; aristae (7) 10–18 (20) mm lg., geniculate curvatae infra geniculum spiraliter curvatae pilis 1.8–2.5 mm lg. tectae, supra geniculum pilis 0.6–1.2 tectae; antherae 1.3–2 mm lg., apice breviter pilosae.

Described from China. Type (Kansu and Tsinghai (Qinghai) border [in region opp. Labrang]. No. 478, leg. I.C. Wu) in Nanking. Isotype in Leningrad. Plate II, fig. 2; map 3.

In grasslands and on rocky slopes; in upper mountain belt.

IIIA. Qinghai: *Nanshan* (in alpine belt of Yuzhno-Tetungsk mountain range, July 25, 1872; Severo-Tetungsk mountain range, 3300 m, July 22; in cent. forest belt of Yuzhno-Tetungsk mountain range, about 2800 m, July 24—1880—Przew.).

IIIB. Tibet: *Weitzan* (in upper Huang He along Razboinik river, 4500–4800 m, July 14, 1884—Przew; Yangtze basin, along Khicha river in Nyamtso district, 4700 m, July 12, 1900—Lad.).

General distribution: China (Cent., South-West.).

Note: One of the specimens mentioned above collected by N.M. Przewalsky in the central forest belt of Yuzhno-Tetungsk mountain range, besides its very large size, differs from type specimens of *P. dichotoma* in very small (3.2–4 mm long, not 4–5.3 mm long) spikelets and highly scabrous branches of panicles due to dense spinules (not subglabrous). Evidently, it belongs to a distinct in very low mountain race, justifying its classification at least as a variety—*P. dichotoma* var. *roshevitsiana* Tzvel. var. nova.

—A varietate typica panisulae rachide et ramis valde scabris et glumis brevioribus vix tinctis differt. Typus: "China, prov. Kansu, in regione media sylvarum jugi Austro-Tetungensis ca. 2800 m, 24 VII (5 VIII) 1880, N. Przewalski" in Herb. Inst. Bot. Acad. Bot. Acad. Sci. URSS—LE conservatur.

P. **junatovii** Grub. in Bot. mater. Gerb. Bot. inst. AN SSSR, 17 (1955) 3, Grubov, Consp. fl. MNR (1955) 61; Malyshev, Vysokogorn. fl. Vost. Sayana [Alpine Flora of Eastern Sayan] (1965) 56.—*Stipa mongholica* f. *minor* Kryl. Fl. Zap. Sib. 2 (1928) 167.

Described from Mongolia (Hang.). Type in Leningrad.

In grasslands and on rocky slopes; in upper (barren) mountain belt.

Found in border region of Nor. Mongolia (Hang.); its occurrence in hills of Mong. Altay is possible.

General distribution: West. Sib. (Altay: Chui river basin), East. Sib. (East. Sayan), Nor. Mong. (Hang.).

3. P. **mongholica** (Turcz. ex Trin.) Griseb. in Ledeb. Fl. Ross. 4 (1852) 447; Roshevitz in Fl. SSSR, 2 (1934) 75; Kitag. Lin. Fl. Mansh. (1939) 89; Fl. Kirgiz. 2 (1950) 44; Grubov, Consp. fl. MNR (1955) 61; Fl. Kazakhst. 1 (1956) 157; Fl. Tadzh. 1 (1957) 432; Keng, Fl. ill. sin., Gram. (1959) 598; Ikonnikov, Opred. rast. Pamira [Key to Plants of Pamir] (1963) 47.—*Stipa mongholica* Turcz. ex Trin. in Bull. Ac. Sci. St.-Pétersb. 1 (1836) 67; Krylov, Fl. Zap. Sib. (1928) 167, excl. var.; Pavlov in Byull. Mosk. obshch. ispyt. prir. 38 (1929) 12.—*S. czekanowskyi* V. Petrov, Fl. Yakut. 1 (1930) 136.—*S. alpina* (Fr. Schmidt) V. Petrov, l.c. 136.—*Lasiagrostis mongholica* (Turcz. ex Trin.) Trin. et. Rupr. in Mém. Ac. Sci. St.-Pétersb., sér. VI, 7, Sci. nat. 5 (1843) 87.—*L. alpina* Fr. Schmidt in Mém. Ac. Sci. St.-Pétersb., sér. VII, 12, 2 (1868) 73.—*Achnatherum mongholicum* (Turcz. ex Trin.) Ohwi in J. Jap. Bot. 17 (1941) 403.—Ic.: Fl. Kazakhst. 1, Plate 10, fig. 5; Fl. Tadzh. 1, Plate 51, figs. 14 and 15; Keng, l.c. fig. 536.

Described from Baikal region (Oki river basin). Type in Leningrad. Map 3.

In grasslands on rocky slopes, in riverine pebble beds and near springs and brooks; in middle and upper mountain belts.

IA. **Mongolia:** *Khobd.* (Kharkiry river valley, July 9; south. head of Kharkiry, July 10—1879—Pot.); *Mong. Alt.* (Khan-Taishiri mountain range).

IIA. Junggar: *Alt. region* (Sarbyuira river of Kran river system, Sept. 12, 1876—Pot.); *Jung. Alt.* (20 km south of Ven'tsyuan', 2810 m, No. 1491, Aug. 14, 1957—Huang); *Tien Shan.*

'IIIC. Pamir (Tagdumbash-Pamir, Pistan gorge in Sarykol mountain range, 4600 m, July 15, 1901—Alekseenko).

General distribution: Jung.-Tarb., Nor. and Cent. Tien Shan, East. Pamir; Middle Asia (Alay), West. Sib. (Altay), East. Sib., Far East (south.), Nor. Mong. (Hubs. region, Hent., Hang.), China (Dunbei).

Note: *P. tibetica* Mez and *P. dichotoma* Keng representing very close ecogeographical races are closely related to this species (*P. mongholica*). The solitary American species of the genus, *P. porteri* (Rydb.) Tzvel. comb. nova [=*Stipa porteri* Rydb. in Bull. Torrey Bot. Club. 32 (1905) 599] is an even less morphologically isolated race of this same genetic group. This species, found only in the hills of Colorado state, is very similar in many respects to *P. mongholica* var. *minutiflora* Titov ex Roshev. (=*Lasiagrostis alpina* Fr. Schmidt, l.c.), a variety distributed in East. Siberia, with relatively small spikelets and short awns.

4. P. pelliotii (Danguy) Grub. Consp. fl. MNR (1955) 62; Hanelt u. Davazamc in Feddes repert. 70, 1–3 (1965) 12.—*Stipa pelliotii* Danguy in Lecomte, Not. Syst. 2 (1912) 167.—*S. mongholica* auct. non Turcz. ex Trin.: Norlindh, Fl. mong. steppe, 1 (1949) 63.

Described from Nor. kashgar. Type in Paris. Plate II, Fig. 1; map 3.

On dry rocky and rubble slopes, rocks; up to midmountain belt.

IA. Mongolia: *Gobi-Alt., East. Gobi, West. Gobi* (Noin somon, hillocky area on road to Chonoin-Bom area, Aug. 18; Bain-Undur somon, 1 km nor. of Burkhantu-Bulak spring on road to pass, Aug. 23—1948—Grub.); *Alash. Gobi* (steppe at southern foot of Tostu mountain range, Aug. 4, 1886—Pot.; "Wen-Tsun-Hai-Tze, prov. Ning-Hsia, June 19, 1922, leg. Söderbom"—Norlindh, l.c.; Obotu-Khar hillocky area south of Tostu-Nur mountain range, July 28, 1943—Yun.; 35 km south-west of In'chuan' town, June 10; south. margin of Yaburai hill 13 km west-nor.-west of Yaburaiyan'chi settlement, June 10; 70 km south-east of Divusum village, nor. rim of Bain-Nor hill, June 11; Bayan-Khoto region 8 km south-east of Bain-Ul hill, June 13; 60 km south of Bayan-Khoto, June 14; 90 km south-west of Inchuan' town, Chintunai gorge, June 18—1958—Petr.); *Khesi.*

IB. Kashgar: *Nor., East.* (near Chukur settlement, Aug. 28, 1929—Pop.; Kuruktag mountain range 35 km south-west of Toksun, Sept. 16, 1957—Yun.; Turfan region, No 5623, June 15, 1958— Lee and Chu; Kiziltag mountain range 25 km south-west of Kyumysh, Sept. 9, 1959—Petr.).

IIA. Junggar: *Jung. Alt.* (Dzhair mountain range 25 km west of Aktam settlement along road to Chuguchak, Aug. 3, 1957—Yun.); *Tien Shan* (near Nan'shan'kou village, June 7, 1877—Pot.; between Barkul' lake and Khami town, May 12, 1879—Przew.; Urumchinki river basin 18–20 km south of Urumchi along road to Davanchin-Daba, June 2, 1957—Yun.); *Jung. Gobi* (west.: 25 km nor.-nor.-east of Orkhu settlement along road from Karamai to Kosh-Tologoi, June 22, 1957—Yun.; between Tien Shan-Laoba station and Myaoergou, No. 2354, Aug. 3, 1957—Huang; east.: Adzhi–Bogdo mountain range, south-west. slope, Toli-Bulak area, June 16, 1877—Pot.; nor. slope of Baga-Khabtak-Nuru mountain range, 1500–1800 m, Sept. 14; Bulugun river meander, south. slope of Barangiin-Khara-Nuru mountain range, Sept. 20—1948, Grub.; between Tsitai and Beidashan [Baitak-Bogdo], 1000 m, No. 5191, Sept. 27; near

Nom, No. 2247, Sept. 29; 20 km south of Beidashan, No. 2443, Sept.—30—1957—Huang; 112–115 km south of Ertai along road to Guchen July, 16, 1959—Yun.).

IIIA. Qinghai: *Nanshan* (high foothills of Altyntag mountain range 3 km south of Aksai settlement, 2330 m, Aug. 2, 1958—Petr.).

General distribution: endemic.

Note: This endemic Central Asian species is not only morphologically the most isolated, but also the most primitive among the species of the genus *Ptilagrostis*.

5. **P. tibetica** (Mez) Tzvel. comb. nova.—*Stipa tibetica* Mez in Feddes repert. 17 (1921) 207.—*S. monholica* auct. non Turcz. ex Trin.: Hemsley, Fl. Tibet (1902) 202; Pampanini, Fl. Carac. (1930) 71; Bor, Grasses Burma, Ceyl., Ind. and Pakist. (1960) 645.

Described from West. Himalayas. Type in Berlin. Isotype in Leningrad. Map 3.

In grasslands and on rocky slopes; in upper mountain belt.

IIIB. Tibet: *South.* ("Guge, 5000 m, 1848, leg. Strachey and Witterbottom"—Hemsley, l.c.).

General distribution: Himalayas.

23. Stipa L.
Sp. pl. (1753) 78.

1. Awn 1–2.7 cm long, scabrous or pilose in lower part and invariably scabrous in upper; glumes up to 1.5 cm long and acute or attenuate-acuminate at tip .... 2.

+   Awn more than 3.5 cm long, pilose or scabrous throughout length or only in upper part; glumes attenuate usually into long membranous cusp at tip .... 8.

2. Awn easily breaking at their base along joint, 1.8–2.6 cm long, genuflexed twice, pilose in lower part (up to second flexure) and gradually narrowing towards upper part from 2–3.5 to 1 mm. Leaf ligules of vegetative shoots (1.5) 2–4 (6) mm long; leaf blades 0.3–0.7 mm in diameter, scabrous below .... 3.

+   Awn not breaking along joint, genuflexed once or twice, scabrous or with up to 1.2 mm long hairs in lower part .... 4.

3. Plant 30–50 cm tall. Panicles 10–20 cm long; spreading at anthesis, with long, often undulating branches; spikelets 7.5–10 mm long; lemma 5.5–8 mm long ....
    ....22. **S. penicillata** Hand.-Mazz.

+   Plant 10–30 cm tall. Panicles 3–12 cm long, invariably compressed, with relatively few spikelets on short branches; spikelets 6–8 mm long; lemma 4–6 mm long ....
    .... 33. **S. subsessiliflora** (Rupr.) Roshev.

4. Awn 1.5–2.5 cm long, genuflexed twice, pilose in lower part (up to first flexure) and gradually narrowing towards top from 1.2 to

0.6 mm; scabrous above first flexure; spikelets 9–11 mm long; lemma 6.5–8.5 mm long, pilose in lower half with longitudinal rows of hairs above. Leaf·blades 0.3–0.7 mm in diameter, glabrous below; ligules 0.2–0.5 mm long .... 1. **S. aliena** Keng.

+ Awn genuflexed once or twice, scabrous in lower part or with up to 0.7 mm long hairs; lemma uniformly covered with hairs .... 5.

5. Panicles 10–30 cm long, with fairly long, scabrous branches and with numerous 5–10 mm long hairs; awn usually scabrous in lower part, genuflexed once or twice. Plants 30–100 cm tall; leaf blades longitudinally folded or flat, 1–4 mm broad, glabrous below; ligules 0.2–1 mm long .... 6.

+ Panicles 3–10 cm long, invariably compressed (subspicate), with few (usually up to 15) 11–15 mm long spikelets on short branches, usually scabrous only in upper part. Lower part of awn highly puberulent. Plants 15–40 cm tall; leaf blades longitudinally folded, 0.3–0.7 mm in diameter, scabrous below (outer surface); ligules 2.5 mm long .... 7.

6. Panicles usually very lax; spikelets 5–6.5 mm long; lemma 4.5–6 mm long, with 10–14 mm long awn; anthers glabrous ....
.... 20. **S. nakaii** Honda.

+ Panicles usually fairly dense; spikelets 7–10 mm long; lemma 6–7.5 mm long with 14–21 mm long awn; anthers with tufts of hairs at tip .... 32. **S. sibirica** (L.) Lam.

7. Lemma 6–7.5 mm long with 10–13 mm long awn, genuflexed once
....**25. S. purpurascens** Hitchc.

+ Lemma 6.5–8.5 mm long with 15–20 mm long awn, genuflexed twice .... 27. **S. regeliana** Hack.

8(1). Awn scabrous throughout length due to spinules, less often uniformly covered with very short (up to 0.8 mm long) stiff hairs
.... 8.

+ Awn scabrous throughout length or only in upper part, with soft more than 1 mm long **hairs** .... 17.

9. Awn covered with very short stiff hairs (0.3–0.8 mm long) throughout length, genuflexed once or twice. Leaf blades 0.3–0.8 mm in diameter, smooth below; leaf ligules of vegetative shoots 0.1–0.4 mm long .... 10.

+ Awn usually scabrous throughout length due to spinules, genuflexed twice .... 11.

10. Awn 8–11 cm long, genuflexed once; lemma 8.5–10 mm long; glumes 22–28 mm long .... 8. **S. consanguinea** Trin. et Rupr.

+ Awn 3.6–6 cm long, genuflexed twice; lemma 6–7.5 mm long; glumes 9–15 mm long .... 28. **S. richteriana** Kar. et Kir.

11. Awn 4–6.5 cm long; glumes 9–15 mm long. Leaf blades 0.6–1.2 mm in diameter .... 12.

+ Awn more than 10 cm long; glumes 17–40 mm long .... 13.

12. Lemma 4.6–5.5 mm long; leaf blades glabrous in lower outer part; leaf ligules of vegetative shoots 0.2–0.5 mm long .... 4. **S. bungeana** Trin.

+ Lemma 8.5–10 mm long; leaf blades somewhat scabrous in lower part; leaf ligules of vegetative shoots 0.4–1 mm long .... 24. **S. przewalskyi** Roshev.

13. Lemma 15–17 mm long with crown of hairs at base of awn; awn 18–27 cm long; glumes 3–4 cm long. Leaf blades glabrous along outer surface .... 11. **S. grandis** P. Smirn.

+ Lemma 9–13.5 mm long; awn 10–18 cm long; glumes 1.8–3.3 cm long .... 14.

14. Lemma 10.5–12 mm long; awn without hairs or spinules at base. Leaf blades 0.5–1.3 mm in diameter, outer surface slightly scabrous due to very short spinules or acute tubercles; inner surface (from top) with very short hairs intermixed with very long ones; leaf ligules of vegetative shoots 0.6–2.5 mm long .... 5. **S. capillata** L.

+ Lemma with crown of short hairs and spinules at base of awn, very rarely (in some specimens of *S. sareptana* Beck.) without crown of hairs. Leaf blades 0.3–0.7 mm in diameter; leaf ligules of vegetative shoots 0.1–0.5 mm long .... 15.

15. Blades of all or most leaves of vegetative shoots scabrous along outer surface due to scattered setaceous spinules; inner surface (from top) with very short hairs. Lemma 9–11 mm long .... 31. **S. sareptana** Beck.

+ Leaf blades with glabrous outer surface (from below) .... 16.

16. Lemma 12–14 mm long; awn 13–18 cm long, from lower part (up to first flexure) 3–4.5 cm long. Leaf blades on inner surface (from top) with very short hairs intermixed with very long ones .... 2. **S. baicalensis** Roshev.

+ Lemma 9–11.5 mm long; awn 10–15 cm long, from lower part (up to first flexure) 2–3 cm long. Leaf blades on inner surface (from top) with very short hairs, without very long ones .... 16. **S. krylovii** Roshev.

17(8). Awn may be glabrous in lower part, smooth but usually scabrous due to short spinules .... 18.

+ Awn pilose throughout length .... 26.

18. Awn genuflexed once, 6–13 cm long. Leaf ligules of vegetative shoots 0.2–0.6 mm long, densely covered with hairs much longer than ligules .... 19.

+   Awn genulexed twice, 10–35 cm long ....21.

19.   Leaf blades glabrous along outer surface (from below). Glumes 2.5–3.5 cm long; lemma 9–11 mm long, without crown of hairs at base of awn; awn 9.5–13 cm long, with up to 6.5 mm long hairs .... 15. **S. klemenzii** Roshev.

+   Leaf blades usually scabrous along outer surface due to scattered spinules. Glumes 1.8–2.7 cm long; lemma 7–9.5 mm long; awn 6–9.5 cm long ....20.

20.   Lemma 7–9 mm long, without crown of hairs at base, less often with stray hairs; awn 6–8 cm long, with up to 3 mm, less often 5 mm long hairs .... 10. **S. gobica** Roshev.

+   Lemma 7.5–9.5 mm long, with crown of hairs at base of awn; awn 7–9.5 cm long, with up to 6 mm long hairs .... 34. **S. tianschanica** Roshev.

21.   Awn 10–20 cm long, with 2.5–3 mm long hairs in upper part; lemma 8–14 mm long. Leaf blades 0.3–0.7 mm in diameter, scabrous in lower part ....22.

+   Awn 15–40 cm long, with 4–6 mm long hairs in upper part. Leaf blades glabrous or usually scabrous in lower part ....23.

22.   Awn scabrous in lower part. Leaf ligules of vegetative shoots 1–5 mm long .... 13. **S. hohenackeriana** Trin. et Rupr.

+   Awn glabrous in lower part. Leaf ligules of vegetative shoots up to 0.2 mm long .... 17. **S. lessingiana** Trin. et Rupr.

23.   Lemma 12–16 mm long with three dorsal bands of long hairs separated from each other almost right from base, middle one longer than half of lemma; awn 15–26 cm long. Leaf blades 0.4–1 mm in diameter, outer surface strongly scabrous due to dense tuberculate spinules ....24.

+   Lemma 16–19 mm long with three dorsal bands of hairs fused below to considerable extent, middle one shorter than half of lemma; awn 25–40 cm long. Leaf ligules of vegetative shoots 0.7–3 mm long ....25.

24.   Leaf ligules of vegetative shoots (0.3) 0.5–1(1.2) mm long. Lemma 14–16 mm long; awn 20–26 cm long... .... 14. **S. kirghisorum** P. Smirn.

+   Leaf ligules of vegetative shoots (1.5) 2–5 (6) mm long. Lemma 12–15 mm long; awn 15–20 cm long ....18. **S. macroglossa** P. Smirn.

25.   Leaf blades 0.7–1.5 mm in diameter, outer surface glabrous (from lower part) or slightly scabrous due to scattered tuberculate spinules. Lemma with longitudinal marginal band of hairs, short of reaching base of awn by 4 to 6 mm ....23. **S. pennata** L.

+ Leaf blades 0.5–1 mm in diameter, outer surface strongly scabrous due to tuberculate spinules intermixed with long spinules. Lemma with longitudinal marginal band of hairs reaching base of awn or short of reaching by not more than 1.5 mm .... 30. **S. rubens** P. Smirn.

26(17). Awn genuflexed once, with hairs in upper arcuate part, shortening in direction of tip of awn from 3–6 to 1.5–3 mm, and hairs 1–2.5 mm long in flexured part. Leaf ligules of vegetative shoots up to 0.5 mm long, densely covered with hairs .... 27.

+ Awn genuflexed twice .... 29.

27. Panicles quite lax with interrupted spikelets; glumes 1.4–1.9 cm long; lemma 6–7.5 mm long, uniformly pilose almost throughout surface; awn 4.5–6.5 cm long, with lower twisted part 0.4–0.7 cm long, 8–10 times shorter than upper part .... 19. **S. mongolorum** Tzvel.

+ Panicles very dense, with close-packed spikelets; glumes 1.8–3 cm long; lemma 7.5–12 mm long, with distinct longitudinal bands of hairs in upper half; lower twisted part of awn 4–6 times shorter than upper part .... 28.

28. Leaf blades glabrous along outer surface. Lemma 9–12 mm long; awn 8–14 cm long, with up to 7 mm long hairs .... 7. **S. caucasica** Schmalh.

+ Leaf blades usually scabrous along outer surface due to dispersed spinules, less often glabrous. Lemma 7.5–9 mm long; awn 4.5–7 cm long, with up to 5 mm long hairs .... 9. **S. glareosa** P. Smirn.

29(26). Awn 10–20 cm long; glumes 1.8–3 cm long .... 30.

+ Awn 4.5–9 cm long .... 31.

30. Hairs in upper part of awn 2–3 mm long; lemma 10–12 mm long. Leaf ligules of vegetative shoots 1–5 mm long; leaf blades usually scabrous along outer surface .... 6. **S. caspia** C. Koch.

+ Hairs in upper part of awn 3.5–4.5 mm long; lemma 8.5–10.5 mm long. Leaf ligules of vegetative shoots 0.4–1 mm long; leaf blades highly scabrous along outer surface due to long spinules .... 12. **S. himalaica** Roshev.

31. Hairs of awn not longer than 1.6 mm, glumes 8–16 mm long; panicles emerging from sheath of upper cauline leaf with interrupted spikelets. Leaf blades 0.4–1 mm in diameter, outer surface glabrous or slightly scabrous .... 32.

+ Hairs in upper part of awn 2.3–3.5 mm long; glumes 12–25 mm long. Leaf blades 0.3–0.7 mm in diameter .... 33.

32. Awn in upper part with 1–1.5 mm long hairs, in lower part with 0.5–0.8 mm long hairs; lemma 4.5–7 mm long; glumes 8–12 mm

long. Leaf ligules of vegetative shoots 0.2–0.4 mm long ....
.... 3. S. breviflora Griseb.

+ Lower and middle part of awn with 1–1.6 mm long hairs, in
upper part with hairs gradually shortening from 0.7 to 0.3 mm
towards tip of awn; lemma 6.5–8 mm long; glumes 12–16 mm
long. Leaf ligules of vegetative shoots 0.5–1.5 mm long ....
.... 29. S. roborovskyi Roshev.

33. Panicles compressed and dense, usually emerging slightly from
sheath of upper cauline leaf; lemma 6.2–8 mm long; awn 4.5–7 cm
long, with hairs 0.5–0.8 mm long in lower part and 2.5–4 mm in
upper. Leaf ligules of vegetative shoots 1–3 mm long; leaf blades
strongly scabrous along outer surface .... 21. S. orientalis Trin.

+ Panicles mostly spreading, with long, usually undulating
branches; lemma 8–13 mm long; awn 6–9 cm long, with hairs 1–
1.5 mm long in lower part and 2–3 mm in upper. Leaf ligules of
vegetative shoots 0.5–1.3 mm long; leaf blades somewhat scabrous
along outer surface ....26. S. purpurea Griseb.

1. **S. aliena** Keng in Sunyatsenia, 6 (1941) 74; ej. Fl. ill. sin., Gram.
(1959) 604—Ic.: Keng, l.c. (1959) fig. 543.

Described from China (Gansu province, Syakhe region). Type in
Nanking.

On rocky slopes and rocks; in middle and upper mountain belts.

IIIA. Qinghai: *Nanshan* (pass between Uvei and Lan'chzhou town, 2820 m, Oct. 9,
1957—Yun.; 108 km west of Xining and 6 km west of Daudankhe village, 3400 m,
Aug. 5, 1959—Petr.).
General distribution: China (Nor.-West., Cent., South-West.).

2. **S. baicalensis** Roshev. in Izv. Glavn. bot. sada SSSR, 28 (1929) 380
(excl. var. *macrocarpa* Roshev.); Roshev. in Fl. SSSR, 2 (1934) 110; Kitag.
Lin. Fl. Mansh. (1939) 94; Grubov, Consp. fl. MNR (1955) 62; Keng, Fl. ill.
sin., Gram. (1959) 610.—*S. attenuata* P. Smirn. in Byull. Mosk. obshch.
ispyt. prir. 45 (1936) 110, diagn. ross. in clavi; Krylov, Fl. Zap. Sib. 12, 1
(1961) 3090.—Ic.: Fl. SSSR, 2, Plate 7, figs, 40–50; Keng, l.c. fig. 551.

Described from Transbaikal. Lectotype (Barguzinsk road, Tataurovo
station,"Belyi Kamen", rocks near ferry station across Senenga river, No.
2441, Aug. 6, 1913, G. Poplavskaya et al.) in Leningrad.

In steppes, on rocky slopes and rocks, in sand; up to lower mountain
belt.

IA. Mongolia: *Cis-Hing.*, *Cent. Khalkha* (midcourse of Kerulen, 1899—Pal.;
Khubagin Khundei region 20–25 km west-south-west of Bain-Ula somon, Aug. 28,
1949—Yun.); *East. Mong.* (common in nor. part; up to Darigangi in south); *East. Gobi*
(Buterin-Obo area 36 km south-east of Khara-Airik somon on road to Sain-Shandu,
Aug. 27, 1940—Yun.).

General distribution: West. Sib. (Altay), East. Sib. (south), Far East (south), Nor. Mong. (Hent., Hang., Mong.-Daur.), China (Dunbei).

Note: L.P. Sergievskaya (Krylov, l.c. 3091), referring to P.A. Smirnov, opined that the name *S. baicalensis* Roshev. cannot be adopted since the author of this species included in it (under the name *S. baicalensis* var. *macrocarpa* Roshev., l.c. 380) the specimens of *S. grandis* P. Smirn. This view cannot be accepted as almost all the specimens cited by R. Roshevitz in the first description belong to *S. baicalensis* and not to *S. grandis.*

oS. borysthenica Klok. ex Prokud. in Fl. Tauriae, 1, 4 (1951) 25—*S. pennata* L. var. *sabulosa* (Pacz.) Tzvel.; Gubanov, Consp. Fl. Outer Mong. (1996) 24.

IA. Mongolia. *East Mong.*

3. S. breviflora Griseb. in Nachr. Gesellsch. Wissensch. u. Univ. Goett. 3 (1868) 82; Roshev. in Fl. SSSR, 2 (1934) 91; Persson in Bot. notiser (1938) 274; Ching in Contribs, U.S. Nat. Herb. 28 (1941) 598; Norlindh, Fl. mong. steppe, 1 (1949) 63; Fl. Kirgiz. 2 (1950) 57; Grubov, Consp. fl. MNR (1955) 62; Keng, Fl. ill. sin., Gram. (1959) 606; Bor, Grasses Burma, Ceyl., Ind. and Pakist. (1960) 643.—*S. aliciae* Kanitz in Széchenyi, Wissensch. Ergebn. 2 (1898) 736; Forbes and Hemsley, Index Fl. Sin. 3(1904) 383.—Ic.: Kanitz, l.c. Plate 7, fig. 2 (sub *S. aliciae*); Keng, l.c. fig. 545.

Described from the Himalayas. Type in Berlin. Isotype in Leningrad. Map 5.

On rocky, rubble and melkozem slopes of hills, and on rocks; up to upper mountain belt.

IA. Mongolia: *Mong. Alt.* (pass through east. extremity of Mong. Altay on road from Bain-Tsagan somon to Tszakha, Aug. 20, 1943—Yun.); *East. Mong.* (Ourato, May 1866—David; Khutokhe river valley near Khukh-Khoto town, June 15, 1884—Pot.; "40 km ad occid. versus ab Khadain-Sume, June 19, 1931, leg. Eriksson"—Norlindh, l.c.); *Gobi-Alt., East. Gobi* (Shabarakh Usu, Outer Mongolia, ca. 1200 m, 1925—Chaney; "Khujirtu-gol, June 2 and July 4, 1927, leg. Hummel; Ikhen-gungo, June 4–6, 1931, leg. Mühlenweg"—Norlindh, l.c.; south. slope of east. extremity of Del'ger-Hangay, Aug. 1, 1931—Ik.-Gal.; nor. foothill of Datsin'shan' mountain range 120 km nor.-west of Uchuan' town; 30 km nor. of Baotou town, June 4, 1958—Petr.; near Bailinmyao, 1959—Ivan.); *Alash. Gobi* (90 km south-west of Chzhunvei, June 30, 1957; 50 km from In'chuan' on road to Bayan-Khoto, 1800 m, June 10; 35 km south-west of In'chuan' town, June 10; Bain-Nor hills 70 km south-east of Bayan-Khoto town, June 11; Bain-Ula hills 10 km south-east of Bayan-Khoto town, June 13—1958—Petr.); *Khesi.*

IB. Kashgar: *Nor.* (nor. slope of Karateke mountain range, 2000–2300 m, June 6, 1889—Rob.; Muzart river valley 4 km below Yangimallya settlement, 2150 m, Sept. 11; right flank of Taushkan-Darya river valley 30 km south-west of Uchturfan, 1700 m, Sept. 17, 1958—Yun.); *West. South* (nor. slope of Russky mountain range, Linchu river estuary, 2500–2800 m, May 28, 1885—Przew.; upper course of Chira-Darya river, 15–17 km south of Ukhu village, Sary-Bulun region, May 19, 1959—Yun.; Cherchen district, 4 km north of Achan village, 2720 m, No. 9393, June 2; same site, 3 km west of Karamulak village, No. 9471, June 7—1959—Lee et al.); *East.* (nor.-east of Turfan on road to Shilaotszy village, 2400 m, No. 5666, June 16; Bogdoshan' foothill near nor. rim of Turfan basin, 1700 m, No. 5747, June 19—1958—Lee and Chu).

IIA. Junggar: *Tarbag.* (Khobuk river valley on road from Karamai to Shara-Sume, 1300m, No. 10565, June 24, 1959—Lee et al.; nor.-east. trails of Saur mountain range, near Kheisangou village on road from Burchum to Karamai, July 10, 1959—Yun.); *Jung. Alt.* (Dzhair mountain range 43 km from Aktam settlement on road to Chuguchak, Aug. 3, 1957—Yun.; nor. margin of Barlyk mountain range on road from Toli to Durbul'dzhin, No. 983, Aug. 5, 1957—Huang); *Tien Shan* (70 km from Dashitou village along road from Muleikhe to Khami town, Oct. 2, 1957—Huang; Ulyasutai-Chulak river valley 8 km above Balinte settlement along road from Karashar to B. Yuldus, Aug. 1, 1958—Yun.; on Urumchi-Karashar road, 1260 m, No. 5955, 1958—Lee and Chu); *Jung. Gobi* (west.: between Duke and Meoergou, 1000 m, No. 366, Aug. 3, 1957—Huang).

IIIA. Qinghai: *Nanshan, Amdo* (along Churmyn river, 3200 m, May 3, 1880—Przew.; 20 km west of Gunkhe, 2980 m, Aug. 6, 1959—Petr.).

IIIB. Tibet: *Chang Tang* (nor. slope of Russky mountain range, 3300–4100 m, Karasai region in Aksu river upper courses, July 3, 1890—Rob., valley of Tiznaf river 25 km from Akkëz pass toward Kyude, May 31; 25–27 km above Kyude settlement in Tiznaf river upper courses, June 1—1959—Yun.); *Weitzan* (nor. slope of Burkhan-Budda mountain range, Khatu gorge, 3500 m, July 8, 1901—Lad.).

IIIC. Pamir (near Pas-Rabat settlement, July 3, 1909—Divn.).

General distribution: Cent. Tien Shan; China (Nor., Nor.-West., South-West.), Himalayas.

4. S. bungeana Trin. in Bunge, Enum. pl. China bor. (1835) 144; Franch. Pl. David. 1 (1884) 330; Forbes and Hemsley, Index Fl. Sin. 3 (1904) 381; Roshev. in Fl. SSSR, 2 (1934) 106; Fl. Kirgiz. 2 (1950) 63; Keng, Fl. ill. sin. Gram. (1959) 606; Bor, Grasses Burma, Ceyl., Ind. and Pakist. (1960) 643. —Ic.: Keng, l.c. fig. 546.

Described from Nor. China. Type in Leningrad.

On rocky and melkozem slopes, in riverine pebble beds; up to midmountain belt.

IA. Mongolia: *East. Mong.* (Ourato, May 1866—David; Baga-Edzhin-Khoro region, Aug. 6, 1884—Pot.); *East. Gobi* (Khara-Narin-Ula hills nor.-west of Huang He river, May 9, 1872—Przew.); *Alash-Gobi* (Alashan mountain range, Tszosto gorge, May 16, 1908—Czet.; 35 km south-west of In'chuan' town, June 10; 25 km south of Bayan-Khoto town, June 10; Divusumu village 70 km south-east of Bayan-Khoto, June 11—1958—Petr.); *Khesi* (100 km nor.-west of Lan'chzhou on road to Uvei, Oct. 9, 1957—Yun.; 23 km east of Yunchan town, June 28, 1958—Petr.).

IB. Kashgar: *Nor.* (west. part of Bai basin, Muzart river along road from Alty-Gumbez to Kurgan, 1800 m, Sept. 6, 1958—Yun.).

IIIA. Qinghai: *Nanshan* (in forest zone of Yuzhno-Tetungsk mountain range, July 13, 1880—Przew.; Loukhu-Shan' mountain range, July 17, 1908—Czet.; 60 km south-east of Chzhan"e town, 2200 m, July 12, 1958; 33 km west of Xining, 2450 m, Aug. 5, 1959—Petr.); *Amdo* (Syansibei mountain range, 2300 m, May 16, 1880—Przew.).

IIIB. Tibet: *South.* ("Vicinity of Lhasa, 4300 m, leg. Richardson"—Bor, l.c.).

General distribution: Cent. Tien Shan; China (Nor. -West., Cent., South-West.).

5. S. capillata L. Sp. pl., ed. 2 (1762) 116; ? Simpson in J. Linn. Soc. London (Bot.) 41 (1913) 451; Krylov, Fl. Zap. Sib. 2 (1928) 179, p.p.; Roshev. in Fl. SSSR, 2 (1934) 109; Fl. Kirgiz. 2 (1950) 51; Fl. Kazakhst. 1 (1956) 155; Fl. Tadzh. 1 (1957) 430; Keng, Fl. ill. sin., Gram. (1959) 608;

Bor, Grasses Burma, Ceyl., Ind. and Pakist. (1960) 644.—Ic.: Keng, l.c. fig. 549.

Described from Cent. Europe. Type in London (Linn.).

In steppes, on rocky slopes and rocks; characteristic species of sheep's fescue *Stipa capillata, Cleistogenes Stipa capillata* and wormwood-*Stipa capillata* steppes; up to midmountain belt.

IA. **Mongolia:** *Mong. Alt.* (Bain somon, midmountain belt near Khalyun, Aug. 24, 1943—Yun.); *Bas. lakes* (Khirgis somon, 5–6 km nor.-nor.-east of Main-Khutul' region, July 23, 1945—Yun.).

IIA. **Junggar:** *Alt. region, Jung. Alt.* (45 km south-west of Toli, 1800 m, No. 1403, Aug. 8; 25 km nor. of Toli, No. 1453, Aug. 9—1957, Huang); *Tien Shan, Zaisan* (left bank of Ch. Irtysh 18–20 km south-west of Burchum settlement on road to Zimunai, July 10, 1959—Yun.); *Balkh.-Alak.* (Chuguchak basin 12 to 14 km east of Durbul-'dzhin on road to Temirtam, Aug. 6, 1957—Yun.).

General distribution: Aral-Casp., Balkh. region, Jung.-Tarb., Nor. and Cent. Tien Shan; Cent. and South. Europe, Mediterr., Balk.-Asia Minor, Near East, Caucasus, Middle Asia, West. Sib. (south.), East. Sib. (south.), Himalayas (west., Kashmir).

Note: The herbarium of the Botanical Institute of the Academy of Sciences, USSR, contains specimens from East. Tien Shan (near Danu village, No. 1432, July 16, 1957—Huang) which differ altogether externally from other specimens of this species. At a height 7–15 cm, they have very dense panicles, 6.5–7.5 mm long lemma and 3.5–4.5 cm long awn. They are very similar to species *S. densiflora* P. Smirn. (=*S. densa* P. Smirn.) described from East. Siberia (Abakari river valley), the only difference being the absence (in the former) of crown of hairs at the base of the awn and leaf blades somewhat scabrous (not smooth) along the outer surface. However, the sharp external difference of these specimens from *S. capillata* was only the result of disease and it is highly possible that the solitary type specimen of *S. densiflora* preserved in Tomsk (no other specimen of this species is known) does not also represent an independent species but a specimen of *S. krylovii* Roshev. infected by a similar disease.

All Eurasian species closely related to *S. capillata* are represented in Cent. Asia. Among them, *S. sareptana* Beck. and *S. krylovii* are vicarious ecogeographical races and very closely related; *S. capillata* and *S. grandis* P. Smirn. are more isolated while *S. baicalensis* Roshev. is probably a hybrid: *S. grandis* × *S. krylovii*.

6. **S. caspia** C. Koch in Linnaea, 21 (1848) 440.—*S. arabica* β. *szowitsiana* Trin. in Mém. Ac. Sci. St.-Pétersb., sér. VI, 7, Sci. nat. 5 (1843) 77.—*S. szowitsiana* (Trin.) Griseb. in Ledeb. Fl. Ross. 4 (1852) 450.—*S. szowitsiana* Trin. in Hohen. in Bull. soc. natur. Moscou, 11, 3 (1838) 243, nom. nud.; Roshev. in Fl. SSSR, 2 (1934) 91; Fl. Kirgiz. 2 (1950) 58; Fl. Kazakhst. 1 (1956) 148; Fl. Tadzh. 1(1957) 424; Bor, Grasses Burma, Ceyl., Ind. and Pakist. (1960) 647; Lavrenko and Nikol'skaya in Bot. zhurn. 50 (1965) 1424, map 7.

Described from East. Transcaucasus. Type in Berlin.

On rocky and rubble slopes, in steppes; up to midmountain belt.

IIA. **Junggar:** *Alt. region* (80 km from Burchum on road to Shara-Sume, July 5, 1959—Yun.); *Dzhark.* (near Kul'dzha, May 13, 1877—A. Reg.); *Balkh.-Alak.*

(Chuguchak basin, 12–14 km east of Durbul'dzhin on road to Temirtam, Aug. 6, 1957—Yun.).

General distribution: Aral-Casp., Balkh. region, Jung.-Tarb., Nor. Tien Shan; Near East, Caucasus (south. and east.), Middle Asia, Himalayas (west., Kashmir).

Note: This species belongs to a small group that is closely related to Near Eastern *S. arabica* Trin. et Rupr. (priority name) and is sometimes merged with it.

7. **S. caucasica** Schmalh. in Ber. Deutsche Bot. Gesellsch. 10 (1892) 293; Roshev. in Fl. SSSR, 2 (1934) 89; Persson in Bot. notiser (1938) 274; Fl. Kirgiz. (1950) 53; Fl. Kazakhst. 1 (1956) 146; Fl. Tadzh. 1 (1957) 421; Bor, Grasses Burma, Ceyl., Ind. and Pakist. (1960) 644; Ikonnikov, Opred. rast. Pamira [Key to Plants of Pamir] (1963) 47.—*S. orientalis* var. *grandiflora* Rupr. in Osten-Sacken and Rupr. Sert. Tiansch. (1869) 35.—*S. bella* Drob. in Fedders repert. 21 (1925) 37.—Ic.: Fl. SSSR, 2, Plate 7, figs 1–6; Fl. Tadzh. 1, Plate 52, figs. 1–4.

Described from Nor. Caucasus. Type in Kiev.

On rocky and rubble slopes, rocks in pebble beds and talus; up to upper mountain belt.

IB. **Kashgar:** *Nor.* (near Khalangao village in Khomote region, No. 7691, Aug. 11, 1958—Lee and Chu); *West., South.* (in Cherchen district, 3400 m, No. 9434, June 5, 1959—Lee et al.).

IIA. **Junggar:** *Tarb.* (Khobuk river valley, July 20, 1914—Sap.); *Jung. Alt.* (near Ven'tsyuan', 2450 m, No. 1457, Aug. 14, 1957—Huang); *Tien Shan.*

IIIC. Pamir.

General distribution: Aral-Casp., Balkh. region, Jung.-Tarb., Nor. and Cent. Tien Shan, East. Pam.; Near East, Caucasus (south. and east.), Middle Asia, Himalayas (west.).

Note: Two quite distinct varieties of this highly polymorphic species are found in Cent. Asia. One of them—*S. caucasica* var. *desertorum* (Roshev.) Tzvel. Comb. nova [=*S. caucasica* f. *desertorum* Roshev. in Fl. Aziatsk. Rossii, 16 (1916) 142]—is characterised by small size of entire plant and awns and occupies a somewhat intermediate position between type specimens of this species and related *S. glareosa* P. Smirn. Most Central Asian specimens of *S. caucasica* belong to this variety. The other variety—*S. caucasica* var. *major* Drob. [in Feddes repert. 21 (1925) 37]—on the contrary, is distinguished by very large plant size, very broad (0.9–1.5 mm in diam.) leaf blades, 11–12 mm long lemma and 10–14 cm long awn. It is widely distributed in the hills of Middle Asia. Within Chinese Central Asia, it has been reported from only one site: East. Tien Shan, Ili river valley 42 km east of bridge on Kash along road to Ziekty, Aug. 29, 1957—Yun.

8. **S. consanguinea** Trin. et Rupr. in Mém. Ac. Sci. St.-Pétersb., sér. VI, 7, Sci. nat. 5 (1843) 78; Krylov, Fl. Zap. Sib. 1 (1928) 177; Roshev. in Fl. SSSR, 2 (1934) 104; ? Bor, Grasses Burma, Ceyl., Ind. and Pakist. (1960) 644.

Described from Altay (Chui river valley). Type in Leningrad.

On rocky and rubble slopes and rocks; in lower and middle mountain belts.

IA. **Mongolia:** *Khobd.* (mud cones on rise toward Ureg-Nur lake, Sept. 14, 1931—Shukh.).

IIA. Junggar: *Tien Shan* (Khanga river valley 25 km nor.-west of Balinte settlement along road from Karashar to Yuldus, Aug. 1, 1958—Yun.; Khotun-Sumbule district, near Lotogao, 2550 m, No. 6263, Aug. 1, 1958—Lee and Chu).

General distribution: West. Sib. (Altay), ? Himalayas (Sikkim).

9. S. glareosa P. Smirn. in Byull. Mosk. obshch. isp. prir. 38 (1929) 13; id. in Feddes repert. 26 (1929) 266; Roshev. in Fl. SSSR, 2 (1934) 89; Fl. Kirgiz. 2 (1950) 54; Grubov, Consp. fl. MNR (1955) 62; Fl. Tadzh. 1 (1957) 423; Keng, Fl. ill. sin., Gram. (1959) 602; Krylov, Fl. Zap. Sib. 12, 1 (1961) 3087; Ikonnikov, Opred. rast. Pamira [Key to Plants of Pamir] (1963) 47; Hanelt u. Davazamc in Feddes repert. 70, 1–3 (1965) 12.—*S. orientalis* var. *trichoglossa* Hack. in Vidensk. Middel. naturh. Foren. Kjobenhavn (1903) 164; Danguy in Bull. Mus. nat. hist. natur. 6 (1911) 7.—*S. orientalis* var. *humilior* Kryl. Fl. Zap. Sib. 2 (1928) 168.—*S. orientalis* auct. non Trin.: Franch. Pl. David. 1 (1884) 330.—*S. caucasica* auct. non Schmalh.: Krylov, l.c. 169; Norlindh, Fl. mong. steppe, 1 (1949) 64; Hanelt u. Davazamc, l.c. 12.—Ic.: Keng, l.c. fig. 540.

Described from Mongolia. Type in Moscow. Isotype in Leningrad. Plate II, fig. 4; map 4.

Characteristic species of different varieties of chee grass barren steppes on rubble trails of hills, in intermontane basins and in pebbly plains, sometimes in thin sand. Chee grass-saltwort (with *Anabasis brevifolia* C.A. Mey., *Sympegma regelii* Bunge and others), wormwood-chee grass (with *Artemisia xerophytica* Krasch., *A. kaschgarica* Krasch. and others), onion bulb-chee grass (with *Allium polyrrhizum* Turcz.), and chee grass-*Cleistogenes* (with species of *Cleistogenes* Keng) are particularly widely distributed on barren steppes; scattered on rocky and stony slopes of hills and mud cones, in hammadas, in pebble beds of gorges; up to upper mountain belt.

IA. **Mongolia:** *Khobd., Mong. Alt., Cent. Khalkha* (8–10 km nor. of west. extremity of Del'ger-Hangay mountain range along old road from Ulan Bator to Dalan-Dzadagad, July 17, 1943—Yun.; hillocky area 60 km nor. of Del'ger-Tsogto somon, June 15, 1950—Kal.); *East. Mong.* (25 km nor.-east of Argaleul hill, Sept. 1, 1931—Pob.; Ongon somon, 4–5 km from Derisun-Khuduk on road to Baishintu, May 18, 1944; Khongor somon, 10 km south-east of Gurban-Khulusutuin-Bulak, Sept. 17, 1949—Yun.); *Bas. lakes, Val. lakes* (among several citations: in basin of Orok-Nur lake, Sept. 7, 1924, No. 169—Pavl., type!); *Gobi-Alt., East. Gobi, West. Gobi* (Bain-Gobi somon, south. foothills of Tsagan-Bogdo mountain range, Aug. 1; Bor-Undur somon, Lamyn-Tooroin-Khuduk, Aug. 7; 20 km south of Saikhan-Bulak towards Atas-Bogdo, Aug. 16—1943, Yun.; Noin somon, along road to Chonoin-Bom area, Aug. 18, 1948—Grub.); *Alash. Gobi, Khesi.*

IB. **Kashgar:** *Nor., West., East.*

IC. **Qaidam:** *plain.* (14 km nor. of Tsagan-Usu settlement, Oct. 14, 1959—Petr.).

IIA. **Junggar:** *Alt. region, Tarb., Jung. Alt., Tien Shan, Jung. Gobi, Zaisan* (left bank of Ch. Irtysh 16–17 km west of Burchum settlement along road to Zimunai; same site, 25–27 km south-west of Burchum settlement, July 10, 1959—Yun.).

IIIA. Qinghai: *Nanshan* (nor. foothill of Altyntag mountain range 3 km south of Aksai settlement, 2330 m; pass through Altyntag 24 km south of Aksai settlement, 3460 m, Aug. 2, 1958—Petr.).

IIIB. Tibet: *Chang Tang* (10–12 km west of Shakhidully along road to Kirgiz–Dzhangil pass, June 3, 1959—Yun.).

IIIC. Pamir.

General distribution: Balkh. region (south-east.), Cent. Tien Shan, East. Pamir; Middle Asia (Alay), West. Sib. (Altay, Chui steppe), Nor. Mong. (Hang.), China (Nor.).

Note: Compared to the related species *S. caucasica* Schmalh., this species is a more eastern ecogeographical race existing under very severe climatic conditions. In many regions of its distribution range (especially in East. Gobi and Tien Shan foothills), specimens are found with leaf blades that are smooth along outer surface, tending in this respect towards *S. caucasica*, but with smaller lemma and awn.

10. **S. gobica** Roshev. in Bot. mater. Gerb. Glavn. bot. sada RSFSR 5 (1924) 13; Pavlov in Byull. Mosk. obshch. ispyt. prir. 38 (1929) 12; Norlindh, Fl. mong. steppe, 1 (1949) 66, excl. var.; Grubov, Consp. fl. MNR (1955) 62; Keng, Fl. ill. sin., Gram. (1959) 602; Hanelt u. Davazamc in Feddes repert. 70, 1–3 (1965) 13, p.p.—*S. potaninii* Roshev. l.c.—*S. sinomongholica* Ohwi in J. Jap. Bot. 19 (1943) 168.—Ic.: Keng, l.c. fig. 541.

Described from Mongolia. Type in Leningrad. Plate II, fig. 5.

On rocky and rubble slopes of hills and mud cones, on rocks, in hammadas and pebble beds of gorges. As in the case of *S. glareosa* P. Smirn., this species is often characteristic of chee grass barren steppes, widely distributed on rubble trails of hills, in intermontane basins and pebbly plains, sometimes in sand; up to lower mountain belt.

IA. Mongolia: *Mong. Alt.* (Tsitsirin-Gol river, June 28, 1877—Pot.; Tukhumyin-Khundei region in Khubchin-Nuru hills west of Adzhi-Bogdo mountain range, Aug. 3, 1947—Yun.); *Cent. Khalkha* (Choiren-Ula on Ulan Bator—Sain-Shanda highway, July 7, 1941—Yun.); *East. Mong.* ("7.5 km ad bor.-orient. Khadain-sume, 26 VI 1936, leg. Eriksson"—Norlindh, l.c.; nor. foothills of Datsin'shan' hill 10 km nor. of Khukh-Khoto town, 1300 m, June 4, 1957—Petr.); *Bas. lakes* (25 km west of Tsagan-Oloma along road to Khobdo, Aug. 28, 1944—Yun.); *Val. lakes* (on Tuin-Gol coast, below its confluence with Sharagol'dzhut, July 8–9, 1893—Klem., lectotype!; Khaileste creek valley near Lamyn-Khita on Ongiin-Gol river, July 21, 1926—Lis.; east. rim of Tuilin-Tala plain, Aug. 26, 1943—Yun.); *Gobi-Alt., East. Gobi, West. Gobi* (15–20 km south-east of Atas-Bogdo, Aug. 11, 1943—Yun.); *Alash. Gobi* (Kobden-Obotu region north of Gashun-Nur lake, Aug. 2, 1886—Pot.; 50 km along road from In'chuan' in Bayan-Khoto, 1800 m, June 10; Divusumu village 70 km south-east of Bayan-Khoto, June 11; Chintunai gorge 90 km south-west of In'chuan' town, June 18, 1958—Petr.); *Khesi* (15 km nor. of Yunchan town, June 28, 1958—Petr.).

IIA. Junggar: *Jung. Gobi* (east.: Beidashan' [Baitak-Bogdo] hills, No. 2372, Sept. 27, 1957—Huang; south.: 112–115 km south of Ertai on Guchen road, July 16, 1959—Yun.).

General distribution: Nor. Mong. (Hang.: west of Arbai-Khere), China (Nor.).

Note: *S. gobica* Roshev., *S. tianschanica* Roshev. and *S. klemenzii* Roshev. represent vicarious ecogeographical races that are closely related. Lower awns in type *S. potaninii* Roshev. (anticline between Nemegetu and Tsomtso mountain ranges, Aug. 20, 1886—

Pot.) are strongly scabrous (not glabrous or subglabrous) but evidently this is of no significant taxonomic importance. In the variety S. *gobica* var. *pubescens* Hanelt et Davazamc (l.c. 14), the sheath of the cauline leaves is short but densely pilose, a characteristic likewise observed among some specimens of closely related species, e.g. *S. glareosa* P. Smirn. and *S. caucasica* Schmalh.

11. **S. grandis** P. Smirn. in Byull. Mosk. obshch. ispyt. prir. 38 (1929) 15; id. in Feddes repert. 26 (1929) 267; Roshev. in Fl. SSSR, 2 (1934) 109; Norlindh, Fl. mong. steppe, 1(1949) 68; Grubov, consp. fl. MNR (1955) 62; Keng, Fl. ill. sin., Gram. (1959) 612.—*S. baicalensis* var. *macrocarpa* Roshev. in Izv. Glavn. bot. sada SSSR, 28 (1929) 380.—Ic.: Fl. SSSR, 2, Plate 7, Figs. 27–39; Keng, l.c. fig. 552.

Described from Nor. Mongolia. Type (steppe between Tamir river and Tola-Bil'chir area, No. 179, July 23, 1924—Pavl.). in Moscow. Isotype in Leningrad.

In steppes, on rocky slopes and rocks, and in sand and pebble beds; up to lower mountain belt.

IA. Mongolia: *Cis-Hing., Cent. Khalkha, East. Mong.*
General distribution: East. Sib. (south.), Far East (south-west.), Nor. Mong. (Hang., Mong.-Daur.), China (Dunbei, Nor.).

12. **S. himalaica** Roshev. in Bot. mater. Gerb. Glavn. bot. sada RSFSR, 5 (1924) 11; Pampanini, Fl. Carac. (1930) 71; Persson in Bot. notiser (1938) 274; Bor, Grasses Burma, Ceyl., Ind. and Pakist. (1960) 644.

Described from the Himalayas. Lectotype (Gilgit Exped., SO of Hindukush, coll. Giles) in Leningrad.

On rocky slopes and rocks; in upper mountain belt.

IB. Kashgar: *West.* (dist. Yarkand, Jerzil, 2800 m, July 11, 1931—Persson).
General distribution: Himalayas (West., Kashmir).

13. **S. hohenackeriana** Trin. et Rupr. in Mém. Ac. Sci. St.-Pétersb., sér. VI, 7, Sci. nat. 5(1843) 80; Roshev. in Fl. SSSR, 2 (1934) 92; Fl. Kirgiz. 2 (1950) 58; Fl. Kazakhst. 1 (1956) 149; Fl. Tadzh. 1 (1957) 426; Krylov, Fl. Zap. Sib. 12, 1 (1961) 3088; Lavrenko and Nikol'skaya in Bot. zhurn. SSSR, 50 (1965) 1423, map 6.—*S. subbarbata* Keller in Bot.-geogr. issled. v Zaisansk. u. 2 (1912) 53.

Described from Transcaucasus. Type in Leningrad.

On rocky, rubble and clayey slopes; up to midmountain belt.

IIA. Junggar: *Alt. region* (left bank of Kran river 3–4 km south of Shara-Sume; 10 km south of Shara-Sume along road to Shipati, July 7, 1959—Yun.).
General distribution: Aral-Casp., Balkh. region, Jung.-Tarb., Nor. Tien Shan; Near East, Caucasus (South. and East. Transcaucasus), Middle Asia.

14. **S. kirghisorum** P. Smirn. in Feddes repert. 21 (1925) 231; Roshev. in Fl. SSSR, 2 (1934) 95; Fl. Kirgiz. 2 (1950) 59; Fl. Kazakhst. 1 (1956) 150; Fl. Tadzh. 1(1957) 429; Krylov, Fl. Zap. Sib. 12, 1 (1961) 3089; Lavrenko and Nikol'skaya in Bot. zhurn. 50 (1965) 1419, map 1.

Described from East. Kazakhstan. Type in Leningrad.

In steppes, on rocky and rubble slopes; up to midmountain belt.

IIA. **Junggar: Alt. region** (in Altay hills, No. 10325, May 28, 1959—Lee et al.; left bank of Kran river, 15 km south-south-east of Shara-Sume along road to Shipati, July 7, 1959—Yun.); *Tarb.* (Mai-Kapchagai hill, June 6, 1914—Schisch); *Jung. Alt.* (Dzhair mountain range, crossing on Shikho-Chuguchak road, No. 1091, June 8, 1957—Huang); *Tien Shan* (Kuitun river basin, right bank of Bain-Gol creek south of Tushantszy settlement, June 29, 1957—Yun.; water divide between Manas and Danu-Gol on Nyutsyuan'tsze-Shikhodze road, No. 532, July 15, 1957—Huang; east. part of Ketmen' mountain range 3–4 km beyond Sarbushin settlement along Kul'dzha-Kyzyl-Kure road, July 23; Manas river basin 3 km from Nyutsyuan'tsze settlement, July 24—1957, Yun.); *Jung. Gobi* (South. Altay, May 1–20, 1877—Pot.; Baitak-Bogdo mountain range, Takhiltu-Ula, Ulyastu-Gola left creek, 2000 m, Sept. 17; same site, upper part of Ulyastu-Gola gorge 7 km from its estuary, Sept. 18—1948, Grub.).

General distribution: Aral-Casp., Balkh. region, Jung.-Tarb., Nor. and Cent. Tien Shan; Middle Asia (east), West. Sib. (south).

15. **S. klemenzii** Roshev. in Bot. mater. Gerb. Glavn. bot. sada RSFSR, 5 (1924) 12; Pavlov in Byull. Mosk. obshch. ispyt. prir. 38 (1929) 12; Grubov, Consp. fl. MNR (1955) 63; Popov, Fl. Sredn. Sib. 1 (1957) 84.—*S. gobica* var. *klemenzii* (Roshev.) Norlindh, Fl. mong. steppe, 1 (1949) 66; Hanelt u. Davazamc in Feddes repert. 70, 1–3 (1965) 13.—*S. gobica* auct. non Roshev.: Popov, l.c.

Described from Mongolia. Lectotype in Leningrad.

On rubble and rocky slopes, in pebble beds and sandy steppes; up to lower mountain belt.

IA. **Mongolia:** *Khobd.* (between the border of Siberia and Kobdo town, 1870—Kalning; Khuityn-Khutyl' crossing south of Tsagan-Nur on road to Kobdo, Aug. 4, 1945—Yun.); *Cent. Khalkha* (on Kharukhi river bank, July 25, 1895—Klem. lectotype!; 60 km south-west of Ulan Bator along road to Ubur Hangay, July 18, 1949—Yun.; plain between mud cones 120 km south of Ulan Bator, Aug. 11, 1950—Kal.); *East. Mong.* ("2.5 km ad orient. Sunit, 19 VI 1934, leg. Eriksson "—Norlindh, l.c.; 18–20 km west-nor.—west of Tamtsag-Bulak, Aug. 14, 1949—Yun.); *Val. lakes* (on right bank of Sharagol'dzhut river, July 5, 1893—Klem.; 35 km nor.-east of east. rim of Artsa-Bogdo mountain range, Oct. 28, 1940—Yun.; 15 km nor.-west of Dzhinsetu somon, June 30, 1941—Tsatsenkin); *Gobi-Alt.* (west. extremity of Dzun-Saikhan near Dalan-Dzadagad-Bain-Dalai road; crossing between Dundu- and Dzun-Saikhan; plain on foothill nor. of Gurban-Saikhan, July 22, 1943—Yun.; south. mountain trails of Baga-Bogdo mountain range, Ermen-Tologoi region, Sept. 18, 1943—Yun.; south. slope of Gurban-Saikhan, July 19, 1950—Kal.); *East. Gobi.*

General distribution: East. Sib. (south.: Dzhida and Borzya basins), Nor. Mong. (Hang., Mong.-Daur.).

Note: Compared to *S. gobica* Roshev., this species is a relatively more mesophilic ecogeographical race and very closely related.

16. **S. krylovii** Roshev. in Izv. Glavn. bot. sada SSSR, 28 (1929) 379 and in Fl. SSSR, 2 (1934) 112; Fl. Kirgiz. 2 (1950) 52; Grubov, Consp. fl. MNR (1955) 63; Keng, Fl. ill. sin., Gram. (1959) 610.—*S. capillata* var. *coronata*

Roshev. in Fl. Aziatsk. Rossii, 12 (1916) 168; Krylov, Fl. Zap. Sib. 2 (1928) 180.—*S. decipiens* P. Smirn. in Byull. Mosk. obshch. ispyt. prir. 45 (1936) 110, diagn. ross. in clavi.—*S. capillata* auct. non. L.: Franch. Pl. David. 1 (1884) 330; Forbes and Hemsley, Index Fl. Sin. 3 (1904) 381; Hao in Engler's Bot. Jahrb. 68 (1938) 583.—*S. baicalensis* auct. non Roshev.; Norlindh, Fl. mong. steppe, 1 (1949) 67, p.p.—Ic.: Keng, l.c. fig. 550.

Described from Altay. Type in Leningrad.

In steppes, on rocky and rubble slopes and rocks, in sand; characteristic of different varieties of stipa steppes; up to midmountain belt.

IA. **Mongolia:** *Khobd., Mong. Alt., Cis-Hing.* (Khuntu somon 5 km west of Toge-Gol, Aug. 7, 1949—Yun.); *Cent. Khalkha, East. Mong., Bas. lakes, Val. lakes, Gobi-Alt., East. Gobi, Alash. Gobi* (in Alashan hills, June 25, 1873—Przew.).

IB. **Kashgar:** Nor. (val. of Muzart river, 2200 m, No. 8321, Sept. 11, 1958—Lee and Chu; nor.-west. slope of Sogdyn-Tau mountain range 10–12 km south-east of Akchit hydropower plant at Kokshaal-Darya, 2900 m, Sept. 19, 1958—Yun.).

IIA. **Junggar:** *Tarb., Jung. Alt.* (Maili-Barlyk mountain range 10–15 km south-west of Karaganda-Daban crossing on road to Junggar outlets, Aug. 14; Shuvutin-Daban crossing on road from Borotal to Sairam-Nur lake, Aug. 18—1957, Yun.); *Tien Shan* (60 km west of Bain-Bulak village, 2540 m, No. 6381, Aug. 2; near Bain-Bulak village, 2560 m, No. 6409, Aug. 6—1958, Lee and Chu; B. Yuldus basin 4–5 km south-west of Bain-Bulak on right bank of Khaidyk-Gol, Aug. 10; Muzart river valley 2–3 km below Sazlik region along road to Oi-Terek, Sept. 9—1958, Yun.; Dzin'kho village 10 km south of Sairam-Nur lake, 1900 m, Aug. 31, 1959—Petr.).

IIIA. **Qinghai:** *Nanshan* (Yuzhno-Tetungsk mountain range, June 20, 1880—Przew.; 60 km south-east of Chzhan"e town, 2200 m, July 12, 1958—Petr.; Tetung river valley near stud farm, 2800 m, July 20, 1958—Dolgushin; east coast of Kukunor lake, 3200 m, Aug. 5, 1959—Petr.); *Amdo* ("auf dem Flusses Kia-po-kia-ho, Aug. 20, 1930; auf dem Ming-ke-schan, Aug. 25, 1930"—Hao, l.c.; 25 km east of Gunkhe town, 3100 m, Aug. 7, 1959—Petr.).

IIIB. **Tibet:** *Weitzan* (Mekong basin, Tszachu river, 3600–4000 m, Sept. 1900—Lad.).

General distribution: Jung.-Tarb., Nor. and Cent. Tien Shan; Middle Asia (Alay), West. Sib. (south), East. Sib. (south.), Nor. Mong. (Hent., Hang., Mong.-Daur.), China (Dunbei, Nor., Nor.-West.).

17. S. lessingiana Trin. et Rupr. in Mém. Ac. Sci. St.-Pétersb., sér. VII, 7, Sci. nat. 5 (1843) 79; Krylov, Fl. Zap. Sib. 2 (1928) 176; Roshev. in Fl. SSSR, 2 (1934) 93; Fl. Kirgiz. 2 (1950) 58; Fl. Kazakhst. 1 (1956) 149; Fl. Tadzh. 1 (1957) 427; Lavrenko and Nikol'skaya in Bot. zhurn. 50 (1965) 1423, map 5.

Described from South. Urals. Type in Leningrad.

In steppes, on rocky and rubble slopes; up to midmountain belt.

IIA. **Junggar:** *Jung. Gobi* (south.: left bank of Manas 20–23 km west of Saeda state farm, June 13, 1957—Yun.); *Tien Shan* (basin of Manas river 3 km east of Nyutsyuan'tsze village, July 24, 1957—Yun.; same site, 1 km east of Nyutsyuan'tsze village, No. 610, July 24, 1957—Huang).

General distribution: Aral-Casp., Balkh. region, Jung.-Tarb., Nor. and Cent. Tien Shan; Europe (south-east.), Near East (nor. Iran), Caucasus (excluding West. Transcaucasus), Middle Asia (hilly), West. Sib. (south).

68

18. **S. macroglossa** P. Smirn. in Bot. mater. Gerb. Glavn. bot. sada RSFSR, 5 (1924) 47; id. in Feddes repert. 21 (1925) 234; Roshev. in Fl. SSSR, '2 (1934) 94; Fl. Kirgiz. 2 (1950) 58; Fl. Kazakhst. 1 (1956) 149; Krylov, Fl. Zap. Sib. 12, 1 (1961) 2088; Lavrenko and Nikol'skaya in Bot. zhurn. 50 (1965) 1421, map 2.

Described from Cent. Kazakhstan (Sarysu river). Type in Leningrad. On rocky and rubble slopes; up to midmountain belt.

IB. **Kashgar:** *Nor.* (lower right opening of Lyangar in Muzart river valley before its discharge into Bai basin, Sept. 12, 1958—Yun.).

IIA. **Junggar:** *Tien Shan* (Aktyube settlement near Kul'dzha, May 13, 1877—A. Reg.; basin of Kuitun river, Bain-Gol creek gorge south of Tushandzy settlement, June 29; same site, in midportion of creek, June 29, 1957—Yun.); *Jung. Gobi* (near Tien Shan-Laoba station on Shikho-Karamai road, No. 971, June 20, 1957—Huang).

General distribution: Aral-Casp., Balkh. region, Jung.-Tarb., Nor. Tien Shan; Middle Asia (West. Tien Shan), West. Sib. (south.: hillocky area).

19. **S. mongolorum** Tzvel. sp. nova.—Planta perennis, dense caespitosa, 15–40 cm alta; culmi in nodis dense, sed brevissime pilosi; vaginae laeves vel sublaeves, glabrae vel margine pilosae; ligulae ad 0.3 mm lg., sat dense pilosae; laminae 0.3–0.7 mm in diam., extus (subtus) laeves vel sublaeves, intus (supra) minute pubescentes. Paniculae 8–20 cm lg., sat laxae, saepe subdiffusae; glumae 14–19 mm lg., 3–5-nerviae, apice longissime acutatae; lemmata 6–7.5 mm lg., sat dense pilosa, sine striis longitudinalibus distinctis, callo ca. 1.5 mm lg.; aristae 4.5–6.5 cm lg.,unigeniculatae, segmento infimo torto, 0.4–0.7 cm lg., pilis ad 2 mm lg. tecto, supero 4–6 cm lg., pilis 2–3.5 mm lg. tecto.

Typus: Mongolia, Erdeni somon, praed. Borocha-Tala ad austr. versus Dzamyn-Ude, in steppa sicca, 16 VI 1941, A. Junatov. In Herb. Inst. Acad. Sci. URSS—LE conservatur.

Affinitas. Haec species a speciebus proximis—S. glareosae P. Smirn. et S. caucasicae Schmalh. paniculis multo laxioribus, glumis brevioribus et aristae segmento infimo breviore (quam segmentum superum in octies-decies brevior) differt.

Described from Mongolia. Type in Leningrad. Plate II, Fig. 3.

In arid steppes, on rocky and rubble slopes, in sand and pebble beds; up to lower mountain belt.

IA. **Mongolia:** *Mong. Alt.* (west. bank of lake Tonkhil'-Nur, mud cone trails, July 16, 1947—Yun.); *Cent. Khalkha* (8–10 km nor. of west. extremity of Delger-Hangay mountain range along old road from Ulan Bator to Dalan-Dzadagad, July 17, 1943—Yun.).; *East. Gobi* (Shabarakh Usu, ca. 1200 m, 1925—Chaney; Undur-Shili somon, slope towards Undegiin-Toirim lowland near Kholbo-Obo, June 2; Khara-Airik somon, upland plain 20–23 km south-east of Buterin-Obo along. Ulan Bator–Sain-Shanda road, Aug. 27; Sain-Usu basin 40 km nor.-east of Sain-Shanda, Aug. 28; 20–25 km nor.-nor.-east of Sain-Shanda, Aug. 28; Tel'ulan-Shanda region 16 km nor.-east of Sain-Shanda, Aug. 31—1940, Yun.: 30 km south-west of Undur-Shili somon, June 2; Erdeni somon, Borokha-Tala region south of Dzamyn-Ude, June 16, type!; Ulan-Baderkhu somon, west. rim of Borokha-Tala region, June 17—1941, Yun.).

General distribution: Endemic.

Note: This species, although apparently resembling *S. glareosa* P. Smirn. and *S. caucasica* Schmalh., is quite distinct and reveals no significant genetic affinity with them. It is quite possible, however, that, it may have been a hybrid in the past of *S. glareosa* P. Smirn. x *S. breviflora* Griseb. or *S. glareosa* P. Smirn. x *Ptilagrostis pelliotii* (Danguy) Grub.

20. **S. nakaii** Honda in Rep. First. Sci. Exped. Manch., sect. IV, 4 (1936) 104; Kitag, Lin. Fl. Mansh. (1939) 94.—*S. roerichii* Keng in J. Wash. Ac. Sci. 28, 7(1938) 307.—*Achnatherum nakaii* (Honda) Tateoka ex Keng, Fl. ill. sin., Gram. (1959) 591.—Ic.: Keng. l.c. (1959) fig. 527.

Described from Dunbei (near Chaoyan town). Type in Kyoto.

On rocky slopes and rocks; in lower mountain belt.

IA. **Mongolia:** *East. Gobi* ("Temur Khada, Peiling Miao, Suiyan Prov., 1500 m, Roerich Exped., Nos. 518 and 905, July 26 and Aug. 24, 1935, leg. Keng"—Keng, l.c. [1938]).

General distribution: China (Dunbei, Nor., Nor.-West.).

Note: Along with *S. sibirica* (L.) Lam., this species belongs to a very primitive group of species of the genus *Stipa* L., which is very close to the genus *Achnatherum* Beauv.

21. **S. orientalis** Trin. in Ledeb. Fl. alt. 1 (1829) 83; ? Deasy, In Tibet and Chin. Turk. (1901) 399; ? Hemsley, Fl. Tibet (1902) 202; Krylov, Fl. Zap. Sib. 2 (1928) 168, p.p.; Pavlov in Byull. Mosk. obshch. ispyt. prir. 38 (1929) 14; ? Pampanini, Fl. Carac. (1930) 71; Roshev. in Fl. SSSR, 2 (1934) 90; Fl. Kirgiz. 2 (1950) 57; Grubov, Consp. fl. MNR (1955) 63; Fl. Kazakhst. 1 (1956) 147; Fl. Tadzh. 1 (1957) 423; Bor, Grasses Burma, Ceyl., Ind. and Pakist. (1960) 645; Ikonnikov, Opred. rast. Pamira [Key to Plants of Pamir] (1963) 49.—Ic.: Ledeb. Ic. pl. fl. ross. 3 (1831) Tab. 223; Fl. Tadzh. 1, Plate 52, figs. 5–7.

Described from Altay. Type in Leningrad.

On rocky, rubble and melkozem slopes and pebble beds; sometimes characteristic of stipa desert steppes; up to upper mountain belt.

IA. **Mongolia:** *Khobd.* (Ulyasty river valley near Achit-Nur lake, Sept. 9, 1931—Bar.); *Mong. Alt.* (near Khalyun, Aug. 24, 1943—Yun.); Bidzhi-Gol river gorge south of Tamchi-Daba crossing, July 17, 1947—Yun.; *Val lakes* (along Baidarik river, July 11, 1924—Gorbunova); *West. Gobi* (Atas-Bogdo mountain range, Aug. 12, 1943—Yun.).

IB. **Kashgar:** *West.*

IIA. **Junggar:** *Alt. region* (20 km nor.-west of Shara-Sume, July 7; 68 km nor.-nor.-west of Ertai settlement near Urungu along road to Koktogoi, June 14, 1959—Yun.); *Tarb., Jung. Alt., Tien Shan, Jung. Gobi* (nor.: nor. of Ertai, No. 10356, June 3, 1959—Lee et al.; 47 km east of Burchum along road to Shara-Sume, July 5, 1959—Yun.; west.: on south-east. margin of Dzhair mountain range 8–10 km west of Aktam settlement along road to Chuguchak, Aug. 3, 1957—Yun.; east : Baitak-Bogdo mountain range, Takhiltu-Ula, left creek of Ulyastu-Gola, about 2000 m, Sept. 17, 1958—Grub.; 60 km south of Ertai along road to Guchen, July 16, 1959—Yun.).

General distribution: Aral-Casp. (Ulutau hills), Balkh. region, Jung. Tarb., Nor. and Cent. Tien Shan, East. Pam.; Middle Asia (east.), West. Sib. (Altay), East. Sib. (Sayans), Nor. Mong. (Hang.), Himalayas (west., Kashmir).

22. **S. penicillata** Hand.-Mazz. in Oesterr. bot. Z. 85 (1936) 226.—*S. laxiflora* Keng in Sunyatsenia, 6 (1941) 73; ej. Fl. ill. sin., Gram. (1959) 603. —Ic.: Keng, l.c. (1959) fig. 542.

Described from Qinghai. Type in Vienna. Map 5.

In sand and pebble beds of river and lake valleys, on rocky slopes and talus; in middle and upper mountain belts.

IIA. **Junggar:** *Tien Shan* (south of Pin'-Daban crossing along road from Urumchi to Karashar, 1900 m, No. 5932, July 21; Luitsigen village in Khomote region, 3100 m, No. 7199, Aug. 10—1958, Lee and Chu).

IIIA. **Qinghai:** *Nanshan* (south-east. bank of Kukunor lake, Aug. 28, 1908—Czet.; "Kukunor, gegen Lombutong, 21 IX 1918, leg. Licent"—Hand.-Mazz. l.c., type!); *Amdo* (Dulan-Khit monastery, 3600 m, Aug. 8, 1901—Lad.).

General distribution: China (Cent., South-West.).

Note: This species, Pamir-Tien Shan *S. subsessiliflora* (Rupr.) Roshev. and Himalayan *S. basiplumosa* Munro ex Hook. f., represent closely related ecogeographical races. *S. penicillata* is the oldest and evidently ancestral to the other two. The Chinese specialist on grasses, Gen I-li [Keng, l.c. (1941) 72] considered *S. penicillata* a much later synonym for *S. chingii* Hitchc. described from Gansu province (Syakhe district) [in Proc. Biol. Soc. Wash. 43 (1930) 94]. However, the characteristics cited in the diagnosis for *S. penicillata* indisputably show the absence of a close relationship between this species and *S. chingii* and places its similarity to *S. laxiflora* Keng described by Gen from Syakhe district in doubt.

23. **S. pennata** L. Sp. pl. (1753) 78, s.s.; Lavrenko and Nikol' skaya in Bot. zhurn. 50 (1965) 1421 map 3.—*S. joannis* Čelak. in Oesterr. bot. Z. 34 (1884) 318; Krylov, Fl. Zap. Sib. 2 (1928) 172; Roshev. in Fl. SSSR, 2 (1934) 96; Fl. Kirgiz. 2 (1950) 59; Fl. Kazakhst. 1 (1956) 151.

Described from Cent. Europe. Type in London (Linn.).

In sandy steppes, on rocky and rubble slopes, in sand; up to midmountain belt.

IIA. **Junggar:** *Alt. region* (in upper Ch. Irtysh west of Koktogoi, 1470 m, No. 10381, June 6, 1959-Lee et al.; 20 km nor.-west of Shara-Sume, July 7, 1959-Yun.), *Zaisan* (left bank of Ch. Irtysh 17–18 km south-west of Burchum along road to Zimunai, July 10, 1959—Yun.).

General distribution: Aral-Casp., Balkh. region, Jung.-Tarb., Nor. Tien Shan; Europe (cent. and east.), Balk.-Asia Minor, Middle Asia (nor.), West. Sib. (south.), East. Sib. (south), Nor. Mong. (Mong.-Daur.: Iro river valley).

24. **S. przewalskyi** Roshev. in Bot. mater. Gerb. Glavn. bot. sada RSFSR, 1, 6 (1920) 3; Keng, Fl. ill. sin., Gram. (1959) 608.—Ic.: Keng, l.c. fig. 548.

Described from Qinghai. Lectotype in Leningrad.

On rocky slopes and rocks; in lower and middle mountain belts.

IA. Mongolia: *East. Mong.* (Sartchy [Salachi], No. 2659—David).

IIIA. Qinghai: *Nanshan* (nor. slope of Yuzhno-Tetungsk mountain range, Aug. 17, 1872—Przew., lectotype!; Tetung river valley near stud farm, 2960 m, Aug. 20, 1958— Dolgushin; 66 km west of Xining, 2800 m, July 5; 24 km south of Xining, 2650 m, Aug. 4; 33 km west of Xining, 2950 m, Aug. 5—1959, Petr.).

General distribution: China (Nor., Nor.-West., Cent.)

25. S. purpurascens Hitchc. in Proc. Biol. Soc. Wash. 43 (1930) 95; Ching in Contribs. U.S. Nat. Herb. 28 (1941) 598.—*Achnatherum purpurascens* (Hitchc.) Keng, Clav. Gram. prim. sin. (1957) 213; ej. Fl. ill. sin., Gram. (1959) 596.—Ic.: Keng, l.c. (1959) fig. 535.

Described from Qinghai. Type in Washington.

IIIA. Qinghai: *Nanshan* ("South of Sining in the La-Che-Tze Mountains, 3350– 3900 m, No. 686, leg. Ching"—Hitchcock, l.c., type!).

General distribution: China (Nor.-West., Cent.).

Note: This species is evidently closely related to the Pamir-Tien Shan *S. regeliana* Hack.

26. S. purpurea Griseb. in Nachr. Gesellsch. Wissensch. u. Univ. Goettingen, 3 (1868) 82; Hemsley in J. Linn. Soc. London (Bot.) 30 (1894) 120; Deasy, In Tibet and Chin. Turk. (1901) 404; Hemsley, Fl. Tibet (1902) 202; Pilger in Hedin, S. Tibet, 6, 3 (1922) 92; Keng, Fl. ill. sin., Gram. (1959) 605; Bor, Grasses Burma, Ceyl., Ind. and Pakist. (1960) 645.—*S. semenowii* Krasn. in Scripta Horti Univ. Petrop. 2, 1 (1889) 22.—*S. pilgeriana* Hao in Engler's Bot. Jahrb. 68 (1938) 583.—*Lasiagrostis tremula* Rupr. in Osten-Sacken and Rupr. Sert. Tiansh. (1869) 35.—*Ptilagrostis purpurea* (Griseb.) Roshev. in Fl. SSSR, 2 (1934) 76; Fl. Kirgiz. 2 (1950) 44; Fl. Kazakhst. 1 (1956) 157; Fl.Tadzh. 1(1957) 433; Ikonnikov, Opred. rast. Pamira [Key to Plants of Pamir] (1963) 46.—Ic.: Fl. SSSR, 2, Plate 6, figs 38–42; Fl. Tadzh. 1, Plate 51, figs. 12 and 13.

Described from the Himalayas. Type in London (K). Isotype in Leningrad. Map 4.

On hill plateaus, rocky and rubble slopes, terraces of river valleys, in sand and pebble beds; one of the characteristic species of false-stipa high-altitude steppes; in upper mountain belt.

IIA. Junggar: *Tien Shan.*

IIIA. Qinghai: *Nanshan* (valley of Sharagol'dzhin river, Paidza-Tologoi region, 3600 m, July 11, 1894—Rob.); *Amdo* (Dulan-Khit monastery, 3400 m, Aug. 12, 1901—Lad.; "Ming-ke-schan im Tsi-gi-gan-ba Gebiete, 3900 m, Aug. 25, 1930"—Hao, l.c.).

IIIB. Tibet: *Chang Tang* (nor. foothills of Przewalsky mountain range, up to 5000 m, Aug. 24, 1890—Rob.; "North. Tibet, 35° 48', 82° 19', 5200 m, Aug. 16, 1898"— Deasy, l.c.; "East. Tibet, 5127 m, Aug. 15, 1901, leg. Hedin"—Pilger, l.c.; in Cherchen district, 4060 m, No. 9400, June 2; same site, No. 9414, June 4—1959, Lee et al.); *Weitzan* (south. bank of Orin-Nur lake, 4500 m, July 6, 1884—Przew.; Yangtze basin, valley of Ichu river and Rkhombomtso lake, 4400 m, Aug. 1, 1900—Lad.); *South.* ("Tibet, 5500 m, leg. Thorold; Tisum, 5000 m, leg. Strachey and Witterbottom"— Hemsley, l.c.).

IIIC. **Pamir** (crossing of Argalyk river gorge in Kizil-Bazar, 3500 m, July 13, 1941; upper course of Lapst river, July 20, 1942; Karadzhilga river, lower course of its tributary Gon-Areka, 4000–4500 m, July 22, 1942—Serp.).

General distribution: Nor. and Cent. Tien Shan, East. Pam.; Middle Asia (Alay), China (South-West.), Himalayas.

Note: The two specimens of V.I. Roborowsky mentioned above were collected in their highly characteristic ecological environment of high-altitude sand steppe. Their perceptible difference from other specimens of *S. purpurea* justified their recognition at least as a distinct variety : *S. purpurea* var. *arenosa* Tzvel. var. nova.—A. varietate typica glumis 17–22 (non 12–17) mm lg., lemmatis 12–13 (non 8–11) mm lg. et aristis 8–9 (non 6–8.5) cm lg. differt. Typus: "Tibet bor.-occid., praemontium bor. jugi Przevalskii, ad 5000 m, in steppa arenosa, 24 VIII 1890, V. Roborovski" in Herb. Inst. Bot. Ac. Sci. URSS—LE conservatur. In addition to differences recorded in diagnosis with respect to the size of glumes and awns of spikelets, this variant differs from type specimens of the species also in that yellowish anthers do not emerge from the lemma.

27. **S. regeliana** Hack. in Sitzber. Akad. Wissensch. Wien, 89 (1884) 130; Roshev. in Fl. SSSR, 2 (1934) 84; Fl. Kirgiz. 2 (1950) 60; Fl. Kazakhst. 1 (1956) 145; Fl. Tadzh. 1 (1957) 417; Bor, Grasses Burma, Ceyl., Ind. and Pakist. (1960) 646; Ikonnikov, Opred. rast. Pamira [Key to Plants of Pamir] (1963) 49. —**Ic.:** Fl. SSSR, 2, Plate 6, figs. 27–32; Fl. Kazakhst. 1, Plate 10, fig. 10.

Described from Cent. Tien Shan (Muzart pass). Type in Vienna. Map 2.
In grasslands and on rocky slopes; in upper mountain belt.

IIA. **Junggar:** *Jung. Alt.* (south. slope of Jung. Alatau below Koketau pass, July 21, 1909—Lipsky; 20 km south of Ven'tsyuan', 2870 m, No. 1492, Aug. 14, 1957—Huang); *Tien Shan*.

General distribution: Jung.-Tarb., Nor. and Cent. Tien Shan, East. Pam.; Himalayas (west., Kashmir).

28. **S. richteriana** Kar. et Kir. in Bull. Soc. natur. Moscou, 14 (1841) 862; Krylov, Fl. Zap. Sib. 2 (1928) 178; Roshev. in Fl. SSSR, 2 (1934) 90; Fl. Kazakhst. 1 (1956) 147; Lavrenko and Nikol'skaya in Bot. zhurn. 50 (1965) 1424, map 8.—*S. woroninii* Krasn. in scripta Horti Univ. Petrop. 2, 1 (1889) 22.—**Ic.:** Fl. Kazakhst. 1, Plate 10, fig. 16.

Described from East. Kazakhstan (Arganaty hills). Type in Leningrad.
On rocky and rubble slopes; up to midmountain belt.

IIA. **Junggar:** *Alt. region* (in upper Ch. Irtysh near Koktogoi, No. 10409, June 7, 1959—Yun.); *Tien Shan* (Manas river valley 15–20 km beyond Manas town, No. 644, June 6, 1957—Huang); *Jung. Gobi* (south.: water divide between Danu-Gol and Manas rivers 26 km from Nyutsyuan'tsze along road to Shikhotsze, Nos. 532 and 534, July 15; near Gan'khedza village on Urumchi-Guchen road, Nos. 2293 and 5133, Sept. 24—1957, Huang).

General distribution: Aral-Casp., Balkh. region, Jung-Tarb., Nor. Tien Shan; Middle Asia (West. Tien Shan, Kyzylkums).

29. **S. roborovskyi** Roshev. in Bot. mater. Gerb. Glavn. bot. sada RSFSR, 1, 6 (1920) 1.—*S. basiplumosa* var. *longearistata* Munro ex Hook. f. Fl. Brit. Ind. 7 (1896) 229.

Described from Nor. Tibet. Type in Leningrad.

In grasslands, on rocky slopes, rocks and in pebble beds; in upper mountain belt.

IIIB. Tibet: *Chang Tang* (nor. slope of Russky mountain range, Kara-Sai region in upper Aksu river, 3300–4200 m, July 3, 1890—Rob., type!; Kerin river basin around Polur settlement, May 11, 1959—Yun.; 4 km south of Polur settlement, 3600 m, No. 64, May 11, 1959—Lee et al.).

IIIC. Pamir (Goo-Dzhiro river, 4500–5500 m, July 27, Kulan-Aryk area between Zaz settlement and Tash-Ui river 3500–3800 m, July 29—1942, Serp.).

General distribution: Himalayas (west.).

30. **S. rubens** P. Smirn. in Feddes repert. 21 (1925) 231; Krylov, Fl. Zap. Sib. 2 (1928) 174; Fl. Kazakhst. 1 (1957) 154; Lavrenko and Nikol'skaya in Bot. zhurn. 50 (1965) 1422, map 4.—*S. zalesskii* auct. non Wilensky: Roshev. in Fl. SSSR, 2 (1934) 102, p.p.

Described from Nor. Kazakhstan (Akmolinsk region). Type in Leningrad.

In stipa-mixed grass steppes (sometimes, their characteristic species), on rocky and rubble slopes; up to midmountain belt.

IIIA. Junggar: *Junggar Alt.* (Dzhair mountain range, Dzhair pass on Toli-Otu road, Aug. 9, 1957—Yun.).

General distribution: Aral-Casp., Balkh. region, Jung.-Tarb., Nor. Tien Shan; Europe (south-east), West. Sib. (south.), East. Sib. (south-west.).

Note: This species is very close to *S. zalesskii* Wilensky (= *S. rubentiformis* P. Smirn.) occurring in the southern European USSR and the two are commonly grouped together.

31. **S. sareptana** Beck. in Bull. Soc. natur. Moscou, 57 (1882) 52; Roshev. in Fl. SSSR, 2 (1934) 111, Fl. Kirgiz. 2 (1950) 52; Fl. Kazakhst. 1(1956) 155; Fl. Tadzh. 1 (1957) 430; Krylov, Fl. Zap. Sib. 12, 1 (1961) 3090. —*S. capillata* var. *sareptana* (Beck.) Schmalh. Fl. Sredn. i Yuzhn. Rossii [Flora of Central and Southern Russia], 2 (1897) 595; Krylov, l.c. 2 (1928) 180.—Ic.: Fl. Kazakhst. 1, Plate 10, fig. 13.

Described from region along Volga (Krasnoarmeisk-Sarepta bay region). Type in Leningrad.

In arid steppes, on solonetz, rubble and rocky slopes; characteristic of different types of stipa desert steppes; up to midmountain belt.

IA. Mongolia: *Mong. Alt.* (Khalyun region, July 24, 1943; between Adzhi-Bogdo mountain range and Altay somon, Aug. 10, 1947; valley of Bidzhi-Gol river 5 km beyond its discharge onto the foothill plain, Aug. 10, 1947—Yun); *Bas. Lakes* (Lamyn-Khuduk collective near nor. margin of Khan-Khukhei mountain range, July 24, 1945—Yun.); *Val. lakes* (east. margin of Guilin-Tala plain, Aug. 26, 1943—Yun.); *Gobi-Alt.* (nor. slope of Ikhe-Bogdo mountain range, Bityuten-Ama creek, Sept. 12, 1943—Yun.; east. slopes of Khatsabgiin-Khara-Uly 20 km south-east of Bain-Undur somon, July 26; Bain-Undur-Nuru mountain range 16 km from Bain-Undur somon on road to Bain-Tsagan near crossing, July 26—1948, Grub.).

IB. kashgar: *Nor.* (nor.-west. slopes of Sogdyn-Tau mountain range 10–12 km south-east of Akchit hydrothermal station at Kokshal-Darya, 2500 m, Sept. 19, 1958—Yun.).

IIA. Junggar: *Alt. region, Jung. Alt., Tien Shan, Jung. Gobi, Zaisan* (left bank of Ch. Irtysh 25–27 km south-west of Burchum settlement on road to Zimunai, July 10, 1959—Yun.).

General distribution: Aral-Casp., Balkh. region, Jung.-Tarb., Nor. and Cent. Tien Shan; Europe (south-east.), Caucasus (Cis-Caucasus), Middle Asia, West. Sib. (south), Nor. Mong. (Hang.: water divide of Chulutu and Khanui rivers, Aug. 31, 1944—Yun.).

32. S. sibirica (L.) Lam. Tab. Encycl. méth. 1(1791) 158; Franch. Pl. David. 1(1884) 330;? Hemsley, Fl. Tibet (1902) 203; Forbes and Hemsley, Index Fl. Sin. 3 (1904) 383, p.p.; Krylov, Fl. Zap. Sib. 2 (1928) 166; Pavlov in Byull. Mosk. obshch. ispyt. prir. 38 (1929) 12; Roshev. in Fl. SSSR, 2 (1934) 85; Hao in Engler's Bot. Jahrb. 68 (1938) 584; Kitag. Lin. Fl. Mansh. (1939) 95; Ching in Contribs, U.S. Nat. Herb. 28 (1941) 598; Keng in Sunyatsenia, 6, 1 (1941) 72; Norlindh, Fl. mong. steppe, 1 (1949) 62; Grubov, Consp. fl. MNR (1955) 63; Fl. Kazakhst. 1 (1956) 145; Bor, Grasses Burma, Ceyl., Ind. and Pakist. (1960) 646.—*S. brandisii* Mez in Feddes repert. 17 (1921) 207.—*Avena sibirica* L. Sp. pl. (1753) 79.—*Achnatherum sibiricum* (L.) Keng, Clav. Gram. Prim. Sin. (1957) 212; ej. Fl. ill. sin., Gram. (1959) 590.—Ic.: Gmel. Fl. sib. 1 (1748) Tab. 22; Fl. SSSR, 2, Plate 6, figs. 22–26; Keng, l.c. (1959), fig. 525.

Described from South. Siberia. Type (specimen selected by Gmelin) in Leningrad.

In cereal-mixed grass steppes, arid meadows, forest glades, and on rocky slopes and rocks; sometimes in river valley sand; up to midmountain belt.

IA. Mongolia: *Khobd.* (Obgor-Ula hill in mid-course of Tsagan-Nurin-Gola, Aug. 2, 1945—Yun.); *Cis-Hing.* (near Yaksha railway station, Aug. 18, 1902—Litw.); *Cent. Khalkha, East. Mong., Bas. lakes* (left bank of Kharkiry river, Sept. 4, 1931—Bar.; Borig-Del' sand 4–5 km from Baga-Nur lake, July 25; 3 km south of Ulangom on road to Kobdo, July 27—1945, Yun.; 40 km south of Ulyasutai on road to Tsagan-Olom, Shurugin-Gol region, Aug. 17, 1947—Yun.); *Val. lakes* (Orok-Nor, No. 308—Chaney); *Gobi-Alt., East. Gobi, Alash. Gobi* ("Chen-Mu-Kuan, western foothills of the Ho-Lan-Shan"—Ching, l.c.); *Ordos* (sand dunes south of Narin-Gol river, Aug. 30, 1884—Pot.).

IIA. Junggar: *Jung. Alt.* (Dzhair mountain range, 1–1.5 km nor. of Otu settlement on road to Chuguchak, Aug. 4, 1957—Yun.; Borotala river valley, 15 km east of Taldy settlement, 1920 m, No. 2073, Aug. 26, 1957—Huang); *Tien Shan* (south of Barkul' lake, 1800 m, No. 2225, Oct. 27, 1957—Huang; on Karashar-Urumchi road, No. 6213, July 22; near Lotogou village, 2550 m, No. 6280, Aug. 1; near Bortu village, No. 6990, Aug. 3—1958, Lee and Chu; valley of Khanga river 25 km nor.-west of Balinte settlement on Karashar-Yuldus road, Aug. 1, 1958—Yun.).

IIIA. Qinghai: *Nanshan* (Yuzhno-Tetungsk mountain range, 2500 m, July 18, 1872; same site, July 14, 1880—Przew.; "auf dem östlichen Nan-Schan, 2800–3000 m, 3 VIII 1930"—Hao, l.c.).

? IIIB. Tibet: *South.* ("Tisum, 5000 m, 1848, leg. Strachey and Witterbottom"—Hemsley, l.c.).

General distribution: Balkh. region, Jung.-Terb., Nor. Tien Shan; Caucasus (Dagestan), West. Sib. (south-east), East. Sib. (south), Far East (south.), Nor. Mong. (Hent., Hang., Mong.-Daur.), China (Dunbei, Nor., Nor.-West., Cent.), Himalayas (possibly a special species —*S. brandisii* Mez), Korea, Japan.

33. S. subsessiliflora (Rupr.) Roshev. in Fl. Aziatsk. Rossii, 12 (1916) 128.—*Lasiagrostis subsessiliflora* Rupr. in Osten-Sacken and Rupr. Sert. Tiansch. (1869) 35. —*Ptilagrostis subsessiliflora* (Rupr.) Roshev. in Fl. SSSR, 2 (1934) 74; Fl. Kirgiz. 2 (1950) 43; Fl. Kazakhst. 1 (1956) 156; Fl. Tadzh. 1 (1957) 432; Ikonnikov, Opred. rast. Pamira [Key to Plants of Pamir] (1963) 47.—**Ic.:** Fl. SSSR, 2, Plate 6, figs. 33–37; Fl. Tadzh. 1, Plate 51, figs. 10 and 11.

Described from Cent. Tien Shan. Type in Leningrad. Map 5.

In steppes, on rubble and rocky slopes; one of the characteristic species of high-altitude false stipa steppes; usually confined to highly elevated intermontane troughs and watershed uplands; in upper mountain belt.

IB. Kashgar: *West.* (Moin river midcourse, Aug. 23, 1913—Schisch. and Genina; Akkëz-Daban pass 110 km south of Kargalyk along Tibet highway, June 5, 1959—Yun.).

IIA. Junggar: *Tien Shan* (M. Yuldus basin, Alashan area, 2900 m, No. 6248, Aug. 2; same site, Ulyasutai area, 2500 m, No. 6341, Aug. 2; B. Yuldus basin, 12 km south-west of Bain-Bulak village, 2570 m, No. 6423, Aug. 9; same site, 2 km south-west of Bain-Bulak, 2650 m, No. 6455, Aug. 9—1958, Lee and Chu).

IIIA. Qinghai: *Nanshan* (Altyntag mountain range 15 km south of Aksai settlement, 2800 m, Aug. 2, 1958—Petr.).

IIIC. Pamir (upper course of Lapst river, July 20, 1942—Serp.; 10–12 km south of Ulug-Rabat crossing on Tashkurgan-Kashgar road, 4300 m, June 14, 1959—Yun.).

General distribution: Jung.-Tarb., Nor. and Cent. Tien Shan, East. Pam.; Middle Asia (Alay), West. Sib. (south-west. Altay).

Note: R. Roshevitz placed this species and *S. purpurea* Griseb. erroneously (Fl. SSSR, 2) in the genus *Ptilagrostis* Griseb. although they are typical species of *Stipa* L. in all respects. A very closely related species—*S. basiplumosa* Munro ex Hook. f.—is found in the Himalayas.

34. S. tianschanica Roshev. in Fl. Aziatsk. Rossii, 12 (1916) 149 and in Fl. SSSR, 2 (1934) 88; Fl. Kazakhst. 1 (1956) 146.—**Ic.:** Fl. Aziatsk. Rossii, 12, Plate 10, fig. 3.

Described from Cent. Tien Shan (Ak-Shiiryak river). Type in Leningrad.

On rocky and rubble slopes, rocks and in pebble beds; in middle and upper mountain belts.

IB. Kashgar: *Nor.* (valley of Taushkan-Darya river 30–35 km, south-west of Uchturfan oasis, Sept. 17, 1958—Yun.); *West.* (Kyzylsu river basin between Shur-Bulak and Kandzhugan villages, July 4, 1929—Pop.; 58–60 km west-nor.-west of Kashgar along road to Ulugchat, June 17; Baikurt settlement 83 km nor.-west of Kashgar, June 19—1959, Yun.; 61 km nor.-west of Kashgar along road to Ulugchat, 2170 m, No. 9375, June 17; between Ulugchat and Baikurt, 2100 m, No. 9692, June

19; same site, 3000 m, No. 9713, June 20; between Kashgar and Baikurt, 2300 m, No. 9763, June 21—1959, Lee et al.).

IIA. Junggar: *Tien Shan* (in Bogdoshan' hills near Turfan, 1700 m, No. 5744, June 19; Khomote district, along road to timber factory from Bort, 2160 m, No. 6983, Aug. 3—1958, Lee and Chu; valley of Khanga river 25 km nor.-west of Balinte settlement on Karashar-B. Yuldus road, Aug. 1; valley of Muzart river upper course, Sazlik region, Sept. 8—1958, Yun.).

IIIA. Qinghai: *Nanshan* (Altyntag mountain range 15 km south of Aksai settlement, 2800 m, Aug. 2, 1958—Petr.).

IIIC. Pamir (Tagdumbash-Pamir, in Kara-Chukur valley near Beik river, July 18, 1901—Alekseenko).

General distribution: Cent. Tien Shan.

Note: This species, very close to *S. gobica* Roshev., relative to the latter is a more western and higher altitude ecogeographical race.

24. Piptatherum Beauv.
Ess. Agrost. (1812) 17.

1. Glumes 5.5–7 mm long, lanceolate, with relatively narrow membranous margin, pink-violet at tip, with lateral ribs reaching up to middle; lemma 4–4.7 mm long, broadly lanceolate, pilose throughout surface during anthesis; anthers 2.3–3 mm long .... 1. **P. munroi** (Stapf) Mez.

+ Glumes 6.2–8.5 mm long, broadly lanceolate, with very broad membranous margin, light green, with lateral ribs perceptible only in lower fourth to third; lemma 3.2–4.5 mm long, lanceolate-ovate, glabrous or subglabrous in lower half during anthesis; anthers 4–5.5 mm long ....2. **P. songaricum** (Trin. et Rupr.) Roshev.

1. **P. munroi** (Stapf) Mez in Feddes repert. 17 (1921) 212; Roshev. in Bot. mater. Gerb. Bot. inst. AN SSSR, 14 (1951) 125.—*Oryzopsis munroi* Stapf in Hook. f. Fl. Brit. Ind. 7 (1897) 644; Pampanini, Fl. Carac. (1930) 72; Hand.-Mazz. Symb. Sin. 7, 5 (1936) 1295; Walker in Contribs, U.S. Nat. Herb. 28 (1941) 597; Keng, Fl. ill. sin., Gram. (1959) 582; Bor, Grasses Burma, Ceyl., Ind. and Pakist. (1960) 640.—**Ic.:** Keng, l.c. fig. 517.

Described from West. Himalayas. Type in London (K).

On rocky slopes, rocks and talus; in middle and upper mountain belts.

IIIA. Qinghai: *Nanshan* ("La-Chi-Tsu-Shan, No. 717, leg. Ching"—Walker, l.c.).

IIIB. Tibet: *Weitzan* (Yangtze basin, Kabchzha-Kamba village on Khichu river, 4000 m, July 20, 1900—Lad.).

General distribution: China (Centr., South-West.) Himalayas.

2. **P. songaricum** (Trin. et Rupr.) Roshev. in Fl. Kirgiz. 2 (1950) 65 and in Bot. mater. Gerb. Bot. inst. AN SSSR, 14 (1951) 106; Fl. Kazakhst. 1(1956)

159. —*Urachne songarica* Trin. et Rupr. in Mém. Ac. Sci. St.-Pétersb., sér. VI, 7, 2, Sci. nat. 5 (1842) 15.—*Oryzopsis songarica* (Trin. et Rupr.) B. Fedtsch. Rast. Turkest. (1915) 94; Roshev. in Fl. SSSR, 2 (1934) 115.—*O. asiatica* Mez in Feddes repert. 17 (1921) 210.—*O. tianschanica* Drob. et Vved. vo Fl. Uzbek 1(1941) 188 and 537.—Ic.: Fl. Kazakhst. 1, Plate 11, fig. 6.

Described from Altay. Lectotype ("Altai, in rupestribus montium Kurtshum, Arkaul et Dolen-Kara, V 1826, leg. Ledebour et Meyer") in Leningrad.

On rocky slopes, rocks and talus; up to midmountain belt.

IIA. **Junggar:** *Alt. region* (near Koktogoi village, 1200 m, No. 1040, June 8, 1959— Lee et al.; 3–4 km south of Shara-Sume along road to Shipati, left bank of Kran river, July 7, 1959—Yun.); *Tien Shan* (Sarybulak village along Suidun river, 1300–2000 m, April 24; Almaty gorge, near Kul'dzha, 1300–1700 m, April 26—1878, A. Reg.; Irenkhabirga mountain range, Taldy gorge, 1300 m, May 15, 1879—A. Reg.; Urumchinki river valley 10–12 km beyond Urumchi town, June 2, 1957—Yun.; Ili valley 42 km east of bridge on Kash river upward along Kunges, Aug. 29, 1958— Yun.); *Jung. Gobi* (nor. slope of Baitak-Bogdo mountain range, above Ulyastu-Gola, Sept. 18, 1948—Grub.); *Zaisan* (Mai-Kapchagai hill, June 6, 1914—Schisch.).

General distribution: Aral-Casp., Balkh. region, Jung.-Tarb., Nor. Tien Shan; Middle Asia (nor. and east.).

25. Milium L.
Sp. pl. (1753) 61, p.p.

1. **M. effusum** L. Sp. pl. (1753) 61; Forbes and Hemsley, Index Fl. Sin. 3 (1904) 383; Krylov, Fl. Zap. Sib. 2 (1928) 183; Pampanini, Fl. Carac. (1930) 72; Roshev. in Fl. SSSR, 2 (1934) 119; Kitag. Lin. Fl. Mansh. (1939) 83; Fl. Kirgiz. 2 (1950) 67; Fl. Kazakhst. 1 (1956) 164; Keng, Fl. ill. sin., Gram. (1950) 67; Bor, Grasses Burma, Ceyl., Ind. and Pakist. (1960) 593.—Ic.: Fl. SSSR, 2, Plate 5, fig. 22; Keng, l.c. fig. 514.

Described from Europe. Type in London (Linn.).

In forests and forest glades, among shrubs and in subalpine meadows; in middle and upper mountain belts.

IIA. **Junggar:** *Alt. region* (in Altay hills, 1800 m, No. 1056, Aug. 18; 120 km nor. of Burchum, 1400 m, No. 3081, Sept. 14—1956, Ching); *Jung. alt.* (Sin'yuan', 2200 m, No. 1121, Aug. 22; 15 km nor.-west of Arasan, No. 1654, Aug. 29—1957, Huang); *Tien Shan.*

IIIA. **Qinghai:** *Nanshan* (on south. slope of Yuzhno-Tetungsk mountain range, July 14, 1872—Przew.).

General distribution: Jung-Tarb., Nor. and Cent. Tien Shan; Europe, Mediterr., Balk.-Asia Minor, Near East, Caucasus, Middle Asia (Alay mountain range), West. Sib., East. Sib. (south.), Far East (south), China (Dunbei, Nor., Nor.-West., Cent.) Himalayas, Korea, Japan, Nor. Amer.

26. Crypsis Ait.

Hort. Kew 1(1789) 48, nom. conserv.

1. Panicles spicate, capitate, often shorter than broad, with greatly shortened and thickened rachis; surrounded at base by two terminal leaves whose blade gradually (without distinct boundary) transformed into sheath below; palea with solitary rib; stamens two .... 1. C. aculeata (L.) Ait.
+ Panicles spicate, shortly cylindrical or ovate, invariably longer than broad, with normally developed slender rachis................. 2.
2. Panicles spicate, shortly cylindrical, less often ovate, usually surrounded at base only in solitary terminal leaf whose blade is distinct from sheath below; palea with two ribs; stamens 3. ....
.... 2. C. schoenoides (L.) Lam.
+ Panicles spicate, ovate, sometimes even broadly ovate but somewhat laterally flattened, surrounded at base by two terminal leaves; blade of one or both these leaves gradually (without distinct boundary) transformed into sheath below; palea with 1 or 2 ribs; stamens 2 or 3. ....3. C. turkestanica Eig.

1. C. aculeata (L.) Ait. Hort. Kew. 1 (1789) 48; Franch. Pl. David. 1 (1884) 331; Forbes and Hemsley, Index Fl. Sin. 3 (1904) 384; Danguy in Bull. Mus. nat. hist. natur. 20 (1914) 146; Krylov, Fl. Zap. Sib. 2 (1928) 184; Pavlov in Byull. Mosk. obshch. ispyt. prir. 38 (1929) 15; Roshev. in Fl. SSSR, 2 (1934) 122; Kitag. Lin. Fl. Mansh. (1938) 70; Norlindh, Fl. mong. steppe, 1(1949) 69; Fl. Kirgiz. 2 (1950) 68; Grubov, Consp. fl. MNR (1955) 63; Fl. Kazakhst. 1 (1956) 165; Fl. Tadzh. 1 (1957) 453; Keng, Fl. ill. sin., Gram. (1959) 577.—Schoenus aculeatus L. Sp. pl. (1753) 42.—Ic.: Fl. SSSR, 2, Plate 8, fig. 1; Tadzh. 1, Plate 55, figs. 1–3; Keng, l.c. fig. 510.

Described from South. Europe. Type in London (Linn.).

In solonchak, saline meadows and along banks of brackish water reservoirs; up to lower mountain belt.

IA. **Mongolia:** Bas. lakes (on Dzabkhyn river in Khirgis-Nur lake region, July 26, 1879—Pot.; Shargain-Gobi, near Ikhe-Gol, Sept. 4; Shargain-Gol river bank between Shargain-Tsagan-Nur lake and Dzak-Obo, Sept. 8—1930; Pob.); Val. lakes (Orok-nur lake bank, Sept. 8, 1924—Pavl.; between Orok-Nur lake and Lamyn-Gegen, 1926—Tug.); Gobi-Alt. (Bain-Tukhum area, Aug. 31, 1931—Ik.-Gal.; same site, July–Aug., 1933—Sim.); West. Gobi (Tseel' somon, Tszakhoi-Tszaram area, Aug. 18, 1943—Yun.); East. Gobi (Baga-Ude in Baishintin lowland, Sept. 15, 1928—L. Shastin); Alash. Gobi (Edzin-Gol river, Butun-Nur, June 16, 1909—Czet.); Ordos (Huang He valley, Aug. 4, 1871—Przew.; Huang He second terrace in Dalat town region, Aug. 10, 1957—Petr.).

IB. **Kashgar:** Nor. (Chadyr-Kul' village between Maralbashi and Aksu, Aug. 6; Sairam village between Kucha and Kurl', Aug. 15—1929, Pop.; around Chiglyk settlement 1 km nor.-west of Mirsali village, No. 8558, Aug. 6; lake near Chiglyk settlement, No. 8598, Aug. 12—1958, Lee and Chu); West. (Yarkand-Darya river valley, 1000 m, June 22, 1889—Rob.; Yandoma village in Kashgar oasis, July 24,

1929—Pop.); *East.* (Bugas village on southern rim of Khami-oasis, Aug. 19, 1895—Rob.; near Toksun settlement, No. 7225, June 9, 1958—Lee and Chu; in Khami settlement zone, No. 10135, Sept. 30, 1959—Lee et al.).

IIA. Junggar: *Tarb.* (Mukur region west of Ulyungur lake, July 30, 1876—Pot.); *Jung. Gobi* ("Bords de l'Irtich, No. 1182, 29 VIII 1895, leg. Chaffanjon"—Danguy, l.c.; Dzheiran-Bulak area 30 km nor.-nor.-east of Karamai settlement along road to Urkho, June 19; Darbaty river valley at its intersection with Karamai-Altay road, June 20—1957, Yun.).

General distribution: Aral-Casp., Balkh. region, Jung.-Tarb., Nor. Tien Shan; Europe (south), Mediterr. Balk.-Asia Minor, Near East, Caucasus, Middle Asia, West. Sib. (south), China (Dunbei, Nor.), Afr. (nor.).

2. C. schoenoides (L.) Lam. Tabl. Encycl. méth. 1(1791) 166; Danguy in Bull. Mus. nat. hist. natur. 20 (1914) 146; Roshev. in Fl. SSSR, 2 (1934) 125; Norlindh, Fl. mong. steppe, 1 (1949) 70; Fl. Kirgiz. 1 (1950) 69; Grubov, Consp. fl. MNR (1955) 63; Fl. Kazakhst. 1 (1956) 166; Fl. Tadzh. 1 (1957) 452; Keng, Fl. ill. sin., Gram. (1959) 577; Bor, Grasses Burma, Ceyl., Ind. and Pakist. (1960) 622.—*Phleum schoenoides* L. Sp. pl. (1753) 60.—*Heleochloa schoenoides* (L.) Host. ex Roem. Collect. (1809) 233; Krylov, Fl. Zap. Sib. 2 (1928) 186.—Ic.: Fl. SSSR, 2, Plate 8, fig. 3; Fl. Tadzh. 1, Plate 55, figs. 4 and 5; Keng, l.c. fig. 509.

Described from South. Europe. Type in London (Linn.).

In solonetz meadows, solonchak, coastal pebble beds and sand, sometimes as weed in irrigated farms; up to lower mountain belt.

IA. **Mongolia:** *East. Mong.* (Dalainur plain south of Gandzhur monastery, 1899—Pal.); *Bas. lakes, East. Gobi* (Inner Mongolia, old flat sand washes, No. 563.—Chaney); *Alash. Gobi* (Dzhargalante area on Edzin-Gol river, June 16, 1909—Czet.; Sanpinkhou village between Bayan—Khoto and Min'tsin on west. margin of Tengeri sand, July 14, 1958—Petr.); *Ordos* (Huang He river second terrace in Dalat town region, Aug. 10, 1957—Petr.).

IB. **Kashgar:** *Nor.* (10 km south-east of Chiglyk settlement, No. 8631, Aug. 17, 1958—Lee and Chu; *West.* (Kashgar oasis, Khan-Aryk village, July 22; same site, Dong-Aryk village, July 22—1929, Pop.); East (Bugas village on southern fringe of Khami oasis, Aug. 22, 1895—Rob.; near Toksun settlement, No. 7225, June 9, 1957—Lee and Chu).

IIA. **Junggar:** *Tien Shan* (nor. of Kul'dzha town, May 3, 1877; lower course of Borotola river, 1000 m, Aug. 19 and 24, 1878; Urtaksary river, 1700 m, Sept. 1880—A. Reg.); *Jung. Gobi* (nor.: crossing of Dyurbel'dzhin on Ch. Irtysh river, July 21; Ch. Irtysh catchment area, Aug. 16—1876, Pot.; "Bords de l'Irtich, 29 VIII 1895, leg. Chaffanjon"—Danguy, l.c.; south.: Mogukhu water reservoir near Savan settlement, No. 1564, June 25, 1957—Huang).

General distribution: Aral-Casp., Balkh. region, Jung.-Tarb., Nor. and Cent. Tien Shan; Europe (south), Mediterr., Balk.-Asia Minor, Near East, Caucasus, Middle Asia, West. Sib. (south), Nor. Mong. (Hang.), China (Dunbei, Nor.), Himalayas (Kashmir), Afr. (nor.); introduced in other countries.

3. C. turkestanica Eig in Agric. Records Inst. Agric. and Nat. Hist. Tel-Aviv, 2 (1929) 206; Roshev. in Fl. SSSR, 2 (1934) 125; Fl. Kirgiz. 2 (1950) 68; Fl. Kazakhst. 1 (1956) 165.—Ic.: Fl. SSSR, 2, Plate 8, fig. 2; Fl. Kirgiz. 2, Plate 11, fig. 3.

Described from Middle Asia. Lectotype (Perovsk bay, Syr Darya district, Katan-Kamys area, June 20, 1914, M. Spiridonov) in Leningrad.

In saline meadows and solonchak, coastal sand and pebble beds; up to lower mountain belt.

IIA. Junggar: *Tarb.* (Mukur area west of Ulyungur lake, July 30, 1876—Pot.). General distribution: Aral-Casp., Balkh. region, Jung.-Tarb.; Caucasus (Casp. region), Middle Asia, West. Sib. (south-east).

27. Phleum L.
Sp. pl. (1753) 59.

1. Panicles spicate, usually elongate-cylindrical, with free branches forming lobes as panicles bend; glumes without long cilia along keels; transformed gradually at tip into short cusp. Leaf blades scabrous on both sides, with greatly thickened marginal ribs .....
.... 1. **Ph. phleoides** (L.) Karst.

+ Panicles spicate, usually shortly cylindrical; their branches almost wholly fused with main rachis and not forming lobes as panicles bend; glumes with long cilia along keels; abruptly transformed into 2–3.5 mm long awn. Leaf blades glabrous or subglabrous below, with slender marginal ribs ....
.... 2. **Ph. alpinum L.**

Section 1. Chilochloa
(Beauv.) Dum.

1. **Ph. phleoides** (L.) Karst. Deutsche Fl. (1880–1883) 374; Pavlov in Byull. Mosk. obshch. ispyt. prir. 38 (1929) 15; Ovczinnikov in Fl. SSSR, 2 (1934) 131; Kitag. Lin. Fl. Mansh. (1939) 85; Norlindh, Fl. mong. steppe, 1 (1949) 70; Fl. Kirgiz. 2 (1950) 70; Grubov, Consp. fl. MNR (1955) 63; Fl. Kazakhst. 1 (1956) 168; Fl. Tadzh. 1 (1957) 397; Keng, Fl. ill. sin., Gram. (1959) 550; Krylov, Fl. Zap. Sib. 12, 1 (1961) 3092.—*Ph. boehmeri* Wib. Prim. Fl. Werth. (1799) 125; Krylov, l.c. 2 (1928) 189.—*Phalaris phleoides* L. Sp. pl. (1753) 55. —Ic.: Fl. SSSR, 2, Plate 9, fig. 3; Keng, l.c. fig. 477.

Described from Europe. Type in London (Linn.).

In meadows, on rocky slopes, rocks and in steppes, forest glades and among shrubs; up to upper mountain belt.

IA. Mongolia: *Mong. Alt.* (Bulugun somon, south. flank of Indertiin-Gol valley, July 24; same site, upper Kharagaitu-Gol, July 24—1947, Yun.); *Cent. Khalkha* (20–25 km nor. of Under Khan, July 25, 1949—Yun.); *Bas. lakes* (Khirgis somon 5–6 km nor.-nor.-east of Main-Khutul' area, July 23; Borig-Del' sand 4–5 km south-east of Baga-Nur lake, July 25—1945, Yun.).

IIA. Junggar: *Alt. region, Jung. Alt., Tien Shan.*

General distribution: Aral-Casp., Balkh. region, Jung.-Tarb., Nor. and Cent. Tien Shan; Europe, Mediterr., Balk.-Asia Minor, Near East, Caucasus, Middle Asia (hilly), West. Sib., East. Sib. (south), Nor. Mong. (Hent., Hang., Mong.-Daur.), China (Nor.); introduced in many other countries of both hemispheres.

Section 2. Phleum

2. **Ph. alpinum** L. Sp. pl. (1753) 59; Forbes and Hemsley, Index Fl. Sin. 3 (1904) 384; Danguy in Bull. Mus. nat. hist. natur. 20 (1914) 146; Krylov, Fl. Zap. Sib. 2 (1928) 188; Ovczinnikov in Fl. SSSR, 2 (1934) 135; Fl. Kirgiz. 2 (1950) 71; Fl. Kazakhst. 1 (1956) 170; Fl. Tadzh. 1 (1957) 400; Keng, Fl. ill. sin., Gram. (1959) 550; Bor, Grasses Burma, Ceyl., Ind. and Pakist. (1960) 402.—*Ph. commutatum* Gaud. in Alpina, 3 (1800) 4 and Agrost. Helv. 1(1811) 40. —Ic.: Fl. SSSR, 2, Plate 9, fig. 11; Fl. Kazakhst. 1, Plate 12, fig. 4; Keng, l.c. fig. 478.

Described from Scandinavian peninsula ("Lapland"). Type in London (Linn.).

In meadows, on rocky slopes and in pebble beds, along banks of water reservoirs; in upper mountain belt.

IA. Mongolia: *Mong. Alt.* ("entre Oulioun-Gur et Kobdo, 2500 m, 7 IX 1895, leg. Chaffanjon"—Danguy, l.c.; summer camp in Bulugun somon in upper Indertiin-Gol, July 24, 1947—Yun.).

IIA. Junggar: *Alt. region* (Oi-Chilik river, Sept. 8, 1876—Pot.; Uichilik valley, July 2, 1908—Sap.; near Tsinkhe, 2200 m, No. 805, Aug. 3; San'chzhuan village in Fuyun' town region, 2500 m, No. 1896, Aug. 17—1956, Ching); *Jung. Alt.* (17 km nor.-west of Ven'tsyuan', 2700 m, Aug. 29; 3 km south of Yakou, 2900 m, No. 1701, Aug. 30—1957, Huang); Tien Shan.

General distribution: Jung.-Tarb., Nor. and Cent. Tien Shan; Arct. (Europ.), Europe (hilly), Balk.-Asia Minor (hilly), Near East (hilly), Caucasus (hilly), Middle Asia (hilly), West. Sib. (Altay), East. Sib. (Sayans), Far East (Kamchatka and Kuril islands), China (Dunbei, Cent., South-West.), Himalayas, Japan, Nor. and South Amer. (Arctic and hills), Afr. (nor.).

Note: This species is sometimes distinguished as *Ph. commitatum* Gaud., restricting the name *Ph. alpinum* L. for the high-altitude European tetraploid and not diploid race. However, *Ph. alpinum* was described not from the Alps but from 'Lapland' and hence this name was first applied to arctic plants. Evidently, it would be appropriate to regard the tetraploid alpine race not as an independent species, but a subspecies— *Ph. alpinum* subsp. *ambiguum* Beck.

28. Alopecurus L.
Sp. pl. (1753) 60.

1. Winter annual 10–40 cm tall, usually with stems ascending geniculately and rooting at lower nodes. Panicles spicate, narrowly cylindrical, 2–7 cm long and 3–5 mm broad; spikelets

2–2.6 mm long; anthers 0.5–0.8 mm long ....
.... 1. **A. aequalis** Sobol.
+ Perennial (10) 20–100 (120) cm tall with erect stems. Panicles spicate, broadly cylindrical, 1–8 cm long and 5–13 mm broad; spikelets 2.5–5.5 mm long; anthers 2–4 mm long............................2.
2. Glumes densely covered with long hairs throughout or (or almost so) outer surface; panicles spicate, shortly cylindrical, usually 1–4 cm long..3.
+ Glumes usually with long hairs only along keel and lateral ribs, less often also along surface between ribs with very short scarcely visible hairs; panicles spicate, on average very long and narrow, usually 2–8 cm long ....4.
3. Plant 10–40 cm tall, confined to bald peaks of hills; spikelets oval, usually violet and glumes convergent at tips; awns exserted from spikelet or reduced and almost not exserted ....
....2. **A. alpinus** Smith.
+ Plant 25–70 cm tall, confined to moderate and low hills and plains; spikelets oblong-oval, usually without violet tinge and glumes with somewhat laterally diverging tips (spikelets slightly "urceolate"); awns invariably welldeveloped, exserted from spikelet ....4. **A. brachystachyus** M.B.
4. Tips of glumes perceptibly deviated laterally, diverging (spikelets "urceolate"); awn arising close to centre or above centre of lemma and usually not or barely exserted from spikelet; less often, strongly exserted (up to 2 mm) from spikelet...............................
....3. **A. arundinaceus** Poir.
+ Tips of glumes not laterally deviated, converging (spikelets not "urceolate"); awn arising below centre of lemma and invariably strongly exserted (by 1.5–3 mm) from spikelet ....
.... 5. **A. pratensis** L.

1. A. aequalis Sobol. Fl. Petropol. (1799) 16; Franch. Pl. David. (1884) 329; Forbes and Hemsley, Index Fl. Sin. 3 (1904) 384; Ovczinnikov in Fl. SSSR, 2 (1934) 158; Fl. Kazakhst. 1 (1956) 174; Fl. Tadzh. 1 (1957) 407; Keng, Fl. ill. sin., Gram. (1959) 574; Bor, Grasses Burma, Ceyl., Ind. and Pakist. (1960) 392.—*A. aristulatus* Michx. Fl. Bor. Amer. 1 (1803) 43.—*A. fulvus* Smith in Smith and Sowerby, Engl. Bot. 21 (1805) Table 1467; Krylov, Fl. Zap. Sib. 2 (1928) 195; Pavlov in Byull. Mosk. obshch. ispyt. prir. 38 (1929) 15.—*A. fulvus* var. *sibiricus* Kryl. Fl. Alt. (1914) 1581 and Fl. Zap. Sib. 2 (1928) 196.—*A. amurensis* Kom. in Izv. Peterb. bot. sada, 16 (1916) 151; id. in Acta Horti Petrop. 20 (1901) 272, nom. nud.; Ovczinnikov, l.c. 158; Kitag. Lin. Fl. Mansh. (1939) 59; Grubov, Consp. fl. MNR (1955) 63.—*A. geniculatus* auct. non L. Fl. Kirgiz. 2 (1950) 75.—**Ic.:** Fl.

SSSR, 2, Plate 10, figs. 1 and 3; Fl. Kazakhst. 1, Plate 12, fig. 5; Keng, l.c. fig. 508.

Described from Leningrad district. Type in Leningrad.

In swamps and along banks of water reservoirs; up to midmountain belt.

IA. **Mongolia: Mong. Alt.** (Bulugun river catchment area at the discharge site of Ulyaste-Gol into it, July 20, 1947—Yun.); *East. Mong.* (Nunlintun' village near Khailar town, No. 685, June 11; near Hailar, No. 778, June 20—1951, Lee et al.).

IIA. **Junggar: Alt. region** (Tsinkhe, Chzhunkhaitsze, No. 1169, Aug. 6; around Koktogoi, 1200 m, Nos. 1812 and 1816, Aug. 13—1956, Ching).

General distribution: Aral-Casp. and Balkh. region (sporadic), Jung.-Tarb., Nor. and Cent. Tien Shan; Europe, Mediterr., Balk.-Asia Minor, Near East, Caucasus, Middle Asia (less frequent), West. and East. Sib., Far East, Nor. Mong., China (Dunbei, Nor., Nor.-West., Cent., East.), Himalayas, Korea, Japan, Nor. Amer.

Note: While in European and American populations of this species awns are exserted from spikelets by not more than 0.6 mm, in Cent. Asia as well as in general over much of the Asian distribution range of *A. aequalis*, populations predominate with awns exserted from spikelets by 0.5–1.5 mm. Specimens with much longer awns of spikelets were described as variety *A. fulvus* var. *sibiricus* Kryl. and as an independent species *A. amurensis* Kom. but, in our view, there is hardly any justification for recognising it as an independent taxonomic entity since awn dimensions do not correlate with any other characteristic; populations with well-developed as well as highly reduced awns are often found in the same region. Thus, among the specimens recorded above from around Hailar town, awns are scarcely exserted from spikelets in No. 778 while their exsertion is 0.7–1.5 mm in No. 685. A very similar picture is noticed with respect to the extent of development of awns in *A. alpinus* Smith.

2. **A. alpinus** Smith, Fl. Brit. 3 (1804) 1386; Krylov, Fl. Zap. Sib. 2 (1928) 193; Ovczinnikov in Fl. SSSR, 2 (1934) 155.—*A. borealis* Trin. Fund. Agrost. (1820) 58; Ovczinnikov, l.c. 155.—*A. alpinus* β. *borealis* (Trin.) Griseb. in Ledeb. Fl. Ross. 4 (1852) 461; Krylov, l.c. 193.—*A. glaucus* β. *altaicus* Griseb. l.c. 462.—*A. alpinus* var. *altaicus* (Griseb.) Kryl. l.c. 193, p.p.—*A. altaicus* (Griseb.) V. Petrov, Fl. Yakut. 1 (1930) 146.—*A. alpinus* subsp. *borealis* (Trin.) Jurtz. in Novosti sist. vyssh. rast. (1965) 307.—? *A. glaucus* auct. non Less. : Danguy in Bull. Mus. nat. hist. natur. 20 (1914) 146.—Ic.: Fl. SSSR, 2, Plate 10, figs. 10 and 11.

Described from Great Britain (Scottish hills). Type in London (K).

In grasslands, on rocky and rubble slopes, in pebble beds and on rocks; in the bald peaks zone of hills.

IA. **Mongolia: Khobd.** (Altyn-Khatasyn area, June 6; Bairimen-Daban pass, 2700 m, June 8; Kharkiry peak, 2700 m, July 10—1879, Pot.); *Mong. Alt., Cent. Khalkha* (around Ikhe-Tukhum-Nur lake, scarps near Ubur-Dzhargalante sources, Aug. 30, 1926—Zam.); *Gobi alt.* (Ikhe-Bogdo mountain range, detritus in upper hill belt, June 29, 1945—Yun.; same site, upper gorge of Narin-Khurimt, 3500 m, July 29, 1948—Grub.).

General distribution: Arct., Europe (Scotland), West. Sib. (nor. and Altay), East. Sib. (hilly), Far East (nor.), Mong. (Fore-Hubs., Hang.).

Note: This circumpolar arctic species, widely distributed on bald peaks of Siberia, reaches the southern boundary of its distribution within Cent. Asia. While in the European Arctic, populations and specimens of this species with highly reduced awns scarcely exserted from spikelets predominate, in the Asian Arctic as well as in the bald peaks belt of Asia, populations and specimens with well-developed awns strongly exserted from spikelets predominate. Specimens with long awns were described as an independent species—*A. borealis* Trin.—but, in our view, should be regarded only as a variety, *A. alpinus* var. *borealis* (Trin.) Griseb. Among the specimens cited above, those with long awns predominate, while among those from Ikhe-Bogdo mountain range (Gobi-Alt.), there are specimens with awns almost completely reduced. Specimens of *A. alpinus* from relatively very low hills are naturally very large in size and recognised as variety *A. alpinus* var. *altaicus* (Griseb.) Kryl. and even as an independent species, *A. altaicus* (Griseb.) V. Petr.

3. **A. arundinaceus** Poir. in Lam. Encycl. méth. 8 (1808) 776; Norlindh, Fl. mong. steppe, 1 (1949) 71; Keng, Fl. ill. sin., Gram. (1959) 572; Bor, Grasses Burma, Ceyl., Ind. and Pakist. (1960) 393.—*A. ventricosus* Pers. Synops. pl. 1 (1805) 80, non Huds. 1778; Krylov, Fl. Zap. Sib. 2 (1928) 194; Pavlov in Byull. Mosk. obshch. ispyt. prir. 38 (1929) 15; Ovczinnikov in Fl. SSSR, 2 (1934) 149; Kitag. Lin. Fl. Mansh. (1939) 60; Fl. Kirgiz. 2 (1950) 72; Grubov, Consp. fl. MNR (1955) 64; Fl. Kazakhst. 1 (1956) 172; Fl. Tadzh. 1 (1957) 405.—*A. ruthenicus* Weinm. in Cat. Sem. Horti Dorpat. (1810) 10; Franch. Pl. David. 1 (1884) 329.—Ic.: Fl. Kazakhst. 1, Plate 12, fig. 6; Keng, l.c. fig. 504.

Described from cultivated specimens of unknown origin. Type in Paris.

In meadows, along banks of water reservoirs, in coastal pebble beds, on rocky slopes, usually in saline soils; up to upper mountain belt.

IA. Mongolia: *Mong. Alt., Cis-Hing.* (Khalkhin-Gol river valley 13 km south-east of Khamar-Daban, Aug. 11, 1949—Yun.; same site, Khalkhin-Gol river catchment area, June 19, 1954—Dashnyam); *Cent. Khalkha, East. Mong., Bas. lakes. Gobi-Alt., East Gobi, Alash. Gobi* (Dynyuanin [Bayan-Khoto] oasis, June 3, 1908—Czet.); *Khesi* (near An'si town, No. 55, July 3, 1956—Ching).

IIA. Junggar: *Alt. region* (around Koktogoi town, No. 2097, June 14, 1956—Ching); *Tarb., Jung. Alt., Tien Shan, Jung. Gobi, Zaisan* (on road from Lasty river to Tumandy river, July 26, 1876—Pot.); *Dzhark.* (Ili river bank west of Kul'dzha town, May, 1877—A. Reg.).

IIIC. Pamir (Tashkurgan, July 25, 1913—Knorring; on west. margin of Tashkurgan, June 13, 1959—Yun.).

General distribution: Aral-Casp., Balkh. region, Jung.-Tarb., Nor. and Cent. Tien Shan; Europe, Mediterr., Balk.-Asia Minor, Near East, Caucasus, Middle Asia, West. Sib., East. Sib. (south), Nor. Mong., China (Dunbei, Nor., Nor.-West., Cent., East.); introduced in the USA and several other countries of both hemispheres.

Note: Specimens with greatly reduced awns scarcely exserted from spikelets predominate in Cent. Asia as also in other parts of the distribution range of *A. arundinaceus*. However, specimens are also found with awns exserted 1–2 mm from spikelet: *A. arundinaceus* var. *exserens* (Griseb.) Marsson, Fl. Neuvorpomm. Rüg. (1869) 555 = *A. ruthenicus* β. *exserens* Griseb. in Ledeb. Fl. Ross. 4 (1853) 464. It is possible that these specimens with long awns represent hybrids, *A. arundinaceus* x *A. pratensis*.

4. **A. brachystachyus** M.B. Fl. taur.-cauc. 3 (1819) 56, in adnot.; Pavlov in Byull. Mosk. obshch. ispyt. prir. 38 (1925) 15; Ovczinnikov in Fl. SSSR, 2 (1934) 145; Kitag. Lin. Fl. Mansh. (1939) 59; Grubov, Consp. fl. MNR (1955) 64; Keng, Fl. ill. sin., Gram. (1959) 572.—Ic.: Fl. SSSR, 2, Plate 10, fig. 8; Keng, l.c. fig. 505.

Described from Transbaikal. Type in Leningrad.

In meadows, coastal pebble beds and along banks of water reservoirs; up to midmountain belt.

IA. **Mongolia:** *Cis-Hing.* (near Yaksha station, June 11, 1902—Litw.; Khalkhin-Gol river, June 10, 1956—Dashnyam); *Cent. Khalkha* (Kerulen valley near Batur-Chzhokhon-Tszasaka, 1899—Pal.; basin of Dzhargalante river, Uste hill at sources of Kharukhe river, Aug. 12, 1925–Krasch. and Zam.); *East. Mong.* (near Manchuria railway station, June 5; close to Kharkhonte railway station, June 7—1902, Litw.; around Manchuria railway station, July 5, 1922—Skvortsov; near Hailar town, No. 596, June 8, 1951—Lee et al., near Shilin-Khoto town, 1960—Ivan.); *Bas.-lakes* (Kobdo river valley near crossing on Kobdo-Ulangom road, Aug. 23, 1944—Yun.); *Gobi-Alt.* (Artsa-Bogdo mountain range, near Kotel'-Usu collective, Aug. 6, 1926—Gus.).

General distribution: East. Sib. (south), Far East (south), Nor. Mong., China (Dunbei, Nor.).

Note: This species is very close to the arctic-alpine (bald peak) species *A. alpinus* Smith and is not always clearly differentiated from it. However, *A. brachystachyus* is a plant of very low hills and is fairly well distributed even on plains.

5. **A. pratensis** L. Sp. pl. 1 (1753) 60; Danguy in Bull. Mus. nat. hist. natur. 20 (1914) 146; Krylov, Fl. Zap. Sib. 2 (1928) 191; Ovczinnikov in Fl. SSSR, 2 (1934) 150; Kitag. Lin. Fl. Mansh. (1955) 60; Fl. Kirgiz. 2 (1950) 75; Grubov, Consp. fl. MNR (1955) 64; Fl. Kazakhst. 1 (1956) 172; Keng, Fl. ill. sin., Gram. (1959) 572.—Ic. : Fl. SSSR, 2, Plate 11, figs. 2 and 3; Fl. Kazakhst. 1, Plate 12, fig. 9; Keng, l.c. fig. 503.

Described from Europe. Type in London (Linn.).

In meadows, riverine pebble beds and on rocky slopes; predominantly in upper and middle mountain belts.

IA. **Mongolia:** *Mong. Alt.* (nor. slope of Khara-Dzarga mountain range near Khairkhan-Duru, Aug. 25, 1930—Pob.; nor. trail of Taishiri-Ula mountain range 7–8 km south-east of Gobi Altay ajmaq centre, July 11, 1945—Yun.); *Cent. Khalkha* ("Vallée du Kéroulen, 25 V 1896"—Danguy, l.c.).

IIA. **Junggar:** *Alt. region* (near Qinhe, No. 961, Aug. 4; near Koktogoi, No. 2221, Aug. 18—1956, Ching); *Tarb.* (Saur mountain range, valley of Karagaitu river in Khobuk river basin, June 23, 1957—Yun); *Jung. Alt.* (20 km south of Ven'tsyuan', 2810 m, No. 1470, Aug. 14; 15 km nor.-west of Ven'tsyuan', No. 1629, Aug. 29—1957, Huang); *Tien Shan.*

General distribution: Aral-Casp. and Balkh. region (hilly zone), Jung.-Tarb., Nor. and Cent. Tien Shan; Arct., Europe, Mediterr., Balk.-Asia-Minor, Near East, Caucasus, Middle Asia (hilly), West. Sib., East. Sib., Nor. Mong., China (Dunbei, Nor.), Nor. Amer.; introduced in some other countries of both hemispheres.

29. Polypogon Desf.
Fl. Atlant. 1 (1798) 66.

1. Perennial, often with procumbent shoots rooting at nodes. Glumes entire or slightly bilobed at tip, with awns shorter than glumes or not more than twice longer; lemma awned .... 2.
+ Annual, without procumbent shoots. Glumes distinctly bilobed at tip, with 2.5–4 times longer awns; lemma awned or not .... 3.
2. Panicles very dense; stalks of spikelets short, longest of them articulated much above their base; palea almost as long as lemma .... 1. **P. fugax** Nees ex Steud.
+ Panicles lax; stalks of spikelets very long, all of them articulated only at their base; palea 1.5 times shorter than lemma .... 2. **P. ivanovae** Tzvel.
3. Plant up to 40 cm tall. Panicles up to 6 cm long; glumes deeply (for 1/4–1/3 their length) bilobed, lobes with long ciliae along margin; lemma not awned .... 3. **P. maritimus** Willd.
+ Plant up to 60 cm tall. Panicles up to 10 cm long; glumes shortly bilobed (for 1/8–1/6 their length) lobes with short ciliae along margin; lemma usually awned (awn easily shedding) .... 4. **P. monspeliensis** (L.) Desf.

Section 1. Nowodworskya
(C. Presl) Tzvel[2].,

1. P. fugax Nees ex Steud. Synops. Pl. Glum. 1 (1854) 184; Bor, Grasses Burma, Ceyl., Ind. and Pakist. (1960) 403.—*P. higagaweri* Steud. l.c. 422; Keng, Fl. ill. sin., Gram. (1959) 553.—*P. demissus* Steud. l.c. 422; Roshev. in Fl. SSSR, 2 (1934) 164; Persson in Bot. notiser (1938) 274; Fl. Kirgiz. 2 (1950) 76; Fl. Kazakhst. 1 (1956) 174; Fl. Tadzh. 1 (1957) 384. —**Ic.:** Fl. SSSR, 2, Plate 12, fig. 4; Fl. Kazakhst. 1, Plate 12, fig. 10.

Described from the Himalayas (Nepal). Type in London (K). Isotype in Leningrad.

Along banks of water reservoirs, in wet sandy sites and coastal pebble beds, often as weed in irrigated fields, around irrigation ditches and in wastelands; up to lower mountain belt.

IB. **Kashgar:** *Nor.* (Keliin-Pichan basin, Chon-Karadzhal settlement, June 22, 1959—Yun.); *West.* ("Bostanterek, ca. 2400 m, 1 VIII 1925"—Persson, l.c.; Upal village in Kashgar oasis, July 10; near Kashgar, July 26—1929, Pop.; "Kashgar, 1330 m, 18 VI 1933"—Persson, l.c.); *East.* (near Toksun town, Sept. 2, 1929—Pop.; Turfan

---

[2]. *Polypogon* sect. Nowodworskya (C. Presl) Tzvel. comb. nova.—*Nowodworskya* C. Presl, Rel. Haenk. 1 (1930) 351. = *Polypogon* β. *Polypogonagrostis* Aschers. et Graebn. Synops. Mitteleur. Fl. 2 (1899) 160.

oasis, No. 5462, May 28; nor.-east of Turfan, No. 5488, June 1; Yarku village in Turfan region, No. 5564, June 9; on Pichan-Nankhu road, No. 6678, June 15; 10 km nor.-east of Toksun, No. 7265, June 15—1958, Lee and Chu).

General distribution: Mediterr., Balk.-Asia Minor, Near East, Caucasus (south. and east. Transcaucasus), Middle Asia, China (Nor., Nor.-West., Cent., East., South-West., South.), Himalayas, Korea, Japan, Indo-Malay., Nor. and South Amer. (possibly introduced).

2. P. ivanovae Tzvel. sp. nova.—Planta perennis, 8–20 cm alta, caespites parvos et laxos formans; culmi geniculato-adscendentes, glabri et laeves; vaginae glabrae, in parte superiore scabriusculae; ligulae 2–4.5 mm lg., dorso scaberrimae, margine cililolatae; laminae 0.8–2.5 mm lt., laxe convolutae vel planae, utrinque scabrae. Paniculae 2.5–7 cm lg. et 0.5–1.5 cm lt., sat laxae, parum roseo-violaceo tinctae, ramis abbreviatis scabris, pedunculis 0.7–2.5 mm lg., scabris, solum ad basin articulationibus donatis; spiculae 2.2–2.8 mm lg., uniflorae; glumae aequilongae et spiculam aequans, scaberrimae, apice obtusatae et in arista recta 0.5–2 mm lg., subito abeuntes; lemmata 1.4–1.8 mm lg., ovata, glabra et laevia, apice aristulato-dentata et in arista recta vel curvatula 2.3–3.5 mm lg. abeuntia; paleae quam lemmata sesqui breviores, glabrae et laeves; antherae 0.5–0.8 mm lg.

Typus: In regione praemontana septentrionali montium Kuenlun, in segetis, 1300–1700 m, 12 VI 1889, W. Roborovski. In Herb. Inst. Bot. Acad. Sci. URSS—LE conservatur.

Affinitas. Haec species a specie proxima: P. fugax Nees ex Steud. paniculis laxioribus, pedunculis longioribus solum ad basin articulationibus dontis et paleis brevioribus differt.

Described from Kashgar. Type in Leningrad.

In field crops; in midmoutain belt.

IB. Kashgar: South. (nor. Kunlun foothills, 1300–1700 m, June 12, 1889—Rob., type!).

General distribution: endemic.

Note: This species was identified for description as a new species by N.A. Ivanova and is named after her.

Section 2. Polypogon

3. P. maritimus Willd. in Neue Schr. Gesellsch. Naturf. Fr. Berlin, 3 (1801) 442; Danguy in Bull. Mus. nat. hist. natur. 17, 6 (1911) and ibid. 20, 3 (1914) 11; Krylov, Fl. Zap. Sib. 2 (1928) 199; Roshev. in Fl. SSSR, 2 (1934) 165; Norlindh, Fl. mong. steppe, 1 (1949) 71; Fl. Kirgiz. 2 (1950) 76; Grubov, Consp. fl. MNR (1955) 64; Fl. Kazakhst. 1 (1956) 175; Fl. Tadzh. 1 (1957) 386.—Ic.: Fl. SSSR, 2, Plate 12, fig. 6; Fl. Kazakhst. 1, Plate 12, fig. 11.

Described from France. Type in Berlin.

In saline meadows and solonchak, wet coastal sand and pebble beds; up to midmountain belt.

IA. Mongolia: *Val. lakes* (Orok-Nur lake, Aug. 4, 1926—Tug.); *Alash. Gobi* (Edzin-Gola valley, Burkhan-Khub, June 12, 1926—Glag.; "Dottoren-namok, No. 1665, 18 IX 1927, leg. Hummel; Bayan-Bogdo, Wentsun-hai-tze, No. 7585, 18 VI, 1929, leg. Söderbom"—Norlindh, l.c.); *Khesi* (along Dankhe river, July 12, 1879—Przew.).

IB. Kashgar: *West.* (Yarkand-Darya, Pshak-Sandy stream, 1300 m, June 13, 1889—Rob.); *East.* (Khami oasis, May 20, 1879—Przew.; "Toksoun, 30 IX, 1907; leg. Vaillant"—Danguy, l.c. [1911]; nor.-east of Toksun settlement, No. 7323, June 19, 1958—Lee and Chu).

IC. Qaidam: *plain* (nor. slope of Burkhan-Budda mountain range, Nomokhun-Gol gorge, Aug. 21, 1884—Przew.); *hilly* (Tuguryuk around Khyrma, 2800 m, July 31, 1901—Lad.).

IIA. Junggar: *Alt. region* (near Ukagou south of Koktogoi, No. 1772, Aug. 11, 1956—Ching); *Tarb.* (Tumandy river, July 27, 1876—Pot.); *Jung. Gobi* ("Konour-Oulen, No. 622, 30 VI, 1895, leg. Chaffanjon"—Danguy, l.c. [1914]; Darbaty river valley at its crossing on Karamai-Altay road, June 20; 3–4 km east of Orkhu settlement on Dyam river, June 21—1957, Yun.); *Zaisan* (right bank of Ch. Irtysh below Burchum river, Sary-Dzhasyk area, June 15, 1914—Schisch.).

General distribution: Aral-Casp., Balkh. region, Jung.-Tarb., Cent. Tien Shan (along Naryn river); Europe (Atlant. and south.), Mediterr., Balk.-Asia Minor, Near East, Caucasus (South. and East. Transcaucasus), Middle Asia, West. Sib. (southwest. Altay); introduced in Nor. Amer. (USA).

4. P. monspeliensis (L.) Desf. Fl. Atlant. 1 (1798) 67; Henderson and Hume, Lahore to Yarkand (1873) 341; Franch. Pl. David. 1 (1884) 332; Kanitz in Széchenyi, Wissensch. Ergebn. 2 (1898) 737; Forbes and Hemsley, Index Fl. Sin. 3 (1904) 386; Danguy in Bull. Mus. nat. hist. natur. 20 (1914) 146; Krylov, Fl. Zap. Sib. 2 (1928) 200; Pampanini, Fl. Carac. (1930) 72; Roshev. in Fl. SSSR, 2 (1934) 164; Persson in Bot. notiser (1938) 274; Fl. Kirgiz. 2 (1950) 79; Grubov, Consp. fl. MNR (1955) 64; Fl. Kazakhst. 1 (1956) 175; Fl. Tadzh. 1 (1957) 385; Keng, Fl. ill. sin., Gram. (1959) 553.—*Alopecurus monspeliensis* L. Sp. pl. (1753) 61.—**Ic.:** Fl. SSSR, 2, Plate 12, Fig. 5; Keng, l.c. fig. 481.

Described from France. Type in London (Linn.).

In saline meadows and solonchak, riverine sand and pebble beds, often as weed in oases, along roadsides and around irrigation ditches; up to midmountain belt.

IA. Mongolia: *East. Mong.* (right bank of Huang He above Hekou town, July 23, 1884—Pot.; 60 km nor. of Dunshen town, valley of Khantaiguan river, Aug. 13, 1957—Petr.); *Bas. lakes* (Shargain-Gobi, Shargain-Gol river bank between Shargain-Tsagan-Nur lake and Dzak-Obo, Sept. 8, 1930—Pob.; same site, valley of Shargain-Gol near Sundultu-Baishing, Sept. 5, 1948—Grub.); *East. Gobi* (near Bailinmyao, 1960—Ivan.); *West. Gobi, Alash. Gobi* (Mukhur-Shara spring south of Tostu-Nur mountain range, Aug. 13, 1948—Grub.); *Ordos*.

IB. Kashgar: *Nor., West., South., East., Takla Makan* (Cherchen oasis, Aug. 3, 1890—Rob.).

IC. Qaidam: *plain* (nor. slope of Burkhan-Budda mountain range, Nomokhun-Gola gorge, Aug. 18, 1884—Przew.).

IIA. Junggar: *Alt. region* (Ukagou near Koktogoi, Aug. 20, 1956—Ching); *Tien Shan* ("entre le Sairam-Nor et l 'Ebi-Nor, 24 VII 1895, leg. Chaffanjon"—Danguy, l.c.; Khaidyk-Gol river, tributary of Tsagan-Usu, 1600 m, Aug. 12, 1893—Rob.); *Jung. Gobi, Dzhark.* (near Suidun town, July 16, 1877—A. Reg.).

IIIB. Tibet: *Weitzan* (nor. slope of Burkhan-Budda mountain range, Khatu gorge, 3500 m, July 25, 1901—Lad.).

IIIC. Pamir (near Tashkurgan town, July 25, 1913—Knorring; Issyksu river estuary, 3100 m, July 19; Pakhpu river gorge, 2700 m, Aug. 2—1942, Serp.).

General distribution: Aral-Casp., Balkh. region, Jung.-Tarb., Nor. and Cent. Tien Shan; Mediterr., Balk.-Asia Minor, Near East, Caucasus, Middle Asia, West. Sib. (far south), China (excluding Dunbei), Himalayas, Korea (south), Japan, Indo-Malay.; introduced weed widely distributed in Americas from USA to Argentina.

30. Agrostis L.
Sp. pl. (1753) 61.

1. Palea 1.5–2 times shorter than lemma; lemma not awned; sometimes dorsally on upper side with very short erect awn not exserted from spikelet .... 2.

+   Palea more than 4 times shorter than lemma, often absent .... 4.

2. Plant 15–50 cm tall, without creeping subsurface shoots but usually with procumbent surface shoots rooting at nodes; stems usually ascending; leaf blades up to 3 mm broad. Panicles 5–12 cm long, usually quite dense, with greatly shortened branches; anthers 0.8–1.3 mm long .... 3. A. stolonifera L.

+   Plants 20–100 cm tall, with creeping subsurface shoots but without procumbent surface shoots; stems usually erect; leaf blades up to 6 mm broad. Panicles 6–25 cm long, lax, with relatively long branches .... 3.

3. Panicles usually small (6–16 cm long), less diffuse with short and thick secondary branches, less deviating from axis of primary branches; spikelets usually 2–3 mm long, greenish or matte pinkish-violet; callus of lemma almost invariably with 0.1 to 0.3 mm long tufts of hairs; palea 1.5 times shorter than lemma; anthers usually 1–1.4 mm long .... 1. A. gigantea Roth.

+   Panicles usually very large (10–25 cm long), widely spreading with very long and slender secondary branches highly deviating from axis of primary branches during anthesis and thereafter; spikelets usually 1.6–2 mm long, pinkish-violet; callus of lemma glabrous or subglabrous; palea 1.5–2 times shorter than lemma; anthers usually 0.7–1 mm long .... 2. A. mongholica Roshev.

4. Anthers 0.4–0.6 mm long; panicles 8–25 cm long, usually diffuse with scabrous branches; spikelets 1.5–2.3 mm long; lemma not awned .... 4. A. clavata Trin.

+   Anthers 0.9–1.5 mm long .... 5.

5.   Panicles 2.5–8 cm long, very dense, with greatly shortened smooth branches; spikelets 2.5–3.5 mm long; lemma not awned. Plant 8–25 cm tall .... **5. A. hugoniana** Rendle.

+   Panicles 5–12 cm long, lax, with fairly long, usually scabrous, less often subglabrous branches; spikelets 1.6–2.8 mm long; lemma with long, genaflexed dorsal awn. Plant 20–60 cm tall .... **6. A. trinii** Turcz.

Section 1. Agrostis

1. **A. gigantea** Roth, Fl. Germ. 1 (1788) 31; Bor, Grasses Burma, Ceyl., Ind. and Pakist. (1960) 387.—*A. alba* auct. non L.: Forbes and Hemsley, Index Fl. Sin. 3 (1904) 389; Krylov, Fl. Zap. Sib. 2 (1928) 206, p.p.; Pampanini, Fl. Carac. (1930) 73; Schischkin in Fl. SSSR, 2 (1934) 183; Fl. Kirgiz. 2 (1950) 83; Fl. Kazakhst. 1 (1956) 179; Fl. Tadzh. 1 (1957) 382; Keng, Fl. ill. Sin., Gram. (1959) 531.—*A. stolonifera* auct. non L.: Norlindh, Fl. mong. steppe, 1 (1949) 72; Ikonnikov, Opred. Rast. Pamira [Key to Plants of Pamir] (1963) 52.—**Ic.:** Fl. Kazakhst. 1, Plate 12, fig. 14; Keng, l.c. fig. 457.

Described from Europe. Type in Berlin.

In meadows, along banks of water reservoirs, in riverine pebble beds, often as weed along roadsides, in field crops and wastelands; up to upper mountain belt.

IA. **Mongolia:** *Khobd.* (valley of Bukhu-Muren river 5–6 km beyond Bukhu-Muren somon, July 31, 1945—Yun.); *Mong. Alt., Cis-Hing.,* East. Mong., Bas. Lakes, Gobi Alt. (Urdzhyum col., Aug. 4, 1886—Pot.); *East. Gobi* (near Bailinmyao, 1960—Ivan.); *Ordos* (20 km west of Dzhasak town, Aug. 16, 1957—Petr.).

IB. **Kashgar:** *Nor., West., South.* (nor. Slope of Kerii mountain range along Gendzhi-Darya river, 2700–4000 m, June 16, 1885—Przew.).

IIA. **Junggar:** *Alt. Region, Tarb.* (on road from Lasty river to Tumandy, July 26, 1876—Pot.); *Jung. Alt., Tien Shan, Jung. Gobi* (St. Kuitun settlement 2–3 km east of Kuitun state farm, June 30; same site, 3–4 km nor. of St. Kuitun settlement, July 6–1957, Yun.); *Dzhark.* (Pilyuchi village near Kul'dzha, June; near Suidun town, July 16—1877, A. Reg.); *Balkh.-Alak.* (near Churchutsu river around Chuguchak, Aug. 1840—Shrenk; Sary-Khulsyn south of Durbul'dzhin, July 22, 1947–Shum.).

IIIC. **Pamir** (Issyksu river estuary, 3100 m, July 19; Pakhpu river gorge, 2700 m, Aug. 2—1942, Serp.).

General distribution: Aral-Casp., Balkh. Region, Jung.-Tarb., Nor. And Cent. Tien Shan, East. Pamir; Arct. (Europ.), Europe, Mediterr., Balk.-Asia Minor, Near East, Caucasus, Middle Asia, West. and East. Sib., Far East (less often, possibly also introduced), Nor. Mong., China, Himalayas, Korea, Japan; introduced in many other countries of both hemispheres.

2. **A. mongolica** Roshev. in: Nor. Mong. 1 (1926) 162 and in Izv. Glavan. bot. Sada SSSR, 28 (1929) 381; Pavlov in Byull. Mosk. Obshch.

ispyt. Prir. 38 (1929) 16; Schischkin in Fl. SSSR, 2 (1934) 184; Kitag. Lin. Fl. Mansh. (1939) 59; Grubov, Consp. fl. MNR (1955) 64; Krylov, Fl. Zap. Sib. 12, 1 (1961) 3093.

Described from Mongolia. Type in Leningrad.

Mostly in saline meadows and solonchak; up to lower mountain belt.

IA. **Mongolia:** *Mong. Alt.* (Khasagtu-Khairkhan hills, near the discharge of Dundu-Tseren-Gol river on to the trails, Sept. 15, 1930—Pob.); Cis-Hing., Cent. Khalkha, East. Mong., Bas. lakes (Shargain-Gobi, near Gol-Ikhe, Sept. 6, 1930—Pob.); *Val. Lakes* (Ongiin-Gol river 30 km below Khushu-Khida, July 17, 1943—Yun.; Ongiin-Gol river near Khushu-Khid monastery, July 14, 1948—Grub.); *Gobi-Alt., East. Gobi* (3 km nor. of Lus somon, Aug. 16, 1950—Lavr.); *West. Gobi* (Tseel' somon, Ikhe-Tszaram area, Aug. 19, 1943—Yun.); *Ordos* (in Huang He valley, July 16, 1871—Przew.; Taitukhai area, Aug. 17, 1884—Pot.; 10 km south-west of Ushin town, Aug. 4, 1957—Petr.).

General distribution: West. Sib. (Altay), East. Sib. (south), Nor. Mong., China (Dunbei, Nor.), Korea (?).

3. A. stolonifera L. Sp. pl. (1753) 62; Kitag. Lin. Fl. Mansh. (1939) 59; Keng, Fl. ill. sin. Gram. (1959) 531; Bor, Grasses Burma, Ceyl., Ind. and Pakist. (1960) 390.—*A. alba* auct. non L.: Krylov, Fl. Zap. Sib. (1928) 206. p.p.—*A. stolonizans* auct. Non Bess.: Schischkin in Fl. SSSR, 2 (1934) 184, p.p.—Ic.: Keng,—l.c. fig. 456.

Described from Europe. Type in London (Linn.).

In wet meadows, marshes and along banks of water reservoirs; up to midmountain belt.

IA. **Mongolia:** *Mong. Alt.* (valley of Buyantu river, Aug. 27; Buyantu camp, Aug. 28—1930, Bar).

General distribution: Aral-Casp., Balkh. Region, Jung.-Tarb., Nor. and Cent. Tien Shan; Arctic (Europ.), Europe, Mediterr., Balk.-Asia Minor, Near East. Caucasus, Middle Asia, West. Sib., East. Sib. (south), China (Dunbei, Nor.), Himalayas, Nor. Amer. (nor.-east.); introduced in many other countries.

Section 2. Trichodium
(Michx.) Dum.

4. A. clavata Trin. in Spreng. Neue Entdeck. 2 (1821) 55; Krylov, Fl. Zap. Sib. 2 (1928) 208; Pavolv in Byull. Mosk. obshch. ispyt. Prir. 38 (1929) 16; Schischkin in Fl. SSSR, 2 (1934) 178; Grubov, Consp. Fl. MNR (1955) 64; Keng, Fl. ill. sin., Gram. (1959) 541.—Ic.: Fl. SSSR, 2, Plate 13, fig. 11; Keng, l.c. fig. 470.

Described from Kamchatka. Type in Leningrad.

In meadows, pebble beds, sand shoals of rivers and forest glades; up to lower mountain belt.

IA. **Mongolia:** *Cis-Hing.* (near Trekhrech'e, No. 1512, July 8; Argun' district, Daulagar village, 700 m, No. 1524, July 13; same site, near Genkhetsyao bridge, No. 1426, July 16—1951, Lee et al.).

General distribution: Europe (Scandinavian peninsula and north-east), West Sib., East. Sib., Far East, Nor. Mong., China (Dunbei, Nor.), Korea, Japan, Nor. Amer.

5. **A. hugoniana** Rendle in Forbes and Hemsley, Index Fl. Sin. 3 (1904) 389; Walker in Contribs, U.S. Nat. Her. 28 (1941) 596; Keng, Fl. ill. Sin., Gram. (1959) 544.—Ic.: Keng, l.c. fig. 472.

Described from China (Shenxi province). Type in London (BM).

In grasslands and on rocky slopes; in upper mountain belt.

IIIA. Qinghai: *Nanshan* ("Ta-Pan-Shan near Sining, No. 672, leg. R.C. Ching"-Walker, l.c.).

General distribution: China (Nor.-West., Cent., South-West.).

A. Kudoi Honda, in Miyabe et Kudo, Fl. Hokk. et Saghal. 2 (1931) 135; Gubanov, Consp Fl. Outer Mong. (1996) 16.—*A. trinii* Turcz., p.p.

IA. Mongolia. *Mong. Alt., Cen. Khalkha.*

6. **A. trinii** Turcz. in Bull. Soc. Natur. Moscou, 29, 1 (1856) 18, in nota; Pavlov in Byull. Mosk. Obshch. ispyt. Prir. 38 (1929) 16; Schischkin in Fl. SSSR, 2 (1934) 175; Kitag. Lin. Fl. Mansh. (1939) 59; Grubov, Consp. fl. MNR (1955) 64; Keng, Fl. ill. sin. Gram. (1959) 538.—*A. canina* subsp. *trinii* (Turcz.) Hult. in Kungl. Sv. Vetensk. Handl., ser. 5, 8, 5 (1962) 114.—*A. canina* auct. non L.: Pavlov, l.c. 16; Grubov, l.c. 64. —Ic.: Fl. SSSR, 2, Plate 13, fig. 5; Keng, l.c. fig. 463.

Described from Baikal area (Irkutsk region). Type in Leningrad.

In meadows, on rocky slopes, in coastal pebble beds, forest glades and meadow steppes; up to upper mountain belt.

IA. Mongolia: *Khobd.* (east. Slope of Ulan-Daba pass on Ulangom-Tsagan-Nuru road, Aug. 29, 1945—Yun.); *Mong. Alt.* (right bank of Buyantu river, July 15, 1925—Klem.; upper course of Dundu-Tumurte river, right tributary of Bulugun river, July 25, 1947—Yun.); *Cis-Hing., Cent. Khalkha, East. Mong.* (near Hailar town, No. 731, June 19, 1951—Lee et al.; same site, 1960—Ivan.).

IIA. Junggar: *Alt. region* (near Qinhe, 2300 m, No. 968, Aug. 4, 1956—Ching).

General distribution: East. Sib., Far East, Nor. Mong., China (Dunbei), Korea, Japan.

Note: Unlike the typical relatively alpine specimens of this species having somewhat scabrous panicle branches, specimens from river valleys in the plains of Cent. Asia and Dunbei have highly scabrous panicle branches and were quite often erroneously identified as *A. canina* L. However, *A. trinii* is readily distinguished from *A. canina* (predominantly European forest species) in the presence of creeping subsurface shoots and very short leaf ligules. *A. trinii* exhibits a closer relationship to the forest-steppe European species *A. vinealis* Schreb.(= *A. syreitschikovii* P. Smirn.).

31. Pentatherum Nábĕl.

In Publ. Fac. Sci. Univ. Masaryk Brno, III (1929) 8; Nevski in Tr. Bot. Inst. AN SSSR, ser. 1, 3 (1936) 146.

1. **P. dshungaricum** Tzvel. sp. nova—Planta perennis, 20–40 cm alta, caespites parvos formans, estolonifera; culmi et vaginae glabri at laeves;

ligulae 1.5–2.5 mm lg. dorso scabriusculae; laminae 1–3 mm lt., planae, glabrae, supra scaberulae, subtus sublaeves. Paniculae 5–10 cm lg., plus minusve contractae, ramis abbreviatis sublaevibus; spiculae 3–4 mm lg., uniflorae, roseoviolaceo tinctae; glumae lanceolatae, glabrae, 1–3-nerviae, carinis scabriusculis; lemmata 2–2.5 mm lg., 5-nervia, pilosa, apice irregulariter dentata, infra medium aristam 3–4.3 mm lg. geniculato-curvatam vel subrectam gerentia; calli pili circa 1/3 lemmatis aequales; paleae quam lemmata sesqui breviores, glabrae et laeves, bicarinatae.; antherae 0.7–0.9 mm lg.

Typus: Jugum Alatau Dshungaricus, 17 km boreali-occidentaliter versus opp. Venjtzjuan (Arasan), 2700 m s.m., No. 1661; 29 VIII 1957, leg. Kuan K.C. In Herb. Inst. Bot. Acad. Sci. URSS—LE conservatur.

Affinitas. A speciebus proximis haec species differt: a *P. pilosulo* (Trin.) Tzvel.—Paniculae ramis sublaevibus et paleis longioribus (non ½ lemmatis brevioribus), a *P. angrenico* Butk.—spiculis majoribus (non 2–2.5 mm lg.) et calli pilis brevioribus.

Described from Junggar Alatau. Type in Leningrad.

In grasslands and on rocky slopes; in upper mountain belt.

IIA. Junggar: *Jung. Alt.* (17 km nor.-west of Ven'tsyuan' [Arasan], 2700 m, No. 1661, Aug. 29, 1957—Huang, type !); *Tien Shan* (between Barchat and Yakou, 2900 m, No. 1679; between Ulastai and Yakou, No. 1968, Aug. 30, 1957—Huang). General distribution: endemic.

Note. Species of the genus *Pentatherum* Nábĕl. are closest to *Agrostis* L. but their inclusion in the latter results in total disappearance of morphological boundaries between genera *Agrostis* and *Calamagrostis* Adans close to it. Thus, like S.A. Nevski (l.c.), we hold *Pentatherum* to be a small independent genus occupying an intermediate position between genera *Agrostis* and *Calamagrostis*. The species described by us differs greatly from the solitary *P. olympicum* (Boiss.) Nábĕl. s.l. (including *Calamagrostis agrostidiformis* Roshev. in it) found in Junggar Alatau of USSR territory. Evidently, *P. dshungaricum* is closest to *P. pilosulum* (Trin.) Tzvel. Comb., nova [=*Agrostis pilosula* Trin. in Mém. Ac. Sci. St.-Pétersb., sér. VI, 6 (1841) 372] found in the hills of northern India and Nepal and differs from the latter mainly in the very long palea. Similarly, *P. angrenicum* Butk. from West. Tien Shan and *P. trichanthum* (Schisch.) Nevski from Armenia (to date these species have been reported only from a single site) are fairly close to *P. dshungaricum*.

32. Calamagrostis Adans.

Fam. pl. 2 (1763) 31.—*Deyeuxia* Clar. in Beauv. Ess. Agrost. (1812) 43.—
Stilpnophleum Nevski in Tr. Bot. inst. AN SSSR, ser. 1, 3 (1936) 143.

1. Rachilla above base of floret not extending or extending in form of very short glabrous outgrowth; lemma usually with 3 (less often 5) ribs; hairs of callus usually distinctly longer than lemma, less often as long. Plants with long creeping subsurface shoots, not forming mats .... 2.

+   Rachilla above base of floret extending in form of fairly long shaft (rudiment of rachilla) covered in fairly long hairs; lemma invariably with 5 ribs; hairs of callus usually shorter than lemma, less often as long .... 8.

2.   Hairs of callus as long (less often almost so) as lemma; lemma with 3–5 ribs; awn terminal (or subterminal) on lemma .... 3.

+   Hairs of callus distinctly (usually by 1.5 times) longer than lemma; lemma almost invariably with 3 ribs; rudiment of rachilla absent or greatly contracted, barely visible even under high magnification .... 4.

3.   Rudiment of rachilla in form of distinctly visible glabrous shaft; awn usually arising from bidentate tip of lemma; panicles lax, slightly diffuse at anthesis. ....
.... 2. **C. alexeenkoana** Litv. ex Roshev.

+   Rudiment of rachilla absent or barely visible even under high magnification; awn usually arising slightly below bidentate tip of lemma; panicles extremely dense, with closely packed spikelets .
.... 22. **C. turkestanica** Hack.

4.   Awn dorsal on lemma near centre, or slightly above centre ...... 5.

+   Awn arising directly from entire or bidentate tip of lemma .... 6.

5.   Spikelets (4) 5–7 (7.5) mm long; upper glume shorter than lower by less than 1 mm .... 18. **C. epigeios** (L.) Roth.

+   Spikelets (7) 7.5–9 (10) mm long; upper glumes 1–1.5 mm shorter than lower .... 20. **C. macrolepis** Litv.

6.   Leaf ligules with spinules and very short hairs dorsally; stems on average very high (50–180 cm) with (2) 3–4 (5) nodes. Panicles greenish or light pinkish-violet, usually quite lax ....
.... 17. **C. dubia** Bunge.

+   Leaf ligules glabrous and smooth or only with spinules drsally; stems on average very low (20–120 cm) with (1) 2–3 (4) nodes. Panicles fairly pinkish-violet in general .... 7.

7.   Panicles 4–12 cm long, very dense, with spikelets closely packed on highly contracted branches; hairs of callus 1.5 or more shorter than lemma. .... 19. **C. hedinii** Pilger.

+   Panicles 10–25 cm long, relatively lax, with very long branches and less closely packed spikelets; hairs of callus 1.5 or more longer than lemma .... 21. **C. pseudophragmites** (Hall. f.) Koel.

8(1) .   Plants with long creeping subsurface shoots, not forming mats; stems with (3) 4–5 (6) nodes, of which upper one located above centre of stem. Panicles lax, with relatively long branches; hairs of callus nearly as long as lemma; awn usually arising close to centre of lemma .... 16. **C. purpurea** (Trin.) Trin.

+   Plants with or without short creeping subsurface shoots, often forming mats; stems with (1) 2–3 (4) nodes, of which upper one located below centre of stem. Panicles quite dense with generally contracted branches, often spicate; hairs of callus usually shorter than lemma, less often as long .... 9.

9.   Awn twisted in lower part, later genuflexed, surpassing tip of lemma by 0.8–2 mm .... 10.

+   Awn not twisted in lower part (or only slightly twisted), almost straight or slightly curved, not surpassing or only barely surpassing by not more than 0.5 mm) tip of lemma. .... 19.

10.   Panicles (1) 1.5–3.5 (4.5) cm long and (1) 1.2–2 (2.3) cm broad, very dense, spicate, cylindrical or ovate (as in species of genus *Alopecurus* L.); spikelets 4.5–8 mm long, usually quite pink or pinkish-violet; awn arising close to base of lemma .... 11.

+   Panicles of different size and form, often, spicate but with branches usually projecting in form of lobes; branches not cylindrical or ovate .... 12.

11.   Plant without creeping subsurface shoots, forming dense mats; glumes with spinules only along ribs .... 4. **C. anthoxanthoides** (Munro) Regel.

+   Plant with short creeping subsurface shoots forming small lax mats; glumes fairly densely covered with spinules over entire outer surface .... 11. **C. przevalskyi** Tzvel.

12.   Awn arising close to centre of lemma or slightly below; palea almost 1.5 times shorter than lemma .... 13.

+   Awn arising close to base of lemma (in their lower one-fourth); palea usually almost as long as lemma, less often not more than ¼ shorter .... 15.

13.   Hairs of callus equal to 2/3–4/5 length of lemma; hairs of relatively short rudiment of rachilla not reaching tip of lemma. Plant with fairly long creeping subsurface shoots .... 1. **C. alaica** Litv.

+   Hairs of callus shorter than 2/3 length of lemma; hairs of long rudiment of rachilla reaching tip of lemma. Plants with short-creeping shoots .... 14.

14.   Plant up to 35 cm tall. Panicles 6–9 cm long, spicate, pink; hairs of callus equal to half length of lemma or slightly longer .... 5. **C. borii** Tzvel.

+   Plant usually very tall (up to 1.5 m). Panicles up to 30 cm long, quite dense but not spicate, greenish or light pinkish-violet; hairs of callus usually slightly shorter than half length of lemma .... 13. **C. scabrescens** Griseb.

15(12). Leaf blades 3–10 mm broad, flat, with upper side altogether glabrous and smooth .... **C. turczaninowii** Litv.

+ Leaf blades in general very narrow, scabrous, with spinules on upper side .... 16.

16. Panicles 10–15 cm long, relatively lax, not spicate; hairs of callus equal to 1/4–1/3 length of lemma .. 15. **C. venusta** (Keng) Tzvel.

+ Panicles 1.5–7 cm long, very dense, spicate; hairs of callus equal to half length of lemma .... 17.

17. Leaf blades green, scattered on upper side with spinules; ligules densely puberulent dorsally .... 3. **C. altaica** Tzvel.

+ Leaf blades greyish-green with upper side densely covered with spinules; ligules glabrous dorsally .... 18.

18. Panicles with branches projecting in the form of lobes. Stems quite scabrous under panicles; spinules on upper side of leaves (as in branches of panicles) very short .... 14. **C. tianschanica** Rupr.

+ Panicles dense and broader than long, nearly cylindrical. Stems usually smooth under panicles; spinules on upper side of leaves (as in branches of panicles) very slender, elongated ....
.... 6. **C. compacta** (Munro ex Hook. f. Hack.)

19(9). Rudiment of rachilla long, roughly 1/2 as long as lemma, densely covered with long hairs, reaching tip of lemma. Greyish-green plants of upland steppes, solonetz and solonchak .... 20.

+ Rudiment of rachilla very short, roughly 1/3 as long as lemma diffusely covered with hairs, usually not reaching tip of lemma .
.... 22.

20. Hairs of callus roughly 1/3 as long as lemma; awn arising in lower fourth of lemma .... 8. **C. kokonorica** (Keng) Tzvel.

+ Hairs of callus 1/2 as long as lemma; awn usually arising at level of lower third of lemma .... 21.

21. Glumes gradually acuminate at tip, usually quite pinkish-violet; panicles with closely packed spikelets, invariably spicate ....
.... 9. **C. macilenta** (Griseb.) Litv.

+ Glumes obtuse or shortly acuminate at tip, usually greenish or light pinkish-violet; panicles with less closely packed spikelets, not always spicate.. 12. **C. salina** Tzvel.

22(19). Palea as long as lemma; hairs of callus 1/2 as long as lemma. Junction between sheath and blade pilose .... **C. obtusata** Trin.

+ Palea 1/5 to 1/3 shorter than lemma; hairs of callus longer than 1/2 of lemma. Junction between sheath and blade glabrous, very rarely pilose .... 23.

23. Glumes subulate or almost subulate-acuminate at tip; hairs of callus usually almost as long as lemma. Greyish-green plant of upland steppes .... **C. sajanensis** Malysch.

+ Glumes shortly acuminate at tip. Green plant .... 24.

24. Glumes 4–6 mm long, slender, with fairly broad membranous margin; hairs of callus usually as long as lemma, less often slightly shorter. Plant of rocky slopes and pebble beds of upper hill belt .... C. lapponica (Wahl.) Hartm.

\+ Glumes (2.5) 3–4.5 (5) mm long, very thick and very dull coloured, with very narrow membranous margin. Plant of marshes and marshy meadows .... 25.

25. Glumes (3.5) 4–4.5 (5) mm long, with fairly large spinules on outer surface; hairs of callus almost as long as lemma or not more than 1/4 shorter. Ligules of upper cauline leaves usually 3–5 mm long .... 7. C. inexpansa A. Gray.

\+ Glumes (2.5) 3–3.5(4) mm long, with smaller spinules on outer surface; hairs of callus usually 2/3–3/4 as long as lemma. Ligules of upper cauline leaves usually 2–3 mm long ....
.... 10. C. neglecta (Ehrh.) Gaertn., Mey. et Scherb.

Section 1. Deyeuxia (Clar.) Dum.

1. C. alaica Litv. in Bot. Mater. Gerb. Glavn. bot. Sada RSFSR, 2 (1921) 122; Roshev. in Fl. SSSR, 2 (1934) 211; Fl. Kirgiz. 2 (1950) 90; Fl. Kazakhst. 1 (1956) 186; Fl. Tadzh. 1 (1957) 370; Tzvelev in Novosti sist. vyssh. rast. (1965) 48.—*C. schugnanica* Litv. l.c. 123; Roshev., l.c. 212; Fl. Kazakhst. 1 (1956) 371.

Described from Middle Asia (Alay valley). Type in Leningrad.

In grasslands, on rocky and clayey slopes and in pebble beds of rivers; in upper mountain belt.

IIIC. Pamir (Tagarma valley, July 25; Karasu river bank, July 26—1909, Divn.). General distribution: Cent. Tien Shan, East. Pam.; Middle Asia (Pamir-Alay).

Note: This species is evidently of hybrid origin—C. *tianschanica* Rupr. × C. *hedinii* Pilg. and, in morphological characteristics, occupies an intermediate position between these assumed parent species. Specimens with extremely narrow (up to 3.5 mm broad) leaf blades and awn arising very low, and described as an independent species, C. *schugnanica* Litv. (l.c.), predominate in Pamir.

2. C. alexeenkoana Litv. ex Roshev. in Fl. SSSR, 2 (1934) 212 and 749. Described from Pamir. Type in Leningrad.

On rocky slopes, talus and in riverine pebble beds; in upper mountain belt.

IIIB. Tibet: *Chang Tang* (Cherchen river basin, 6–7 km south of Isachan settlement, 3400 m, No. 936, June 5, 1959—Lee et al.).

IIIC. *Pamir* (Tagdumbash-Pamir, Kara-Chukur river valley, Tura area in Tekezikerik gorge, 4300 m, July 18, 1901—Alekseenko, type!).

General distribution: endemic.

Note: This species is very close to *C. turkestanica* Hack., differing from it in the presence of rudiment of rachilla, lemma invariably with 5 ribs and slightly more lax panicles. Both species were evidently of hybriod origin in the past from *C. tianschanica* Rupr. × *C. pseudophragmites* (Hall. f.) Koel. s.l. (including *C. hedinii* Pilg.) but, unlike *C. alaica* Litv., largely tend toward the latter of the assumed parent species.

3. **C. altaica Tzvel. sp. Nova.**—Planta perennis, laxe caespitosa, 40–65 cm alta; culmi sat tenues, 2–3-nodis, laeves; foliorum laminae (1.5) 2–4 (5) mm lt., planae, virides, scabridae; ligulae 1.5–4 mm lg., dorso copiose, sed brevissime pilosae. Paniculae 5–7 cm lg. et 0.7–1.4 cm lt. spiciformes, ramulis brevibus (ad 2 cm lg.), scabris, oligospiculatis; spiculae 4–5 mm lg., plus minusve roseo-violaceae; glumae late lanceolatae, paulo inaequales, apice acuminatae, minute scabrae; lemmata glumis breviora, 5-nervia, minute scabra, apice vulgo denticulis 2–4 instructa; arista subbasalis, geniculata, lemmatis apicem 0.8–1.5 mm superans; calli pili circa 1/2 lemmatis longitudinis aequales; rachillae rudimentum 0.8–1.3 mm lg., pilosum; antherae 1.8–2.5 mm lg.; caryopsis ignota.

Typus: China boreali-occidentalis, Altaj Mongolicus, ad ripam dextram fl. Kairty 20 km ad septentriones versus pag. Koktogoj in declivitate stepposa vallis Kuidyn, No. 1462, 15 VII 1959, A. junatov. In Herb. Inst. Bot. Acad. Sci. URSS—LE conservatur.

Affinitas: Haec species *C. macilentae* (Griseb.) Litv. et *C. tianschanicae* Rupr. proxima, sed a specie prima—aristis subbasalibus longioribus valde geniculatis et paleis lemmatis subaequalibus, a secunda—foliis viridbus, ligulis extus copiose (sed brevissime) pilosis et paleis longioribus differt.

Described from Junggar (south. slope of Mongolian Altay). Type in Leningrad. Plate III, fig. 1.

On rocky slopes, in grasslands and riverine pebble beds; in midmountain belt.

IIA. Junggar: *Alt. Region* (in Qinhe region, 1800 m, No. 1235, Aug. 2; between Qinhe and Tsagan, 2100 m, No. 1514, Aug. 8; near Fuyun' town, 1800 m, No. 2055, Aug. 18; near Klamar-Yu coll 1400 m, No. 2498, Aug. 27—1956, Ching; right bank of Kairty river 20 km nor. of Koktogoi town, Kuidun valley, No. 1462, July 15, 1959—Yun., type!).

General distribution: endemic.

Note: This species differs from other Central Asian species of the genus by spicate panicles—*C. tianschanica* Rupr. and *C. macilenta* (Griseb.) Litv.—in lemma and palea of nearly same length and considerably less xeromorphic habit. Moreover, awns are longer and resemble those of the Eurasian forest species *C. arundinacea* (L.) Roth. From yet another relatively close species, *C. pavlovii* Roshev. = *C. krylovii* Reverd.), found in the hilly border regions of the USSR from Trans-Ili Alatau to the Sayans, *C. altaica* differs in small spicate panicles, very short hairs of callus and very long awns.

4. **C. anthoxanthoides** (Munro) Regel in Acta Horti Petrop. 7 (1880) 639; Pilger in Hedin, S. Tibet, 6, 3 (1922) 92; Roshev. in Fl. SSSR, 2 (1934) 209;

Persson in Bot. notiser (1938) 274; Fl. **Kirgiz.** 2 (1950) 8; Fl. **Kazakhst.** 1 (1956) 185; Ikonnikov, Opred. rast. Pamira [Key to Plants of Pamir] (1963) 52.—*Deyeuxia anthoxanthoides* Munro in Henderson and Hume, Lahore to Yarkand (1873) 339.—*Stilpnophleum anthoxanthoides* (Munro) Nevski in Tr. Bot. inst. AN SSSR, ser. I, 3 (1936) 144; Fl. Tadzh. 1 (1957) 372.—Ic.: Munro, l.c. 339; Fl. Tadzh. 1, Plate 44, figs. 8–13.

Described from Tibet. Type in London (K).

On rocky slopes, in grasslands, riverine pebble beds, talus and on rocks; in upper mountain belt.

IIA. **Kashgar:** *Nor.* (Uchturfan region from Karoli village to Bedel' pass, June 27, 1908—Divn.); *West.* ("Bostanterek, ca. 2400 m, 1925"—Persson, l.c.; near Bostanterek area, July 11, 1929—Pop.; Sulu-Sakal valley 25 km east of Irkeshtam, 3000 m, July 26, 1935—Olsuf'ev).

IIIB. **Tibet:** *Chang Tang* ("Sanju Pass above Kitchik Yilak, Aug. 1870, leg. Henderson and Hume"—Munro, l.c. type!).

IIIC. **Pamir** ("Eastern Pamir, Yam-Bulak-Bashi glacier, Mustagh-Ata, 4439 m, Aug. 17, 1894, leg. Hedin"—Pilger, l.c.; Tagdumbash-Pamir, Pistan gorge, 4600 m, July 15, 1901—Alekseenko; Billuli pass, June 13; around Orya-Malo pass, June 18—1909, Divn.; Piyak-Davan pass, about 5000 m, July 10; upper Kaplyk river, 4500–5000 m, July 14; Shor-Luk river gorge; about 5000 m, July 28—1942, Serp.).

General distribution: Cent. Tien Shan, East. Pam.; Middle Asia (Pamir-Alay), Himalayas (west.).

5. **C. borii** Tzvel. mon. novum.—*Deyeuxia rosea* Bor in Kew Bull. (1954) 498 and ej. Grasses Burma, Ceyl., Ind. and Pakist. (1960) 399, non *Calamagrostis rosea* Hack. ex Stuckert, 1905.

Described from Tibet. Type in London (K).

IIIB. **Tibet:** *South* (Lhasa, 3300 m, Sept. 1904—Walton, type!).
General distribution: endemic.

Note: Judging from the type specimen kindly furnished by Kew herbarium, this species is very close to *C. scabrescens* Griseb., probably representing its alpine derivative. Lemma and awn in the type specimen are markedly underdeveloped.

6. **C. compacta** (Munro ex Hook. f.) Hack. in Vidensk. Meddel. naturh. Foren. Kjobenhavn (1903) 167; Tzvelev in Bot. Mater. Gerb. Bot. inst. AN SSSR, 21 (1961) 33 and in Novosti sist. Vyssh. rast. (1965) 43; Ikonnikov, Opred. rast. Pamira [Key to Plants of Pamir] (1963) 52.—*Deyeuxia compacta* Munro ex Hook. f. Fl. Brit. Ind. 7 (1897) 267; ? Hemsley, Fl. Tibet (1902) 203.—*D. holciformis* (Jaub. et Spach) Bor, Grasses Burma, Ceyl., Ind. and Pakist. (1960) 398, p.p. (excl. typo).

Described from West. Himalayas. Lectotype (West. Tibet, 1828, leg. Jacquemont) in London (K).

On rocky slopes, in grasslands and riverine pebble beds; in upper mountain belt.

IIIC. **Pamir** (Tagdumbash-Pamir, Pistan gorge, 4600 m, July 15, 1901—Alekseenko).

General distribution: Cent. Tien Shan, East. Pam.; Middle Asia (Pamir-Alay), Himalayas (west.).

Note: Specimens of this species occupy an intermediate position between such distantly related species as *C. tianschanica* Rupr. and *C. anthoxanthoides* (Munro) Regel and probably represent their hybrids (Tzvelev, l.c., 1965). This view is confirmed by the fact that *C. compacta* was collected together with *C. anthoxanthoides* from the solitary site in Cent. Asia cited above.

7. **C. inexpansa** A. Gray, N. Amer. Gram. and Cyper. 1 (1834) No. 2; Tzvelev in Novosti sist. vyssh. rast. (1965) 28.—*C. stricta* var. *aculeolata* Hack. in Bull. Herb. Boiss. 7 (1899) 652.—*C. neglecta* var. *aculeolata* (Hack.) Miyabe et Kudo in J. Fac. Agric. Hokk. Univ. 26, 2 (1930) 140; Kitag. Lin. Fl. Mansh. (1939) 66.—*C. aculeolata* (Hack.) Ohwi in Acta Phytotax and Geobot. 2 (1933) 278, in adnot.—? *C. neglecta* var. *mongolica* Kitag. l.c. 66, p.p.

Described from Nor. America. Type in London (K). Isotype in Leningrad.

In marshes and marshy meadows.

IA. **Mongolia**: *Cis-Hing.* (Khuntu somon, 5 km west of Toge-Gol river, Aug. 7; 5–7 km south-west of Dzhara-Ul hill, Aug. 8—1949, Yun.).

General distribution: Far East, China (Dunbei), Korea, Japan, Nor. Amer.

Note: In our view, it is possible to combine *C. aculeolata* (Hack.) Ohwi and *C. inexpansa* into a single Amphi-Pacific species, very closely related on the one hand to the widely distributed boreal *C. neglecta* (Ehrh.) Gaertn., Mey. et Schreb. and, on the other, to the arctic-alpine (especially characteristic of bald peaks) *C. lapponica* (Wahl.) Hartm. All three species, among which *C. inexpansa* is evidently the oldest, are interrelated through populations and clones of hybrid origin.

8. **C. kokonorica** (Keng) Tzvel. Comb. nova.—*Deyeuxia kokonorica* Keng, Fl. ill. sin., Gram. (1959) 520, diagn. Sin.—Ic.: Keng, l.c. fig. 447 (type!). — Planta perennis 30–60 cm alta, innovationes subterraneos breviter repentes emittens; laminae 1.5–3.5 mm lt., planae vel laxe convolutae. Paniculae 3.5–8 cm lg., sat densae, sub-spiciformes, ramis scabris valde abbreviatis; glumae 3.5–4.5 mm lg., apice acutiusculae, carinis scabridulis; lemma 3.5–4.5 mm lg., 5-nervium, scabriusculum; calli pili ca. 1/4 lemmatis aequilongi; arista prope lemmatis basin inserta, leviter geniculatim curvata et lemmatis apicem vix superans; rachillae rudimentum magnum, longe et dense pilosum; palea ca. 3/4 lemmatis aequilonga.

Described from Qinghai. Type in Nanking.

IIIA. **Qinghai**: *Nanshan* ("Kukunor lake, No. 5305"—Keng, l.c., type!).
General distribution: endemic.

Note. Judging from the figure of the type cited above (Keng, l.c.), this species is close to *C. macilenta* (Griseb.) Litv. and *C. salina* Tzvel., differing from them mainly in very short hairs of the callus.

C. lapponica (Wahl.) Hartm. Handb. Skand. Fl. (1820) 46; Pavlov in Byull. Mosk. Obshch. ispyt. Prir. 38 (1929) 17; Roshev. in Fl. SSSR, 2 (1934) 219; Grubov, Consp. fl. MNR (1955) 65.—*C. sibirica* V. Petrov, Fl. Yakut. 1 (1930) 203.—*Arundo lapponica* Wahl Fl. Lapp. (1812) 27.

Described from Scandinavia (Lapland). Type in Uppsala.

On rocky slopes, rocks in riverine pebble beds, grasslands, thin forests; in middle and upper mountain belts, especially on bald peaks.

Found in border regions of Nor. Mongolia (Hent., Hang., Khan-Khukhei mountain range) and China (Great Hinggan).

General distribution: Arct., West. Sib. (Altay), East. Sib., Far East, Nor. Mong. (Fore Hubs., Hent., Hang.), China (Dunbei), Korea, Nor. Amer.

9. **C. macilenta** (Griseb.) Litv. in Bot. mater. Gerb. Glavn. bot. sada RSFSR, 2 (1921) 119; Krylov, Fl. Zap. Sib. 2 (1934) 205; Grubov, Consp. fl. MNR (1955) 65.—*C. varia* γ. *macilenta* Griseb. in Ledeb. Fl. Ross. 4 (1852) 427.—*Deyeuxia macilenta* (Griseb.) Keng, Fl. ill. sin., Gram. (1959) 520, quoad nomen.—*Calamagrostis tianschanica* auct. non Rupr.: Pavlov in Byull. Mosk. Obshch. ispyt. Prir. 38 (1929) 17.

Described from Altay. Type in Leningrad. Map 6.

In grasslands, on rocky slopes and in riverine pebble beds in lower and middle mountain belts.

IA. **Mongolia:** *Khobd.* (upper course of Kharkiry river, July 10, 1879—Pot.); *Mong. Alt., Cent. Khalkha* (Kerulen basin, Bain-Gol valley, July 17, 1924—Lis.; water divide between Ubur- and Ara-Dzhirgalante rivers, July 2; Batu-Norbo somon, Khardzanai-Gol river, July 27—1949, Yun.); *Bas. Lakes* (Ulangom district, June 20, 1879—Pot.; Barun-Gol river bank, July 5, 1898—Klem.; Chitsirtsany-Bulak area nor. of Khirgis-Nur lake; Kharkiry river valley 4 km south of Ulangom, July 27, 1945—Yun.); *Gobi-alt.* (Artza-Bogdo, No. 342, 1925—Chaney; Bain-Gobi somon, bank of Tsagan-Gol river, July 27, 1948—Grub.).

IIA. **Junggar:** *Jung. Gobi* (Barkul' lake, No. 223, Sept. 27, 1957—Huang; Tamchi somon, Gun—Tamchi area, Aug. 2, 1947—Yun.).

General distribution: West. Sib. (Altay), East. Sib. (south.), Nor. Mong. (Forhubs., Hent., Hang.).

10. **C. neglecta** (Ehrh.) Gaertn., Mey. et Scherb. Fl. Wett. 1(1799) 94; Krylov, Fl. Zap. Sib. 2 (1928) 218; Pavlov in Byull. Mosk. Obshch. ispyt. prir. 38 (1929) 17; Roshev. in Fl. SSSR, 2 (1934) 215; Kitag. Lin. Fl. Mansh. (1939) 66, excl. var.; ? Norlindh, Fl. mong. steppe, 1 (1949) 77; Grubov, Consp. fl. MNR (1955) 65; Fl. Kazakhst. 1 (1956) 187; Fl. Tadzh. 1 (1957) 371; Ikonnikov, Opred. rast. Pamira [Key to Plants of Pamir] (1963) 52. —*Deyeuxia neglecta* (Ehrh.) Kunth, Rév. Gram 1 (1829) 76; Keng, Fl. ill. Sin., Gram. (1959) 523.—Ic. : Keng, l.c. fig. 448.

Described from Sweden. Type probably in Goettingen. Isotype in Leningrad.

In marshes, marshy meadows, coastal sand and pebble beds; up to upper mountain belt.

IA. **Mongolia:** *Cent. Khalkha* (Dzhirgalante river basin, marsh among sand, Sept. 15, 1925—Krasch. and Zam.), *East. Mong.* ("Doyen, in collo arenoso, June 29, 1934, leg. Eriksson"—Norlindh, l.c.).

IIA. **Junggar:** *Tien Shan* (B. Yuldus 30–35 km south-west of Bain-Bulak settlement, Aug. 10, 1958—Yun.).

General distribution: Cent. Tien Shan, East. Pam.; Arct., Europe, Caucasus (south, Georgia), Middle Asia (West. Pam.) West. Sib., East. Sib., Far East, Nor. Mong. (Fore Hubs., Hent., Hang.), China (Dunbei, Nor.-West.), Korea, Japan, Nor. Amer.

C. obtusata Trin. Gram. Unifl. (1824) 225; Krylov, Fl. Zap. Sib. 2 (1928) 215; Pavlov in Byull. Mosk. Obshch. ispyt. prir. 38 (1929) 17; Roshev. in Fl. SSSR, 2 (1934) 220; Grubov, Consp. fl. MNR (1955) 65.

Described from West. Siberia. Type in Leningrad.

In forests, forest glades, meadows, among shrubs; up to midmountain belt.

Found in the border regions of Nor. Mongolia (Hent., Hang.).

General distribution: Europe (nor.-east), West. Sib., East. Sib., Nor. Mong. (Hent., Hang.).

11. **C. przevalskyi** Tzvel. sp. nova.—Planta perennis, 4–10 cm alta, laxe caespitosa, surculis breviter repentibus emittens; foliorum laminae 1.5–3.5 mm lt., planae, glauco-viridia; foliorum superiorum vaginae valde inflatae, extus scabrae. Paniculae 1.5–2.5 cm lg. et 1–1.5 cm lt., spiciformes, densissimae, ovatae vel breviter cylindricae, roseo-violaceae; glumae 4.5–6.5 mm lg. extus facie tota scabrae; calli pili circa 1/2–1/3 lemmatis longitudinis aequales; arista longa, geniculata, prope lemmatis basin fixa.

Typus: Tibet bor.-orient., jugum Syan' sibei, ad rivulum in fauce, 4000 m, No. 153, 17 V 1880, n. Przevalsky. In Herb. Inst. Bot. Ac. Sci. URSS—LE conservatur.

Affinitas: Haec species a specie proxima—*C. anthoxanthoides* (Munro) Regel glumis facie tota scabris et surculis partim repentibus differt.

Described from Qinghai. Type in Leningrad. Plate III, fig. 2.

IIIA. **Qinghai:** *Amdo* (Syan'sibei mountain range, in gorge of brook, 4000 m, No. 153, May 17, 1880—Przew., type!).

General distribution: endemic.

Note: Typical specimens of this species mostly resemble the specimens of *C. anthoxanthoides* (Munro) Regel but differ from them in the presence of short creeping subsurface shoots and glumes covered with spinules throughout their surface and not just along ribs. This species evidently replaces *C. anthoxanthoides* in Nor.-East. Tibet and Qinghai while in the Himalayan regions adjoining Cent. Asia (Yatung town vicinity and Khakalunga hill in Nor. Sikkim) and perhaps even in South. Tibet, another species of this same genetic affinity is encountered, namely *C. tibetica* (Bor) Tzvel. comb. nova [=*Deyeuxia tibetica* Bor in Kew Bull. (1949) 66; Keng, Fl. ill. sin., Gram. (1959) 518], differing from other related species in pilose glumes and branches of panicles.

C. sajanensis Malysch. in Bot. mater. Gerb. Bot. Inst. AN SSSR, 21 (1961) 452; Tzvelev in Novosti sist. vyssh. rast. (1965) 49.

Described from East. Siberia (Sayans, Tunkinsk valley). Type in Leningrad.

In grasslands, on rocky slopes and in riverine pebble beds; in lower and middle mountain belts.

Found in border regions of Nor. Mongolia (Hang.).

General distribution: East. Sib. (south.).

12. **C. salina** Tzvel. in Novosti sist. vyssh. Rast. (1965) 27.—*Deyeuxia macilenta* (Griseb.) Keng, Fl. ill. sin., Gram. (1959) 520, quoad pl.—Ic.: Keng, l.c. fig. 446.

Described from Transbaikal. Type in Leningrad. Map 6.

On solonchak and in saline meadows.

IA. **Mongolia:** *Cent. Khalkha* (mid. Kerulen, near Chzhovana camp, 1899—Pal.); *East. Mong.* (south-east. margin of Baishintyn-Sume, Orle area, Aug. 17 and 18, 1927—Zam.; Khuntu somon, 18 km south-east of Bain-Tsagan, Aug. 6, 1949—Yun.); *Bas. lakes* (near Dzergin-Nur lake, Aug. 9, 1930—Bar.); *Gobi-Alt.* (valley of Legin-Gol river, Aug. 20, 1927—Simukova).

IB. **Kashgar:** *Nor.* (Keinsk basin in upper course of Kyzyl river, nor.-west of Kucha town, Sept. 2, 1958—Yun.; 5 km nor. of Ven'su on Asku road, Sept. 30, 1958—Lee and Chu).

IC. **Qaidam:** *plain* (south-east of Qaidam, 3000 m, Aug. 29, 1884—Przew.).

IIA. **Junggar:** *Jung. Gobi* (from Tsukhul-Nur lake to Tsagan-Tunge station, Sept. 22, 1930—Bar.; catchment area of Bodonchi river 2-3 km south of Bodonchin-Khure, July 19, 1947—Yun.; Uienchi somon, Borotsonchzhi area, Sept. 13, 1948—Grub.).

General distribution: East. Sib. (south. Trans baikal).

Note: Large rudiment of rachilla densely covered in long hairs is characteristic of this species as well as the related species *C. macilenta* (Griseb.) Litv. and *C. kokonorica* (Keng) Tzvel.

13. **C. scabrescens** Griseb. in Nachr. Gesellsch. Wissensch. u. Univ. Goettingen, 3 (1868) 79; Pampanini, Fl. Carac. (1930) 73.—*Deyeuxia scabrescens* (Griseb.) Munro ex Duthie in Atkins. Gazett. N.W. Ind. (1882) 628; Forbes and Hemsley, Index Fl. Sin. 3 (1905) 395; Keng, Fl. ill. sin., Gram (1959) 509; Bor, Grasses Burma, Ceyl., Ind. and Pakist. (1960) 399. —Ic.: Keng. l.c. fig. 434.

Described from Sikkim. Type in London (K).

In forests and in forest glades, in scrub and meadows; in middle and upper mountain belts.

IIIA. **Qinghai:** *Nanshan* (in forest belt of Yuzhno-Tetungsk mountain range, 2800 m, July 24, 1880—Przew.).

IIIB. **Tibet:** *Weitzan* (Yangtze basin, Darindo region near Chzherku, 3900 m, Aug 8, less 1900—Lad.).

General distribution: China (Cent., South-West.), Himalayas.

14. **C. tianschanica** Rupr. in Osten-Sacken and Rupr. Sert. Tiansch. (1869) 34; Pampanini, Fl. Carac. (1930) 73; Roshev. in Fl. SSSR, 2 (1934)

205, excl. syn.; Fl. Kirgiz. 2 (1950) 89; Fl. Kazakhst. 1 (1956) 185; Fl. Tadzh. 1 (1957) 369; Ikonnikov, Opred. rast. Pamira [Key to Plants of Pamir] (1963) 54.—*C. arundinacea* var. *purpurascens* (R. Br.) Pilger in Hedin, S. Tibet, 6, 3 (1922) 92, quoad pl.—*Deyeuxia tianschanica* (Rupr.) Bor in Kew Bull. (1949) 66.—*D. holciformis* (Jaub. and Spach) Bor, Grasses Burma, Ceyl., Ind. and Pakist. (1960) 398, p.p. (excl. typo). —Ic.: Fl. Kazakhst. 1, Plate 13, fig. 1; Fl. Tadzh. 1, Plate 43, figs. 1–3.

Described from Cent. Tien Shan. Type in Leningrad. Map 6.

On rocky and clayey slopes, in grasslands and riverine pebble beds; in upper mountain belt.

IIA. Junggar: *Jung. Alt.* (Kazan pass, 1878—A. Reg.; 20 km south of Ven'tsyuan', No. 48, July 14, 1957—Huang); *Tien Shan* (Mengote hill, 3300–3700 m, July 4, 1879—A. Reg.; valley of B. Yuldus, Sept. 3, 1893—Rob.; on Urumchi-Karashar highway, 2800 m, No. 6129, July 22, 1958—Lee and Chu).

IIIA. Qinghai: *Nanshan* (Yarkhu region on Sharagol'dzhin, June 13, 1894—Rob.).

IIIB. Tibet: *Chang Tang* (Mudzhik-Atasy mountain range, 4300 m, Aug. 16; Przewalsky mountain range, 4500–5000 m, Sept. 22—1890, Rob.; Raskem village, 1898—Nov.; upper course of Tiznaf river 27 km beyond Kyude village along road to Saryk pass, 3850 m, June 1; 15 km east of Kirgiz-Dzhangil crossing on road to Karakash river [Sinkiang-Tibet highway], 4500 m, June 2—1959, Yun.; same site, 4500 m, June 2, 1959—Lee et al.).

IIIC. Pamir.

General distribution: Nor. and Cent. Tien Shan, East. Pam.; Middle Asia (Pamir-Alay).

C. turczaninowii Litv. in Bot. mater. Gerb. Glavn. bot. sada RSFSR, 2 (1921) 115; Roshev. in Fl. SSSR, 2 (1934) 227; Grubov, Consp. fl. MNR (1955) 66.— *C. korotkyi* Litv. in Spiske rast. Gerb. russk. fl. [List of Plants in the Herbarium of the Russian Flora] 8 (1922) 182; roshev. l.c. 228.

Described from Transbaikal. Type in Leningrad.

In sparse forests, forest glades and grasslands, on rocky slopes and rocks; in middle and upper mountain belts.

Found in border regions of Nor. Mongolia (Hentey hills) and China (Great Hinggan).

General distribution: East. Sib. (south), Far East, Nor. Mong. (Hent.), China (Dunbei).

15. C. venusta (Keng) Tzvel. comb. nova.—*Deyeuxia venusta* Keng in Sunyatsenia, 6, 1 (1941) 66.

Described from Qinghai. Type in Washington.

In meadows and forest glades; in middle and upper mountain belts.

IIIA. Qinghai: *Nanshan* ("at Ta-Hwa near Ping Fan [Yunden], Kansu prov., 2900–3100 m, No. 521a, 17 VII 1923, leg. Ching"—Keng, l.c. type!).

General distribution: endemic.

Note. The author of this species considers it closer to the East. Chinese hill species *C. hupehensis* (Rendle) Tzvel. comb. nova [= *Deyeuxia hupehensis* Rendle in Forbes and Hemsley, Index Fl. Sin. 3 (1904) 394], which in turn belongs to a large group of closely related species grouped around *C. arundinacea* (L.) Roth. However, the original diagnosis of *C. venusta* does not indicate whether this species possesses such a characteristic feature of *C. hupehensis* as glumes with ciliate margin.

Section 2. Calamagrostis

16. C. purpurea (Trin.) Trin. Gram. Unifl. (1824) 219; Tzvelev in Novosti sist. vyssh. rast. (1965) 34.—*C. langsdorffii* (Link) Trin. l.c. 225; Franch. Pl. David. 1 (1884) 332; Forbes and Hemsley, Index Fl. Sin. 3 (1904) 391; Krylöv, Fl . Zap. Sib. 2 (1928) 219; Pavlov in Byull. Mosk. Obshch. ispyt. prir. 38 (1929) 16; Roshev. in Fl. SSSR, 2 (1934) 213; Kitag. Lin. Fl. Mansh. (1939) 66; Grubov, Consp. fl. MNR (1955) 65; Fl. Kazakhst. 1 (1956) 186.—*Arundo purpurea* Trin. in Spreng. Neue Entdeck. 2 (1820) 52.—*A. langsdorffii* Link, Enum. Pl. Hort. Berol. 1 (1821) 74.—*Deyeuxia langsdorffii* (Link) Keng, Fl. ill. sin., Gram. (1959) 518. —Ic.: Trin. l.c. (1824) Table 4, fig. 3; Keng, l.c. fig. 444.

Described from Baikal region. Type in Leningrad.

In marshy meadows, along banks of water reservoirs, in forests and among shrubs; up to upper mountain belt.

IA. **Mongolia:** *Mong. Alt.* (Bulugun somon, upper course of Dundu-Tumurte river, July 25, 1947—Yun.); *Cis-Hing.* (valley of Upper Numuryg river, Aug. 6–7, 1959—Yun.).

IIA. **Junggar:** *Alt. region* (Chzhuntsinkhe river, 2400 m, No. 1473, Aug. 7; Kanas river nor. of Burchum settlement, No. 3045, Sept. 18—1956, Ching; right bank of Kairty river 20 km nor. of Koktogoi, valley of Kuidun river, July 15, 1959—Yun.); *Jung. Alt.* (San'tai village 20 km nor. of Ulastai settlement, No. 3850, Aug. 28, 1957—Huang); *Tien Shan* (Borgatu brook nor. of Kash river valley, 1700–2000 m, July 5; valley of Kash river, 2000–2800 m, July 29—1879, A. Reg.).

General distribution: Jung.-Tarb.; Europe (nor.-east.), West. Sib., East. Sib., Far East, Nor. Mong., China (Altay, Dunbei, Nor., East., South-West.), Korea, Japan, Nor. Amer.

Note. Typical specimens of this highly polymorphous species have relatively small (2.5–4 mm long), quite closely packed spikelets and are found mainly in southern Siberia and, within Cent. Asia, in Cis-Hinggan, Junggar Alatau and Tien Shan. Specimens with larger (3.5–6 mm long) and less closely packed spikelets regarded by us (Tzvelev, l.c. 34) as subspecies *C. purpurea* subsp. *langsdorffii* (Link) Tzvel., are widely distributed in Siberia and North America and are found in Mongolian Altay of Central Asia. The variety "*C. langsdorffii* var. *decipiens* Litv. ex Pavl." with very short callus hairs described from Hentey (Pavlov, l.c. 17) was probably described from hybrid specimens.

Section 3. Pseudophragmites Tzvel.[3]

17. C. dubia Bunge, Reliq. Lehmann. (1852) 348; Roshev. in Fl. SSSR, 2 (1934) 195; Fl. Kirgiz. 2 (1950) 87; Fl. Kazakhst. 1 (1956) 184; Fl. Tadzh. 1 (1957) 364; Tzvelev in Novosti sist. vyssh. rast. (1965) 43.

Described from Middle Asia (Amu Darya valley). Type in Paris.

In sand and pebble beds of river and lake valleys, in tugai and around irrigation ditches; up to lower mountain belt.

IA. **Mongolia:** *East. Mong.* (near Khukh-Khoto town, July 19, 1884—Pot.; valley of Khiengukhe river nor.-west of Kalgan town, July 21, 1934—Kozlov); *East. Gobi* (sandy shore of Batu-Kalka river, Roerich Exped., No. 801, Aug. 11, 1935—Keng); *Alash. Gobi* (near Chzhintasy town, June 22; Taleulyu area in Edzin-Gol valley, June 23 and 24; left bank of Ezin-Gol river opposite Mumin town, June 28; Tufyn area, June 29—1886, Pot.); *Ordos* (in Huang He valley, July 19, 1871—Przew.; Baga-Edzhin-Khoro area, Aug. 6, 1884—Pot.); *Khesi* (near Yunchin town, July 20, 1875—Pias.).

IB. **Kashgar:** *Nor., West., East., Takla-Makan* (along Khotan river, Sept. 15, 1885—Przew.; near Cherchen river 8 km south of Cherchen town, No. 9489, June 9, 1959—Lee et al.).

IIA. **Junggar:** *Jung. Gobi* (south.: 5 km west of Yantszykhai village, No. 290, June 30; 2 km south-west of Ulausu village, No. 1603, July 3; in the region of Gan'khedze settlement, No. 547, Sept. 23—1957, Huang; 2-3 km east of Santokhodze settlement along road to Manas village, July 4, 1957—Yun.); *Dzhark.* (along Ili river close to Chabuchar village, No. 3076, Aug. 5, 1957—Huang).

General distribution: Aral-Casp., Balkh. region; Near East, Caucasus, Middle Asia, China (south. Dunbei, Nor.).

Note. *C. dubia* represents the oldest ecogeographical race genetically affiliated to *C. pseudophragmites* (Hall. f.) Koel. s.l. and is confined to the plains and low mountain regions. This race is evidently ancestral to the extensively distributed, predominantly mountain race *C. pseudophragmites* s.s., the Near Eastern mountain race *C. persica* Boiss. and the high mountain and more xerophilous Central Asian race *C. hedinii* Pilg.

18. C. epigeios (L.) Roth, Tent. Fl. Germ. 1 (1788) 34; Franch. Pl. David. 1 (1884) 331; Forbes and Hemsley, Index Fl. Sin. 3 (1905) 391; Danguy in Bull. Mus. nat. hist. natur. 20 (1914) 146; Krylov, Fl. Zap. Sib. 2 (1928) 224; Pavlov in Byull. Mosk. obshch. ispyt. prir. 38 (1929) 16; Roshev. in Fl. SSSR, 2 (1934) 194, p.p.; Kitag. Lin. Fl. Mansh. (1939) 65; Ching in Contribs, U.S. Nat. Herb. 28 (1941) 596; Norlindh, Fl. mong. steppe, 1 (1949) 74; Fl. Kirgiz. 2 (1950) 87; Grubov, Consp. fl. MNR (1955) 65; Fl. Kazakhst. 1 (1956) 182; Fl. Tadzh. 1 (1957) 364, p.p.; Keng, Fl. ill. sin., Gram. (1959) 528; Bor, Grasses Burma, Ceyl., Ind. and Pakist. (1960) 396.—*C. epigeios* var. *densiflora* Ledeb. Fl. Ross. 4 (1852) 433, p.p.; Franch. l.c. 332; Keng, l.c. 528.—*C. glomerata* Boiss. et Buhse in Mém. Soc. natur. Moscou, 12 (1860) 229.—*C. koibalensis* Reverd. in Sist. zam. Gerb. Tomsk. univ. 1 (1941) 3.—*C. epigeios* subsp. *glomerata* (Boiss. et Buhse) Tzvel. in

---

[3]*Calamagrostis* sect. Pseudophragmites Tzvel. in Novosti sist. vyssh. rast. (1965) 38.

Novosti sist. vyssh. rast. (1965) 41.—*Arundo epigeios* L. Sp. pl. (1753) 81.
—Ic.: Fl. Tadzh. 1, Plate 43, figs. 4–7; Keng, l.c. fig. 455.

Described from Europe. Type in London (Linn.).

In sand and pebble beds of river and lake valleys, on rocky slopes, in steppes and along roadsides; up to upper mountain belt.

IA. **Mongolia:** *Cis-Hing.* (Khalkhin-Gol somon, Verkhn. Numuryg, July 6, 1949—Yun.); *Cent. Khalkha, East. Mong., Bas. lakes, Val. lakes* (valley of Ongiin-Gol river 40–45 km south of Khushu-Khida, Aug. 18, 1945—Yun.); *Gobi-Alt.* (south. slope of Ikhe-Bogdo mountain range, upper creek valley of Narin-Khurimt, June 28, 1945—Yun.); *Alash-Gobi* (Alashan' desert, May 28, 1872—Przew.); *Ordos* (in Huang He valley, July 16, 1871—Przew.; Taitukhai area, Aug. 17, 1884—Pot.; 60 km west of Ushin town, Aug. 2; 10 km south-west of Ushin town, Aug. 4—1957, Petr.).

IIA. **Junggar:** *Alt. Region, Tarb.* (Ulasty river, July 26, 1876—Pot.); *Jung. Alt., Tien Shan, Jung. Gobi, Dzhark.*

IIIA. **Qinghai:** *Nanshan* (in Severo-Tetungsk forest belt, 2500 m, Aug. 1, 1880—Przew.).

General distribution: Aral-Casp., Balkh. region, Jung.-Tarb., Nor. and Cent. Tien Shan; Europe, Mediterr., Balk.-Asia Minor, Near East, Caucasus, Middle Asia, West, Sib., East. Sib. (south), Far East (south), Nor. Mong., China (Dunbei, Nor., Nor.-West., Cent., East.), Himalayas, Korea, Japan; introduced in many other countries.

Note. Two widely distributed ecogeographical races are found within this species. One, comprising the typical specimens of the species, has very lax panicles with large (5.5–7 mm long) spikelets and is found in the forest zone of Eurasia and in the more southern hilly regions. Most Central Asian specimens belong to this race. The second race, established sometimes as the variety *C. epigeios* var. *densiflora* Ledeb. (l.c.), sometimes as the subspecies *C. epigeios* subsp. *glomerata* (Boiss. et Buhse) Tzvel. (l.c.), or even as an independent species *C. glomerata* Boiss. et Buhse (= *C. koibalensis* Reverd. l.c.), has far denser panicles with very small (4–5.5 mm long) spikelets, it is distributed in the steppe regions of Eurasia, including the steppes of southern East. Siberia from where it entered Cent. Asia, i.e., steppes of Cis-Hinggan and East. Mongolia.

19. **C. hedinii** Pilg. in Hedin, S. Tibet, 6, 3 (1922) 93.—*C. littorea* var. *tartarica* Hook. f. Fl. Brit. Ind. 7 (1896) 261.—*C. emodensis* var. *breviseta* Hack. in Pauls. in Vidensk. Meddel. naturh. Foren. Kjobenhavn (1903) 167; Pampanini, Fl. Carac. (1930) 73; Roshev. in Fl. SSSR, 2 (1934) 202; Fl. Tadzh. 1 (1957) 369.—*C. pseudophragmites* var. *tartarica* (Hook. f.) Bor, Grasses Burma, Ceyl. Ind., and Pakist. (1960) 396.—*C. pseudophragmites* subsp. *tartarica* (Hook. f.) Tzvel. in Novosti sist. vyssh. rast. (1965) 42.— C. *turkestanica* auct. non Hack.: Tzvelev in Bot. mater. Gerb. Bot. inst. AN SSSR, 21 (1961) 31, p.p.; Ikonnikov, Opred. rast. pamira [Key to Plants of Pamir] (1963) 52, p.p.

Described from Tibet. Type in Stockholm.

On rocky and rubble slopes, in grasslands, riverine pebble beds and talus; in upper mountain belt.

IIIB. **Tibet:** *Chang Tang* (Russky mountain range, in pebble bed of Aksu river, 4000 m, June 3, 1890—Rob.; "N. Tibet, Kash-Otak, 2916 m, Aug. 3–20, 1894, leg. Hedin"—Pilger, l.c., type!).

General distribution: Cent. Tien Shan, East. Pam.; Middle Asia (Pamir-Alay), Himálayas (west., Kashmir).

Note. This species, very close to C. *pseudophragmites* (Hall. f.) Koel., is commonly found in Pamir within the USSR but evidently is far rarer in the Chinese Pamir.

20. **C. macrolepis** Litv. in Bot. mater. Gerb. Glavn. bot. sada RSFSR, 2 (1921) 125; Tzvelev in Bot. mater. Gerb. Bot. inst. AN SSSR, 21 (1961) 30.— *C. gigantea* Roshev. in Izv. Bot. sada AN SSSR, 30 (1932) 294 and in Fl. SSSR, 2 (1934) 195, non Nutt. 1837; Fl. Kazakhst. 1 (1956) 182; Keng, Fl. ill. sin., Gram. (1959) 528; Bor, Grasses Burma, Ceyl., Ind. and Pakist. (1960) 396. —Ic.: Fl. Kazakhst. 1, Plate 13, fig. 4; Keng, l.c. fig. 454.

Described from West. Pamir (Shugnan). Type in Leningrad.

In coastal sand and pebble beds, saline meadows and along banks of water reservoirs; up to midmountain belt.

IA. **Mongolia:** *Mong Alt.* (valley of Buyantu river, Sept. 18, 1930—Bar.; in the lower courses of Buyantu-Gol river, 1941—Kondratenko); *Cent. Khalkha* (25–30 km along road from Sorgol-Khairkhan hill to Dalan-Dzadagad, July 16, 1943—Yun.; Undur-Ul'dza somon, valley of Ongiin-Gol river, Aug. 26, 1952—Davazhamts); *East. Mong.* (Kulun-Buir-Nur plain, near Gandzhur monastery, Aug. 4, 1899—Pot. and Sold.; between Urshun' river and Gandzhur, 1899—Pal.; near Shilin-Khoto, 1959—Ivan.); *Bas. lakes* (40 km south of Ulyasutai along road to Tsagan-Ol, July 17, 1947—Yun.); *Val. lakes* (left bank of Tuin-Gol river, July 12, 1893—Klem.); *Gobi-Alt., East. Gobi* (Bain-Dzak basin, July 20, 1943—Yun.); *Alash. Gobi.*

IB. **Kashgar:** *Nor., West., East., Takla-Makan* (catchment area of Tarim river between Yuili and Tikeklik villages, Aug. 19, 1958—Yun.; 3 km nor. of Cherchen settlement, No. 9518, June 14, 1959—Lee et al.).

IIA. **Junggar:** *Alt. region* (near Shara-Sume, No. 674, Sept. 6, 1956—Ching); Jung. Alt. (20 km south of Arasan, No. 1479, Aug. 14, 1957—Huang); *Tien Shan, Jung. Gobi, Dzhark.* (Ili river bank south of Kul'dzha, town. May 27, 1877—A. Reg.).

General distribution: Aral-Casp., Balkh. region, Jung.-Tarb., Nor. and Cent. Tien Shan; Caucasus (Caspian sand), Middle Asia, West. Sib. (Altay), East. Sib. (south), Nor. Mong., China (Nor.), Himalayas (west., Kashmir).

Note. This species represents a predominantly desert ecogeographical race related to C. *epigeios* (L.) Roth s.l. and, in this respect, is similar to C. *dubia* Bunge related to C. *pseudophragmites* (Hall. f.) Koel. S.l.

21. **C. pseudophragmites** (Hall. f.) Koel. Descr. Gram. (1802) 106; Krylov, Fl. Zap. Sib. 2 (1928) 226; Pavlov in Byull. Mosk. obshch. isp. prir. 38 (1929) 16; Roshev. in Fl. SSSR, 2 (1934) 196; Kitag. Lin. Fl. Mansh. (1939) 67; Norlindh, Fl. mong. steppe, 1 (1949) 74; Fl. Kirgiz. 2 (1950) 88; Grubov, Consp. fl. MNR (1955) 66; Fl. Kazakhst. 1 (1956) 184; Fl. Tadzh. 1 (1957) 367; Keng, Fl. ill. sin., Gram. (1959) 526; Bor, Grasses Burma, Ceyl., Ind. and Pakist. (1960) 396.—C. *littorea* (Schrad.) DC. Fl. Fr. 5 (1815) 255; Pampanini, Fl. Carac. (1930) 73.—C. *glauca* (M.B.) Trin. in Hohen. Enum. Pl. Prov. Talysh (1837) 14, non Reichb. 1830; Roshev. l.c. 196; Fl. Kirgiz. 2 (1950) 88; Fl. Kazakhst. 1 (1956) 184; Fl. Tadzh. 1 (1957) 366.—*Arundo pseudophragmites* Hall. f. in Roem. Arch. 1, 2 (1796) 10.—*A. littorea* Schrad.

Fl. Germ. 1 (1806) 212.—*A. glauca* M.B. Fl. taur.-cauc. 1 (1808) 79, p.p.—
Calamagrostis onoei auct. non Franch. et Savat.: Franch. Pl. David. 1 (1884)
332.—*C. purpurea* auct. non Trin.: Hao in Engler's Bot. Jahrb. 68 (1938)
583. —Ic.: Norlindh, l.c. fig. 6, a–b, c–d (var. *minor* Meld.), e (var. *stepposa*
Meld.); Keng, l.c. fig. 452.

Described from Switzerland. Type in Geneva.

On rocky slopes, in grasslands, along banks of water reservoirs, in
coastal pebble beds, sand and talus; predominantly in lower and middle
mountain belts.

IA. **Mongolia:** *Mong. Alt., Cent. Khalkha* (Toly meander 120 km south-west of
Ulan-Bator, July 1,1949—Yun.); *Bas. lakes* (40 km south of Ulyasutai along road to
Tsagan-Olom, Aug. 17, 1947—Letov); *Val. lakes* (left bank of Baidarik river, June 20,
1894—Klem.; valley of Ongiin-Gol river 30 km below Khushu-Khida, July 17, 1943—
Yun.); *East. Gobi* ("Hailutan-Gol, 1250 m, No. 1352, Aug. 8; Oboin-Gol, No. 1706,
Oct. 3, 1927, leg. Hummel"—Norlindh, l.c. [var. *minor* Meld.]; near Bailinmyao, 1959—
Ivan.); *Alash. Gobi* (near Dzhintasy village, July 2; between Khoir-Toor and Gantsy-
Dzak on left bank of Edzin-Gol, July 7—1886, Pot.; "near Shire-Sume, distr.
Dundagung, No. 6845, 5 VIII 1927, leg. Söderbom [var. *minor* Meld., typus!]; prope
Khoburin-Nor, No. 1511, 29 VIII 1927, leg. Hummel" [var. *stepposa* Meld., type!]—
Norlindh, l.c. ; near Chzhunvei, near Sabotou village, July 26; 9 km south of Chzhunvei
town, July 29—1957, Petr.; near Bayan-Khoto town, June 14, 1958—Petr.); *Ordos*
(Baga-Edzhin-Khoro area, Aug. 5 and 6, 1884—Pot.; Khaolaitunao lake 25 km south-
east of Otok town, Aug. 1, 1957—Petr.); *Khesi* (near Tszyutsyuan' town, July 2,
1956—Ching).

IB. **Kashgar:** *Nor., West.* (Yarkand-Darya valley 20 km south-east of Yarkand
town, Sept. 25, 1958—Yun.); *South.* (near Karatash river, 2600 m, July 25, 1885—
Przew.); *East.* (Algoi river near Toksun settlement, No. 7283, June 16; on banks of
Tsinshuikhe river near Khomote settlement, No. 7650, Aug. 28—1958, Lee and Chu).

IIA. **Junggar:** *Jung. Alt.* (on Ven'tsyuan'khe river, No. 1438, Aug. 14, 1957—
Huang); *Tien Shan, Jung. Gobi* (nor.: catchment area of Bulugun river, July 28, 1947—
Yun.; between Barbagai and Burchum villages, Sept. 11, 1956—Ching; south.:
midcourse of Manas river, No. 1351, July 12; between Kazvan and Tsynitsyuan'
villages in the vicinity of Savan settlement, No. 1693, July 21—1957, Huang); *Balkh.-
Alak.* (valley of Emel' river, June 20, 1905—Obruchev).

IIIA. **Qinghai** (Yuzhno-Kukunor mountain range, Bain-Gol river near confluence
with Ara-Gol river, 3300 m, June 27, 1894—Rob.; same site, on bank of Usubin-Gol
river, 3500 m, Aug. 16, 1901—Lad.; "Kokonor, Tsi-gi-gan-ba, 3600 m",—Hao, l.c.).

IIIB. **Tibet:** *Weitzan* (nor. slope of Burkhan-Budda mountain range, near Khatu-
Gol river, 3400 m, Aug. 13, 1884—Przew.; Yangtze and Mekong interfluve zone, along
Gochu river, 4100 m, Aug. 23, 1900—Lad.).

IIIC. **Pamir** (valley of Karasu river near Tagarma village, July 25, 1909—Divn.).

General distribution: Aral-Casp., Balkh. region, Jung.-Tarb., Nor. and Cent. Tien
Shan; Europe (south), Mediterr., Balk.-Asia Minor, Near East, Caucasus, Middle Asia
(hilly), West. Sib. (south, along Ob up to Berezovo town), East. Sib. (south), Far East
(near Vladivostok), Nor. Mong., China (Dunbei, Nor., Nor.-West., Cent., East., South-
West.), Himalayas, Korea; very close to species, found in Japan *C. onoei* Franch et
Savat.

Note. Relatively small (up to 40–60 cm high) specimens of this species are common
in Cent. Asia. Variety *C. pseudophragmites* var. *minor* Meld. has evidently been described

from them (in Norlindh, l.c. 75). Another variety described by the same author, *C. pseudophragmites* var. *stepposa* Meld. (in Norlindh, l.c. 77), according to the key has very dense panicles and lemma with five ribs. Judging from these characteristics, it is similar to *C. hedinii* Pilg. and *C. turkestanica* Hack. but the presence of these species in Alashan Gobi from where the variety was described is not very likely. Specimens mentioned above from Yuzhno-Kukunor mountain range (collections of Roborowsky and Ladygin) have relatively small panicles and very large and unequal glumes as in the very close Japanese species *C. onoei* Franch. et Savat. but differ from it in very dense panicles and very narrow (2–3 mm broad) leaf blades usually folded longitudinally. In our opinion, therefore, they may be regarded as a new variety—*C. pseudophragmites* var. *ladyginii* Tzvel. var. nova. Planta 40–60 cm alta; laminae 2–3 mm lt., vulgo plus minusve convolutae; paniculae 8–12 cm lg. et 2–4 cm lt., sat laxae, roseo-violaceae; spiculae 5–9 mm lg.; glumae valde inaequalis. Typus: "Jugum Austro-Kukunoricum, ad ripam fl. Usubin-Gol, 3500 m, No. 437, 16 VIII 1901, V. Ladygin" in Herb. Inst. Bot. Ac. Sci. URSS—LE conservatur.

22. **C. turkestanica** Hack. in Acta Horti Petrop. 26 (1906) 59; Roshev. in Fl. SSSR, 2 (1934) 202; Fl. Kirgiz. 2 (1950) 88; Fl. Tadzh. 1 (1957) 368; Tzvelev in Bot. mater. Gerb. Bot. inst. AN SSSR, 21 (1961) 31, p.p. and in Novosti sist. vyssh. rast. (1965) 48; Ikonnikov, Opred. rast. Pamira [Key to Plants of Pamir] (1963) 52, p.p.— *C. alopecuroides* Roshev. in Bot. mater. Gerb. Glavn. bot. sada RSFSR, 3 (1922) 199 and in Fl. SSSR, 2 (1934) 202; Fl. Kirgiz. 2 (1950) 89.

Described form Middle Asia (Transalay mountain range). Type in Leningrad.

In grasslands, riverine pebble beds, on rocky slopes and talus; in upper mountain belt.

IB. **Kashgar:** *West.* (Sulu-Sakal valley 25 km east of Irkeshtam settlement, July 26, 1935—Olsuf'ev).

IIA. **Junggar:** *Tien Shan* (Khaidyk-Gol river, 2000 m, 1893—Rob.; valley of Muzart river 10–12 km above its discharge into Bai basin, Chokarpa area, Sept. 7; Oi-Terek region in upper right tributary of Muzart, Tuku-Daban, 3100 m, Sept. 10—1958, Yun.; valley of Muzart river nor.-west of Pocheentszy village, No. 8347, Sept. 12, 1958—Lee and Chu).

General distribution: Cent. Tien Shan, East. Pam.; Middle Asia (Alay, Fergana mountain range).

Note. This species is very close to species of section Deyeuxia (Clar.) Dum. in relatively short hairs of callus and, often, five-ribbed lemma; it is likewise close to *C. hedinii* Pilg. as well as *C. alexeenkoana* Litv. ex Roshev. Quite probably, therefore, *C. turkestanica* is of hybrid origin from *C. hedinii* Pilg. x *C. tianschanica* Rupr., preserving more the characteristics of the former of the supposed parent species.

33. **Pappagrostis** Roshev.
in Fl. SSSR, 2 (1934) 230 and 749.

1. **P. pappophorea** (Hack.) Roshev. in Fl. SSSR, 2 (1934) 231.— *Calamagrostis pappophorea* Hack. in Ann. Conserv. et Jard. Bot. Genéve, 7–

8 (1904) 325.—*Stephanachne pappophorea* (Hack.) Keng in Contribs. Biol. Lab. Sci. Soc. China, Bot. 9 (1934) 136 and ej. Fl. ill. sin., Gram. (1959) 587. —Ic.: Keng, l.c. (1959) fig. 521.

Described from Cent. Tien Shan. Type in Geneva. Map 6.

On rocky and clayey slopes, in pebble beds, talus; up to upper mountain belt.

IA. **Mongolia**: *Alash. Gobi* (Bayan-Khoto vicinity, Shuankheishan' hills, Aug. 4, 1958—Petr.); *Khesi* (left bank of Huang He, Sikou village nor. of Lanzhou, Oct. 20, 1884—Pot.: nor. foothill plain of Nanshan mountain range 60 km west of Gaotai town, Aug. 7, 1958—Petr.).

IB. **Kashgar**: *Nor., West.* (along the lateral valley of Tiznaf river, June 5; Ak-Këz pass 110 km south of Kargalyk, upper course of Raskem-Darya, June 5—1959, Yun.); *South.* (Cherchen valley, 2600–3000 m, Aug. 7, 1890—Rob.); *East.*

IIA. **Junggar**: *Tien Shan* (Algoi, 2000–2700 m, Sept. 12, 1879—A. Reg.).

IIIA. **Qinghai**: *Nanshan* (nor. foothills of Altyntag mountain range 3 km south of Aksai, 2330 m, Aug. 2, 1959—Petr.).

IIIB. **Tibet**: *Weitzan* (nor. slope of Burkhan-Budda mountain range, Khatu gorge, 3500 m, June 18 and 25, 1901—Lad.).

IIIC. **Pamir** (Toili-Bulun area on Pas-Rabat river, Aug. 2, 1909—Divn.; upper course of Tiznaf river 20 km south of Kyude near Saryk pass, 3500 m, June 1, 1959— Yun.).

General distribution: Cent. Tien Shan.

Note. The chinese specialist on Gramineae, Gen I-li (Keng, l.c.), has placed this species in the genus *Stephanachne* Keng, established by him. The date of its publication (Aug. 1934) precedes by just one month that of the genus *Pappagrostis* Roshev. (Sept. 1934). However, in the type genus *Stephanachne* described from the hilly regions of Sychuang province, *S. nigrescens* Keng [specimen available in the herbarium of the Botanical Institute of the Academy of Sciences of the USSR (Nor. Sychuang, near Kserntso river, Aug. 6, 1885—Pot.)], differs so much from *Pappagrostis pappophorea* (Hack.) Roshev. that, in our opinion, it would be more correct to consider these two species independent monotypic genera. In particular, *Stephanachne nigrescens*, like the genera grouped around genus *Stipa* L., has three large bract husks on each flower while there are only two in the case of *Pappagrostis pappophorea* as in species of *Calamagrostis* Adans. With which it is very similar in many other respects also.

34. Sinochasea Keng

in J. Wash. Ac. Sci. 48, 4 (1958) 115.

1. **S. trigyna** Keng in J. Wash. Ac. Sci. 48, 4 (1958) 115; Keng, Fl. ill. Sin., Gram. (1959) 525. —Ic.: Keng, l.c. (1958) fig. 1 and l.c. (1959) fig. 451.

Described from Qinghai. Type in Nanking.

On rocky slopes and rocks.

IIIA. **Qinghai**: *Nanshan* ("Hai-yen Hsien, San Chio Shen, Sheep-breeding Station, San Tui Pu, Aug. 21, 1954, leg. P.C. Yeh and W.C. Wang"—Keng, l.c., type!).

General distribution: endemic.

Note. The solitary species of this very recently discovered genus is, in our view, very close to genus *Pappagrostis* Roshev., from which it differs only in the presence of three (not two as is usual) stigmas. This feature has been observed in only one of the relatively close genera, the monotypic Himalayan genus *Pseudodanthonia* Bor et C.E. Hubb.

35. Deschampsia Beauv.
Ess. Agrost. (1812) 91.

1. Glumes as large as spikelet and fairly long; lower glume exceeding adjoining lemma by 1/4 to 1/3; panicles usually fairly dense, less often lax and generally spreading, usually with slightly sinuate rachis and rather scabrous branches, often pendent at tip ....
   .... 2. **D. ivanovae** Tzvel.

+ Glumes shorter than spikelets or as long; lower glume not, or only slightly exceeding adjoining lemma .... 2.

2. Predominantly 30–100 cm tall midmountain plant. Panicles broadly spreading during and after anthesis, usually pyramidal, with fairly long scabrous branches; spikelets on branches relatively interrupted .... 3.

+ Predominantly high-mountain plant 10–70 cm tall. Panicles quite dense, usually compressed, with distinctly shortened, slightly scabrous branches; spikelets on branches usually crowded ...... 4.

3. Panicles pyramidal, not pendent, with strongly scabrous branches; awn usually not, or only slightly exceeding tip of lemma. Leaf blades with prominent ribs on upper surface extremely scabrous due to dense spinules ....
   .... 1. **D. caespitosa** (L.) Beauv.

+ Panicles usually porrect, sometimes pendent at tip, usually with reatively slightly scabrous branches, awn usually exceeding tip of lemma. Leaf blades with less prominent and less scabrous ribs on upper surface .... 5. **D. sukatschewii** (Popl.) Roshev.

4. Panicles 2–8 cm long, very dense, with very short branches bearing closely packed spikelets; generally pinkish-violet; awn generally reaching tip of lemma or slightly exceeding ....
   .... 3. **D. koelerioides** Regel.

+ Panicles 5–12 cm long, very lax, with fairly long branches and less closely packed spikelets, usually light coloured; awn highly reduced, usually not reaching tip of lemma ....
   .... 4. **D. pamirica** Roshev.

1. D. caespitosa (L.) Beauv. Ess. Agrost. (1812) 160; ? Hemsley, Fl. Tibet (1902) 203; Forbes and Hemsley, Index Fl. Sin. 3 (1904) 399, p.p.; Krylov,

Fl. Zap. Sib. 2 (1928) 229, p.p.; Roshev. in Fl. SSSR, 2 (1934) 245; Hao in Engler's Bot. Jahrb. 68 (1938) 585; Fl. Kirgiz. 2 (1950) 91; Grubov, Consp. fl.MNR (1955) 66; p.p.; Fl. Kazakhst. 1 (1956) 190; Fl, Tadzh. 1 (1957) 359; Keng, Fl. ill. sin., Gram. (1959) 484, p.p.; Bor, Grasses Burma, Ceyl., Ind. and Pakist. (1960) 435.—*Aira caespitosa* L. Sp. pl. (1753) 64; Pavlov in Byull. Mosk. obshch. ispyt. prir. 38 (1929) 17.—Ic. : Fl. SSSR, 2, Plate 18, fig. 14; Fl. Kazakhst. 1, Plate 14, fig. 9.

Described from Europe. Type in London (Linn.)

In meadows, riverine pebble beds and along banks of water reservoirs; up to upper mountain belt.

IA. Mongolia: *Mong. Alt.*

IIA. Junggar: *Alt. region, Jung. Alt.* (Toli district, Albaksin hill, No. 2754, Aug. 8, 1957—Huang; flood plain of Argaty-Gol river, Aug. 17,1957—Yun.); Tien Shan.

IIIA. Qinghai: *Nanshan* ("Schalakutu, 3400 m, 1935"—Hao, l.c.; meadow near east coast of Kukunor lake, 3210 m, Aug. 5, 1959—Petr.)

General distribution: Jung. Tarb., Nor. and Cent. Tien Shan; Arct. (Europ.), Europe, Mediterr. (hilly), Balk.-Asia Minor (hilly), Near East (hilly), Caucasus (hilly), Middle Asia (West. Tien Shan, Gissar, Turkestan and Zeravshan mountain ranges), West. Sib., East. Sib. (Sayans), Nor. Mong. (Fore Hubs., Hent., Hang.), Himalayas (?), Nor. Amer.

Note. Closely related to other Central Asian species of the genus.

2. D. ivanovae Tzvel. in Bot. mater. Gerb. Bot. inst. AN SSSR, 21(1961) 49.—*D. litoralis* auct. non Reut.: Hao in Engler's Bot. Jahrb. 68 (1938) 583; Keng, Fl. ill. sin., Gram. (1959) 483.—Ic.: Keng, l.c. fig. 415.

Described from Qinghai. Type in Leningrad. Plate III, fig. 3; map 2.

In forest and coastal meadows, riverine pebble beds, on rocky slopes, along banks of water reservoirs; in middle and upper mountain belts.

IIIA. Qinghai: *Nanshan* (in forest belt of Severo-Tetungsk mountain range, July 28; same site, in alpine belt of mountain range, Aug. 6—1872, Przew., type!; Sharogol 'dzhin river, Paidza-Tologoi area, Sept. 11, 1894—Rob.; valley of Ganshig river, left tributary of Peishikhe river, 3350–3720 m, Aug.; source of Ganshig, 3900–4300 m, Aug. 18—1958, Dolgushin); *Amdo* ("auf dem Ming-ke unweit des Klosters Ta-Schiu-Sze, 1935"—Hao, l.c.).

IIIB. Tibet: *Weitzan* (along Dyaochu river, July 1; in alpine belt of Burkhan-Budda mountain range, 4300–5100 m, Aug. 2, 1884—Przew.; nor. and east. coast of Alyk-Nor lake, 4200 m, May 30; along Khichu river in Nyamtso district, 4600 m, July 11, 1900; nor. slope of Burkhan-Budda mountain range, Khatu gorge, 4000–4700 m, July 12 and 18, 1901—Lad.).

General distribution: China (Nor.-West., South-West.).

Note. Specimens of this species from the very low mountain areas (of forest belt), in habitat very closely resemble D. caespitosa (L.) Beauv. and D. sukatschewii (Popl.) Roshev. and possibly form an independent race.

3. D. koelerioides Regel in Bull. Soc. Natur. Moscou, 41 (1868) 299; Krylov, Fl. Zap. Sib. 2 (1928) 232; Roshev. in Fl. SSSR, 2 (1934) 251; Fl. Kirgiz. 2 (1950) 91; Grubov, Consp. fl. MNR (1955) 66; Fl. Kazakhst. 1

(1956) 190; Fl. Tadzh. 1 (1957) 359, p.p.; Bor, Grasses Burma, Ceyl., Ind. and Pakist. (1960) 435. —Ic.: Fl.SSSR, 2, Plate 18, fig. 10.

Described from Tien Shan (Tersk Alatau). Type in Leningrad.

In grasslands, along banks of rivers and brooks; in upper mountain belt.

IA. **Mongolia:** *Khobd.* (upper course of Kharkiry river, July 10, 1879—Pot.); *Mong. Alt.* (Aksu river or Belaya Kobdo, July 22 and 23, 1909—Sap.).

IIA. **Junggar:** *Alt. Region* (in Altay hills, No. 10735, July 18, 1959—Lee et al.); *Tien Shan.*

IIIC. **Pamir** (near the confluence of Kara-Chukur and Ilyksu rivers, July 16, 1901—Alekseenko; near west. fringe of Tashkurgan town, June 13, 1959—Yun.).

General distribution: Jung.-Tarb., Nor. And Cent. Tien Shan, East. Pam.; Middle Asia (Pamir-Alay), West. Sib. (Altay), East. Sib. (Sayans), Nor. Mong. (Hang.), Himalayas (west.).

4. D. pamirica Roshev. in Fl. SSSR, 2 (1934) 252; Ikonnikov, Opred. rast. Pamira [Key to Plants of Pamir] (1963) 54.—*D. koelerioides* auct. non Regel: Fl.Tadzh. 1 (1957) 359, p.p.

Described from Pamir. Type in Leningrad.

In saline meadows and riverine pebble beds, in upper mountain belt.

IIA. **Junggar:** *Tien Shan* (B. Yuldus, 2700 m, Aug. 2, 1893—Rob.; B. Yuldus trough 30–35 km south-west of Bain-Bulk village, Aug. 10; Muzart river valley upper course, Sazlik area, Sept. 9; same site, upper Tuku-Daban near Oi-Terek area, 3100 m, Sept. 10—1958, Yun.).

General distribution: Cent. Tien Shan, East. Pam.; Middle Asia (West. Pam.).

Note. Compared to *D. koelerioides* Regel, this species is found in the relatively lower mountain belt and more hydrophiloussites. It is usually confined to saline meadows of river and lake valleys.

5. D. sukatschewii (Popl.) Roshev. in Fl. SSSR, 2 (1934) 246.—*D. caespitosa* subsp. *orientalis* Hult. Fl. Kamtsch. 1 (1927) 109.— *Aira sukatschewii* Popl. in Ocherki po fitosots. i fitogeogr. [Essays of Phytosociology and Phytogeography] (1929) 382. *Deschampsia caespitosa* auct. non Beauv.: Grubov, consp. fl. MNR (1955) 66, p.p.; Keng, Fl. ill. sin., Gram. (1959) 484, p.p.—Ic.: fl. SSSR, 2, Plate 18, fig. 15.

Described from Transbaikal. Type in Leningrad.

In meadows, sand and pebble beds of river and lake valleys; up to mid-mountain belt.

IA. **Mongolia:** *Cis-Hing.* (Siguitu district, between Yangetsi and Yoan'li rivers, No. 2606; same site, between Genkhe station and Yangetsi river, No. 2454—1954, Wang); *East. Mong.* (Argun' district, near Genkhetsyao bridge, No. 1180, July 8; same site, Klyuchevaya village, No. 1813, Aug. 6—1951, Lee et al.).

General distribution: Arct. (Asiat.). East. Sib., Far East, Nor. Mong. (Fore Hubs., Hent., Mong.-Daur.), China (Dunbei, Nor., East.), Korea, Japan, Nor. Amer. (west.).

Note. The boundary between the distribution ranges of this predominantly East Asian species and predominantly European species *D. caespitosa* (L.) Beauv. runs through Mongolia. Consequently hybrid transitional forms are abundant here .

36. Trisetum Pers.
Syn. pl. 1 (1805) 97.

1. Plants 40–130 cm tall with short creeping subsurface shoots; stems with 3–5 nodes, slightly scabrous or glabrous under panicle; sheath glabrous; leaf blades 2.5–8 mm broad. Panicles 8–25 cm long, usually spreading and lax, with fairly long scabrous branches; spikelets 6–8 mm long; awn of lemma 5–8 mm long, recurved; anthers 1.8–2.4 mm long .... 4. T. sibiricum Rupr.

+ Plants 6–50 cm tall, without creeping subsurface shoots; stems with 1–3 nodes close set in lower part; sheath glabrous or pilose; leaf blades 0.7–4 mm broad. Panicles 1.5–10 cm long, usually dense, with very short branches; spikelets 4.5–7 mm long; anthers 0.4–1.5 mm long .... 2.

2. Branches of panicles glabrous or slightly scabrous; stems glabrous and smooth under panicle, awn of lemma 4–8 mm long, recurved; anthers 0.8–1.2 mm long .... 1. T. altaicum Roshev.

+ Branches of panicles very short but usually pilose; stems densely puberulent under panicle .... 3.

3. Awn of lemma 0.5–3 mm long, erect, less often suberect; panicles very dense, light green or light greyish-violet; anthers 0.8–1.5 mm long .... 2. T. litvinowii (Domin) Nevski.

+ Awn of lemma 3.5–6 mm long, recurved .... 4.

4. Leaf ligules 0.7–2.5 mm long. Panicles light green, quite lax, usually interrupted .... 3. T. seravschanicum Roshev.

+ Ligules 0.2–0.7 mm long. Panicles usually distinctly pinkish or greyish -violet, dense, less frequently interrupted ....
.... 5. T. spicatum (L.) Richt.

1. **T. altaicum** Roshev. in Bot. mater. Garb. Glavn. bot. sada RSFSR, 3 (1922) 85; Krylov, Fl. Zap. sib. 2 (1928) 235; Pavlov in Byull. Mosk. obshch. ispyt. prir. 38 (1929) 18; Roshev. in Fl. SSSR, 2 (1934) 254; Fl. kirgiz. 2 (1950) 95; Grubov, Consp. fl. MNR (1955) 66; Fl. Kazakhst. 1 (1956) 193.—Ic.: Fl. SSSR, 2, Plate 18, fig. 23; Fl. Kazakhst.1, Plate 14, fig. 10.

Described from Altay. Type in Leningrad.

In meadows, on rocky slopes and rocks; in upper mountain belt.

IIA. **Junggar:** *Alt. Region* (near Shara-Sume, 1700 m, No. 2620, Aug. 29, 1956—Ching); **Tien Shan** (Kyzemchek hills in Sairam-Nur lake region, July 31, 1878—A.

Reg.; 3 km south of Barchat village, No. 1678, Aug. 30; Nilki district, 60 km nor. of Ulastai, No. 1989, Aug. 31—1957, Huang).

General distribution: Jung.-Tarb., Nor. and Cent. Tien Shan; Middle Asia (West. Tien Shan), West. Sib. (Altay), East. Sib. (south), Nor. Mong. (Hent.).

Note. This species is very close to species *T. subalpestre* (Hartm.) Neum. found on barren peaks of East. Siberia and in the northern Scandinavian peninsula.

2. **T. litvinowii (Domin) Nevski** in Tr. Sredneaz. gos. univ., ser. 8b,. 17 (1934) 1; Fl. Kazakhst. 1 (1956) 193; Ikonnikov, Opred. rast. Pamira [Key to Plants of Pamir] (1963) 55.—*Koeleria litvinowii* Domin in Bibl. Bot. 65 (1907) 116; Goncharov in Fl. SSSR, 2 (1934) 325; Fl. Kirgiz. 2 (1950) 111; Keng, Fl. ill. sin., Gram. (1959) 477.—*K. argentea* Griseb. in Nachr. Gesellsch. Wissensch. u. Univ. Goettingen, 3 (1868) 77, non *Trisetum argenteum* Roem. et Schult. (1817); Pampanini, Fl. Carac. (1930) 75; Bor, Grasses Burma, Ceyl., Ind. and Pakist. (1960) 444.—*Koeleria hosseana* Domin in Feddes repert. 10 (1912) 54.—*K. enodis* Keng in Sunyatsenia, 6, 1 (1941) 60.—Ic.: Fl. SSSR, 2, Plate 25, fig. 4; Keng, l.c. fig. 409.

Described from Tien Shan (Chatkal mountain range). Type in Leningrad.

On rocky slopes, in grasslands and pebble beds; in upper mountain belt.

IA. Mongolia: *Mong. Alt.* (Tamchi somon, midportion of Bus-Khairkhan mountain range, July 17; Adzhi-Bogdo mountain range, Burgastyin-Daba pass in upper Indertiin-Gol, Aug. 6—1947, Yun); *Gobi-Alt.* (Ikhe-Bogdo mountain range, upper Narin-Khurimt creek valley, Sept. 6; same site, lower part of Arkatuin-Ama creek valley, Sept. 7—1943,Yun.).

IIA. Junggar: *Tien Shan* (30 km south of Nyutsyuan'tsze village along Ulausu river, 3150 m, No. 316, July 18; between Daban and Danu villages, No. 1990, July 19—1957, Huang; from Bartu to timber factory in Khomote, 2460 m, No. 7029, Aug. 4, 1958—Lee and Chu).

IIIA. Qinghai: *Nanshan* (24 km south of Xining, 2650 m, Aug. 4, 1959—Petr.); *Amdo* (vicinity of Huang He river near Guide town, 2300 m, June 19–23, 1880—Przew.; "on meadows of Htsechu between Labrang and the Yellow River, about 3870 m, July 1926, leg. J.F. Rock"—Keng, l.c. [1941]).

IIIB. Tibet: *Chang Tang* (south of Cherchen, 4190 m, No. 9417, June 4, 1959—Lee et al.); *Weitzan* (on Dychu river, June 4; on Bychu river, right tributary of Dychu, 4600 m, June 19; on southern bank of Orin-Nor lake, 4300 m, July 18—1884, Przew.; nor. bank of Russky lake, 4500 m, June 20; Yangtze river basin near Kambchzha-Kamba village, 4000 m, July 20—1900, Lad.; nor. Slope of Burkhan-Budda mountain range, Khatu gorge, 4300–4700 m, July 18—1901, Lad.; "Quellgebiet des Hoangho, Westrand der Odontala, 4300 m No. 50, Aug. 14, 1906; Tal des Scokoh-tschü, linker Nebenfluss des Jang-tse-kiang, No. 137, 23 VIII 1906, leg. A. Tafel "—Domin, l.c. [1912]); *South.* (Gyangtse, July to Sept., 1904—Walton).

IIIC. Pamir.

General distribution: Jung.-Tarb., Nor. and Cent. Tien Shan, East. Pam.; Middle Asia (West. Tien Shan, Pamir-Alay), Nor. Mong. (Hang.: bank of one of the tributaries of Urdu-Tamir river near Zain-Shabi, July 24, 1926—Ik.-Gal.), Himalayas.

Note. This species occupies an intermediate position between closely related genera *Trisetum* Pers. and *Koeleria* Pers. Many investigators place it in the genus *Koeleria* but considering the totality of characteristics (in particular the length of hairs on the rachilla of the spikelet and the number of anthers), it exhibits considerably closer affinity to *Trisetum spicatum* (L.) Richt. than to any other species of the genus *Koeleria*. Moreover, there is justification to assume that *T. litvinowii* is of hybrid origin—from *Trisetum spicatum* (L.) Richt. x *Koeleria cristata* (L.) Pers.—although this has yet to be confirmed. Typical specimens of *K. litvinowii* with profusely pubescent leaves and generally greyish-violet panicles are confined to the northern and western parts of the distribution range of the species while specimens with glabrous or subglabrous leaves and light green (silvery) panicles can be regarded as a district variety, *T. litvinowii* var. *argentea* (Griseb.) Tzvel. comb. nova (= *Koeleria argentea* Griseb. l.c.), distributed mainly in southern Tibet and in the Himalayas.

3. **T. seravschanicum** Roshev. in Bot. mater. Gerb. Glavn. bot. sada RSFSR, 3 (1922) 88 and in Fl. SSSR, 2 (1934) 225.—*T. virescens* (Regel) Roshev. in Izv. Bot. sada Petra vel. 14, suppl. 2 (1914) 64 and in Fl. SSSR, 2 (1934) 255, non Nees ex Steud. (1854); Fl. Kirgiz. 2 (1950) 95.—*T. fedtschenkoi* Henr. in Blumea, 3, 3 (1940) 425.—*Avena flavescens* β. *virescens* Regel in Bull. Soc. natur. Moscou, 2 (1868) 299.—*A. virescens* (Regel) Regel in Acta Horti Petrop. 7, 2 (1881) 635.— *Trisetum spicatum* auct. non Richt.: Nevski in Tr. Sredneaz. gos. univ., ser. 8b, 17 (1934) 1, p.p.; Fl. Kazakhst. 1 (1956) 192, p.p.; Fl. Tadzh. 1 (1957) 356, p.p.—Ic.: Fl. SSSR, 2, Plate 18, figs. 20 and 21.

Described from Middle Asia (Zeravshan mountain range). Type in Leningrad.

In meadows, on rocky slopes, in sparse forests, along banks of water reservoirs; in middle and upper mountain belts.

IIA. **Junggar:** *Jung. Alt.* (Toli district, 2000 m, No. 1256, Aug. 6; between Syata village and Ven'tsyuan', 2000 m, Nos. 1380 and 1392, Aug. 13—1957, Huang); *Tien Shan* (Bogdo hill, 2700–3000 m, July; Kokkamyr hill, July; Talki gorge, July—1877, A. Reg.; south-east. bank of Sairam-Nur lake, 1878—A. Reg.; south of Nyutsyuan'tsze village, No. 554, July 16; near Danu village, No. 1412, July 16; near Kelisu village, July 17; 5 km south-west of Dzhagastai village, No. 1753, Aug. 8—1957, Huang; nor. foothill of Narat mountain range descending into Tsanma valley, left tributary Kunges, 2150 m, Aug. 7, 1958—Yun.).

General distribution: Nor. and Cent. Tien Shan; Middle Asia (hilly).

Note. Recently, this species has been rather often combined with *T. spicatum* (L.) Richt. but the many persistent morphological characteristics of the specimens as well as their adaptation to just one relatively small region of the extensive distribution range of *T. spicatum*, could well serve as justification for holding *T. seravschanicum* as an independent species and relatively low mountain ecogeographical race.

4. **T. sibiricum** Rupr. in Beitr. Pflanzenk. Russ. Reich. 2 (1845) 65; Krylov, Fl. Zap. Sib. 2 (1928) 234; Pavlov in Byull. Mosk. obshch. ispyt. prir. 38 (1929) 18; Roshev. in Fl. SSSR, 2 (1934) 253; Kitag. Lin. Fl. Mansh. (1939) 95; Fl. Kirgiz. 2 (1950) 95; Grubov, Consp. fl. MNR (1955) 66; Fl. Kazakhst. 1 (1956) 193; Keng, Fl. ill. sin., Gram. (1959) 482.—*T. flavescens* var. *sibiricum* (Rupr.) Ohwi in Bot. Mag. Tokyo, 4 (1931) 192.—Ic.: Fl. SSSR, 2, Plate 18, fig. 24; Keng, l.c. fig. 413.

Described from north European USSR. Type in Leningrad.

In meadows, along fringes of swamps and in sparse (predominantly larch) forests; up to upper mountain belt.

IA. **Mongolia:** *Cis-Hing.* (Khuntu somon 5 km west of Toge-Gol river, Aug. 7, 1949—Yun.); *Cent. Khalkha, East. Mong.* (near Hailar town, 1960—Ivan.); *Bas. lakes* (Khirgis somon, 5–6 km nor.-nor.-east of Mani-Khutul' area, July 23, 1945—Yun.); *Gobi-alt.* (Dzun-Saikhan mountain range, Yalo creek valley, Aug. 26, 1931—(Ik. Gal.).

IIA. **Junggar:** *Jung. Alt., Tien Shan.*

IIIA. **Qinghai:** *Nanshan* (along Yusun-Khatyma river, 3000–3300 m, July 12, 1880—Przew.).

General distribution: Jung.-Tarb., Nor. and Cent. Tien Shan; Arct., Europe (east.), Balk.-Asia Minor, West. Sib., East. Sib., Far East, Nor. Mong., China (Dunbei, Cent., South-West.), Korea, Japan, Nor. Amer. (Aleutian islands and Yukon).

5. **T. spicatum** (L.) Richt. Pl. Eur. 1 (1890) 59; Krylov, Fl. Zap. Sib. 2 (1928) 236; Pavlov in Byull. Mosk. obshch. ispyt. prir. 38 (1929) 18; Pampanini, Fl. Carac. (1930) 73; Rehder in J. Arnold. Arb. 14 (1933) 3; Roshev. in Fl. SSSR, 2 (1934) 255; Walker in Contribs. U.S. Nat. Herb. 28 (1941) 599; Keng in Sunyatsenia, 6, 1 (1941) 60; Fl. Kirgiz. 2 (1950) 92; Grubov, Consp. fl. MNR (1955) 66; Fl. Kazakhst. 1 (1956) 192; Fl. Tadzh. 1 (1957) 356; Ikonnikov in Dokl. AN Tadzh SSR 20 (1957) 56; Bor, Grasses Burma, Ceyl., Ind. and Pakist. (1960) 448.—*T. subspicatum* (L.) Beauv. Ess. Agrost. (1812) 88; Forbes and Hemsley, Index Fl. Sin. 3 (1904) 400; Keng, Fl. ill. Sin., Gram. (1959) 482.—*Aira spicata* L. Sp. pl. (1753) 64.—*A. subspicata* L. Syst. nat., ed. 10, 2 (1759) 873, nom. illeg.; Hemsley, Fl. Tibet (1902) 203.—Ic.: Fl. SSSR, 2, Plate 18, fig. 22; Fl. Tadzh. 1, Plate 42; Keng, l.c. (1959) fig. 414.

Described from Scandinavia (Lapland). Type in Leningrad.

On rocky and rubble slopes, in grasslands, pebble beds, on rocks; in upper mountain belt.

IA. **Mongolia:** *Khobd.* (Tszusylan river, July 1; Kharkiry peak, July 11 and 12—1879, Pot.); *Mong. Alt., Gobi-Alt.* (Ikhe-Bogdo mountain range, in upper creek valley of Narin-Khurimt, Sept. 6, 1943; same site, between Narin-Khurimt and Ketsu creek valleys, June 28, 1945; same site, above Ketsu creek valley, June 29, 1945; same site, upper creek valley of Shimkhid, June 30, 1945—Yun.; same site, head of Narin-Khurimt gorge, 3500 m, July 29, 1948—Grub.); *Khesi* (in region of Suchzhou oasis [Tszyutsyuan'], Aug. 1, 1820—Marten).

IIA. **Junggar:** *Alt. region* (Oi-Chilik river, Sept. 8, 1876—Pot.; Qinhe district, No. 1074, Aug. 4, 1956—Ching); *Jung. Alt.* (30 km west of Ven'tsyuan', No. 2015, Aug. 25, 1957—Huang); *Tien Shan.*

IIIA. **Qinghai:** *Nanshan* (Severo-Tetungsk mountain range, Cherik pass, Aug. 8, 1890—Gr.-Grzh.; Humboldt mountain range, Ulan Bulak, 3700–4000 m, June 5, 1894—Rob.; "Near Sining, No. 672, leg. Ching"—Keng, l.c. 1941; Mon'yuan', valley of Ganshig river, left tributary of Peishikhe river, 3350–3720 m, Aug.; same site, sources of Ganshig river, 3900–4300 m, Aug. 18, 1958—Dolgushin).

IIIB. **Tibet:** *Chang Tang* ("Tibet, 4100 m, leg. Thorold"—Hemsley, l.c.); *Weitzan* (Burkhan Budda mountain range, nor. Slope of Khatu pass, 5000 m, July 12, 1901—Lad.); *South.* ("Balch pass, 5700 m, leg. Strachey and Witterbottom"—Hemsley, l.c.).

IIIC. **Pamir** (Ulug-Tuz gorge in Charlym river basin, May 28, 1909—Divn.; Chiganlyk village on descent from Kok-Mainak pass, July 27, 1913—Knorring; Mia river gorge, 4000 m, July 21, 1941—Serp.; "nor.-west. slope of Kongur, 4800 m, 13–17, VIII 1956"—Ikonnikov, l.c.).

General distribution: Jung.-Tarb., Nor. and Cent. Tien Shan, East. Pam.; Arct., Europe (hilly), Balk.-Asia Minor (hilly), Near East, Caucasus, Middle Asia (hilly), West. Sib. (Altay), East. Sib., Far East, Nor. Mong., China (Dunbei, Nor., Nor.-West., South-West.), Himalayas, Korea, Japan, Nor. America, South America.

Note. Highly polymorphic species within which a series of subspecies and varieties are distinguished. Central Asian specimens belong to subspecies *T. spicatum* subsp. *mongolicum* Hult. [Svensk Bot. Tidskr. 53, 2 (1959) 214] which differs only on average from the predominantly Arctic type subspecies *T. spicatum* subsp. *spicatum* in less uneven glumes and slightly longer anthers. The largest specimens were described as form *T. spicatum* f. *elatior* Kryl. and variety *T. spicatum* var. *robusta* Gamajun., which are scarcely of taxonomic importance.

37. Helictotrichon Bess.

in Schult. and Schult. f. Add. ad Mant. 3 (1827) 526.—
Avenastrum Jessen, Deutschl. Gräs. (1863) 214.

1. Palea glabrous and smooth along keel, sometimes almost smooth (with stray, very short spinules). Half or more of sheath closed; sheath of lower leaves generally rather pilose, less often glabrous; leaf blades 2–8 mm broad, flat, often rather pilose; ligules of cauline leaves 2–7 mm long; ligules of leaves of vegetative shoots 0.2–0.7 mm long; all shoots extravaginal; subsurface shoots short, creeping .... 6. **H. pubescens** (Huds.) Pilg.

+ Palea profusely puberulent along keel. Sheaths laciniate almost up to base .... 2.

2. Leaf blades flat or longitudinally folded with strongly thickened and prominent midrib in form of keel, with highly characteristic thickened cartilaginous whitish margin 0.2–0.6 mm broad; glabrous on upper surface, with scattered very short spinules, without prominent ribs; ligules of cauline leaves 3–10 mm long. Panicles rather compressed, with short branches .... 3.

+ Leaf blades flat or longitudinally folded, sometimes setaceous, with less prominent midrib and without thickened cartilaginous margin, generally rather scabrous or pilose on upper surface, usually with prominent ribs .... 5.

3. Plant 40–100 cm tall, usually not forming mats, with short creeping subsurface shoots; leaf blades flat, 4–12 mm broad. Panicles 12–25 cm long; spikelets 12–22 mm long; hairs in upper part of rachilla segments 2–2.6 mm long; lemma of lowest flowers 11–14 mm long; hairs on callus 1.5–2 mm long ....
.... 8. **H. dahuricum** (Kom.) Kitag.

+   Plants 20–80 cm tall, usually forming mats, without creeping subsurface shoots; leaf blades flat or longitudinally folded, 1.5–4 mm broad. Panicles 5–15 cm long; spikelets 9–15 mm long; hairs in upper part of rachilla segments 0.8–1.3 mm long; lemma of lowest flowers 9–11 mm long; hairs on callus 0.5–0.8 mm long .. .. 4.

4.   Panicles 4–8 cm long, very dense; spikelets brownish-green, with golden tinge .... 7. H. asiaticum (Roshev.) Grossh.

+   Panicles 5–15 cm long, usually very lax; spikelets pale green ....9. H. schellianum (Hack.) Kitag.

5.   Plant 70–140 cm tall, with short creeping subsurface shoots; leaf blades 2–7 mm broad, flat, with faintly visible ribs on upper surface. Panicles 10–20 cm long, narrow, but fairly lax; spikelets 9–13 mm long, usually pale green; rachilla segments in upper part with hairs 1.5–2.2 mm long .... 1. H. altius (Hitchc.) Ohwi.

+   Plants up to 70 cm tall, forming dense mats, without creeping subsurface shoots; leaf blades up to 3 mm broad, usually longitudinally folded, less often flat, with highly prominent ribs on upper surface. Panicles up to 10 cm long .... 6.

6.   Leaf ligules of vegetative shoots 0.8–3 mm long, ligules of upper cauline leaf 0.3–1.5 mm long; leaf blades very long and narrow, invariably longitudinally folded, 0.5–0.9 mm in diameter, with continuous ring of subepidermal mechanical tissue, without sclerenchyma band above central vascular bundle .... 2. H. altaicum Tzvel.

+   All leaf ligules up to 0.6 (0.8) mm long; leaf blades relatively short, longitudinally folded, less often flat, up to 3.5 mm broad, without continuous ring of subepidermal mechanical tissue; invariably with sclerenchyma band above central vascular bundle .... 7.

7.   Panicles very dense, with axis and branchlets throughout length with short but abundant hairs; spikelets 7–9 mm long, rather golden in colour; lemma of lowest flowers 6.5–7.5 mm long; stems under panicle puberulent for small extent; leaf blades scabrous on upper surface .... 5. H. tibeticum (Roshev.) Holub.

-    Panicles very lax, with scabrous branches sometimes puberulent only in upper part; spikelets 8–12 mm long; lemma of lowest flowers 7.5–10 mm long; stems under panicle glabrous, smooth or somewhat scabrous .... 8.

8.   Sheaths of cauline leaves glabrous, less often with few scattered hairs; leaf blades densely puberulent on upper surface; stalks of spikelets in upper part with elongated spinules passing into short hairs; spikelets usually brownish-violet, brightly coloured .... 3. H. mongolicum (Roshev.) Henr.

+ Sheaths of cauline leaves usually short but profusely pilose, less often glabrous; leaf blades scabrous on upper surface due to spinules; stalks of spikelets throughout length only with spinules; spikelets usually not lustrous ....
.... 4. H. tianschanicum (Roshev.) Henr.

Section 1. Helictotrichon

1. H. altius (Hitchc.) Ohwi in J. Jap. Bot. 17 (1941) 440; Keng, Fl. ill. sin., Gram. (1959) 490.—*Avena altior* Hitchc. in Proc. Biol. Soc. Wash. 43 (1930) 96.—Ic.: Keng, l.c. fig. 421.

Described from Qinghai. Type in Washington.

In meadows, sparse forests, among shrubs; in middle and upper mountain belts.

IIIA. **Qinghai:** *Nanshan* ("South of Sining in the La-Che-Tze mountains, 3350–3900 m, No. 716, leg. Ching"—Hitchcock, l.c. type!).

General distribution: China (Cent., South-West.), Himalayas (east.).

Note. This species falls among the more primitive species of the section and genus recently classified as a distinct subgenus, *Archavenastrum* (Vierh.) Holub, but, in our opinion, it is very closely related to the type genus *H. desertorum* (Less.) Nevski. Another new species from Kam upland collected by G.N. Potanin is closer to *H. altius* and some Himalayan species of the genus. We named the new species *H. potaninii* Tzvel. sp. nova.—Planta 25–50 cm alta, laxe caespitosa; culmi et vaginae glabrae et laeves; ligulae 1.5–3.5 mm lg., extus glabrae; laminae 2–4 mm lt., planae, supra plus minusve scabrae et sparse pilosae, subtus glabrae et sublaeves. Paniculae 5–8 cm lg., sat densae, roseo-violaceo tinctae, ramis scabris multo abbreviatis; spiculae 7–9 mm lg., 3–4-florae; rachillae internodia densissime et longe pilosa, in parte superiore pilis 3–4 mm lg. tecta; glumae lanceolatae, superior spiculam subaequilonga, trinervia, inferior subsesqui brevior, uninervia; lemmata lanceolata, scaberula, 7–9.5 mm lg., in parte superiore membranacea et aristulatobidentata; aristae geniculate curvatae, 10–14 mm lg., prope medium lemmatis insertae; paleae lemmatis paulo breviores, carinis copiose sed brevissime pilosis; antherae 1–1.6 mm lg. Typus: "Prov. Szetschuan septentrionalis, in alpe Su-Ye-Schan ad limitem nivis aeternae, 30 VII 1885, G. Potanin". In Herb. Bot. Inst. Ac. Sci. URSS—LE conservatur. A specierum proximis : *H. asperum* (Munro) Bor, *H. virescens* (Nees ex Steud.) Henr., *H. schmidii* (Hook. f.) Henr. et *H. altius* (Hitchc.) Ohwi praesertim rachillae pilis densioribus et longioribus necnon paniculis densis differt.

2. H. altaicum Tzvel. sp. nova.—*Avenastrum desertorum* auct. non podpéra: Roshev. in Fl. SSSR, 2 (1934) 276, p.p.; Fl. Kirgiz. 2 (1950) 100; Grubov, Consp. fl. MNR (1955) 67.—*A. tianschanicum* auct. non Roshev. : Fl. Kazakshst. 1 (1956) 199; fl. Tadzh. 1 (1957) 352.—Haec species *H. desertorum* (Less.) Nevski proxima, sed praesertim ligulis parum brevioribus (ad 3 mm lg.), extus breviter pilosis et margine ciliatis differt.

Typus: "Mongolia borealis, circa lacus Ubsu-Nur, trajectus Ulan-Daban, 11 VI 1879, G. Potanin". In Herb. Inst. Bot. Acad. Sci. URSS—LE conservatur.

Described from Mongolia. Type in Leningrad.

In upland steppes, on rocky slopes and rocks; in midmountain belt.

IA. **Mongolia:** *Khobd.* (Ulan-Daban pass, June 11, 1879—Pot., type!); *Mong. Alt.* (Tamchi somon, nor. Mountain trail of Bus-Khairkhan range, June 17; same site, south. slope of Tamchi-Daba pass, July 16; Bulugun somon, south, flank of Indertiin-Gol valley, July 24—1947, Yun.).

IIA. **Junggar:** *Alt. region,* Tarb. (south. slope of Saur mountain range, karagaity river valley, Bain-Tsagan creek valley, June 23, 1957—Yun.): *Jung. Alt., Tien Shan* (nor. slope of Tien Shan [below koshety-Daban pass], June 10, 1877—Pot.; near Koksu river, 1700 m, May 30, 1878; Irenkhabirga mountain range, Naryn-Gol brook near Tsagan-Usu, 2000–2700 m, June 10, 1879—A. Reg.; near Tekes town, No. 1019, Aug. 18; Barkul' lake, No. 2240, Sept. 28—1957, Huang; east. part of Ketmen' mountain range 3–4 km above Sarbushin settlement on Kul'dzha-Kzyl Kure road, Aug. 23, 1957—Yun.).

General distribution: Jung.-Tarb., Nor. and Cent. Tien Shan; Middle Asia (east.), West. Sib. (Altay), East. Sib. (Sayans).

Note. This species represents one of two ecogeographical races that were formerly combined under the name *H. desertorum* (Less.) Nevski. The other race, *H. desertorum* s.s., described from South. Urals and having very long (in leaves of vegetative shoots 2–7 mm long), glabrous, less often subglabrous, ligules, is widely distributed in steppes on plains and low mountain steppes of Eurasia from Austria and Czechoslovakia to Minusinsk basin; it does not enter Mongolia or China. Grisebach too was aware of the existence of these races but erroneously regarded the Altay specimens collected by him as the type specimen of *H. desertorum* and described the plains race as an independent species, *Avena besseri* Griseb. [in Ledeb. Fl. Ross. 4 (1852) 415]. The same mistake has been repeated recently by other investigators, who have adopted different names for the typical plains race *H. desertorum* : *H. besseri* (Griseb.) Janch. [in Phyton, 5 (1953) 64], *H. desertorum* proles *basalticum* (Podpéra) Holub [in Preslia, 31 (1959) 50], *H. desertorum* var. *planitierum* Pavl. f. [in Bot. mater. Gerb. Bot. inst. AN SSSR, 22 (1963) 72]. The significant differences between Central Asian specimens and typical plains specimens of *H. desertorum* were also recorded by N.A. Ivanova who erroneously combined them with *H. mongolicum* (Roshev.) Henr. as a variety.

3. **H. mongolicum** (Roshev.) Henr. in Blumea, 3, 3 (1940) 431; Ohwi in J. Jap. Bot. 17 (1941) 440; Krylov, Fl. Zap. Sib. 12, 1 (1961) 3098.—*Avena mongolica* Roshev. in Izv. Glavn. bot. sada SSSR, 27 (1928) 96; Pavlov in Byull. Mosk. obshch. ispyt. prir. 38 (1929) 19.—*Avenastrum mongolicum* (Roshev.) Roshev. in Fl. SSSR, 2 (1934) 280; Grubov, Consp. fl. MNR (1955) 67; Fl. Kazakhst. 1 (1956) 198. —Ic.: Fl. SSSR, 2, Plate 21, fig. 13.

Described from Nor. Mongolia (Fore Hubs.). Lectotype (Munku-Sardyk, Aug. 17, 1871, leg. Czekanovsky) in Leningrad. Map 7.

In grasslands, on rocky slopes and in riverine pebble beds; in middle and upper mountain belts.

IA. **Mongolia:** *Khobd.* (Tszusylan river, July 1; south. peak of Kharkira, July 11 and 12—1879, Pot.); *Mong. Alt.* (Taishiri-Ula mountain range, July 1, 1877—Pot.); *Gobi-Alt.* (Ikhe-Bogdo mountain range, Aug. 23, 1926—Tug.; same site, upper creek valley of Ikhe-Khurimt, June 28; nor. slope of Ikhe-Bogdo mountain range, upper creek valley of Shishkhid, June 30; Tsagan-Gol river bank, near Bain-Gobi somon, July 3—

1945, Yun.; south-east. slope of Ikhe-Bogdo mountain range, Narin-Khurimt gorge, 2800 m, July 28, 1948—Grub.).

IIA. Junggar: *Tarb.* (south. slope of Saur mountain range. Karagaitu river valley, Bain-Tsagan creek valley, June 23, 1957—Yun.); *Tien Shan* (Manas river basin, upper course of Danu-Gol river near Se-Daban pass, July 21, 1957—Yun.; same site, 2600 m, No. 2141, July 21; 4 km south of Danu village, 3250 m, No. 444, July 22; along Ulausu river, No. 580, July 23—1957, Huang).

IIIA. Qinghai: *Nanshan* (Xining hills, Myn'dan'sha river, May 22, 1890—Gr.-Grzh.). General distribution: Jung.-Tarb. (Tarbagatai); West. Sib. (south. Altay), East. Sib. (East. Sayan, Barguzinsk mountain range and Sokhondo hill in Borshchevok mountain range), Nor. Mong. (Fore Hubs., Hent., Hang.).

Note. The solitary defective specimen of this species from Qinghai has very broad, often flat leaf blades.

4. **H. tianschanicum** (Roshev.) Henr. in Blumea, 3, 3 (1940) 429.— *Avenastrum tianschanicum* Roshev. in Izv. Bot. sada AN SSSR, 30 (1932) 771; Roshev. in Fl. SSSR, 2(1934) 280; Fl. Kirgiz. 2 (1950) 99. —**Ic.:** Fl. SSSR, 2, Plate 21, fig. 10.

Described from Cent. Tien Shan. Lectotype (Kara-Archa river, Aug. 6, 1912, V. Sapozhnikov and B. Schischkin) in Leningrad. Map 7.

On rocky and clayey slopes and in riverine pebble beds; in middle and upper mountain belts.

IB. Kashgar: *Nor.* (near Uchturfan, Airi gorge, June 2; same site, Karagailik gorge, June 20—1908, Divn.; Sogdyntau mountain range, Taushkan-Darya valley 10–12 km south-east of Akchet hydraulic power station, Sept. 19, 1958—Yun.).

IIA. Junggar: *Tien Shan* (nor. of Turfan, 2600 m, June 23; Khomote dist. between Bartu village and timber factory, 2590 m, No. 1063, Aug. 4—1958, Lee and Chu; valley of Muzart river in its upper course, Lyangar gorge, Sept. 12, 1958—Yun.).

General distribution: Cent. Tien Shan.

Note. The specimens noted above, collected by Lee and Chu, have very broad, often flat leaf blades and sparsely hairy sheath of cauline leaves.

5. **H. tibeticum** (Roshev.) Holub in Preslia, 31, 1 (1959) 50.—*H. tibeticum* (Roshev.) Keng f. in Keng, Fl. ill. sin., Gram. (1959) 496, nom. nud.—*Avena tibetica* Roshev. in Izv. Glavn. bot. sada SSSR, 27 (1928) 98.—*A. suffusca* Hitchc. in Proc. Biol. Soc. Wash. 43 (1930) 95; Walker in Contribs. U.S. Nat. Herb. 28 (1941) 596. —**Ic.:** Keng, l.c. fig. 426.

Described from Tibet. Lectotype in Leningrad. Map 7.

On rocky slopes, in grasslands and among shrubs; in middle and upper mountain belts.

IIIA. Qinghai: *Nanshan* (Severo-Tetungsk mountain range, June 22, 1872; in Nanshan hills, July 1879; between Nanshan and Rako-Gol river, 3300–3700 m, July 10, 1880—Przew.; in Gansu province, Aug. 3, 1890—Marten; "La-Chang-Kou near Sining, 3000-3300 m; La-Che-Tze Mountains, 3350–3900 m; at Ta-Hwa near Pinfan, 2900-3100 m, July 17—1923, leg. Ching"—Hitchcock, l.c.; Ganshiga river valley, left tributary of Peishikhe river, 3350–3920 m, 1958—Dolgushin).

IIIB. Tibet: *Weitzan* (near Dyaochu river, No. 284, June 29, 1884—Przew., lectotype!; Yangtze river basin, near Nko-Gun river entering Rkhombo-Mtso lake, 4400 m, Aug. 6, 1900; nor. slope of Burkhan-Budda mountain range, Khatu gorge, 4000–4700 m, Aug. 17, 1901—Lad.).

General distribution: China (South-West.).

Note. Typical specimens of this species have sheaths of cauline leaves and leaf blades with short but dense hairs on upper surface while specimens from Qinghai have glabrous or subglabrous sheaths and leaf blades scabrous on upper surface. This provides a basis for distinguishing the latter as variety *H. tibeticum* var. *suffuscum* (Hitchc.) Tzvel. comb. nova (=*Avena suffusca* Hitchc. l.c.).

Section 2. Pubescentes
(Rouy) Tzvel.[4]

6. **H. pubescens** (Huds.) Pilg. in Feddes repert. 45 (1938) 6; Krylov, Fl. Zap. Sib. 12, 1 (1961) 3099.—*Avena pubescens* Huds. Fl. Angl. (1762) 42; Krylov, l.c. 2 (1928) 242; Pavlov in Byull. Mosk. obshch. ispyt. prir. 38 (1929) 18.—*Avenastrum pubescens* (Huds.) Opiz, Seznam Rostl. Kvét. české (1852) 20; Jessen, Deutschl. Gräs. (1863) 53; Roshev. in Fl. SSSR, 2 (1934) 275; Fl. Kirgiz. 2 (1950) 99; Grubov, Consp. fl. MNR (1955) 67; Fl. Kazakhst. 1 (1956) 197.—*Avenochloa pubescens* (Huds.) Holub in Acta Horti Bot. Prag. (1962) 84. —**Ic.**: Fl. SSSR, 2, Plate 21, fig. 7; Fl. Kazakhst. 1, Plate 14, fig. 1.

Described from England. Type in London (K).

In meadows and forest glades; in middle and upper mountain belts.

IA. Mongolia: *Mong. Alt.* (2 km west of Bulugun somon on road to Kharagaitu-Khutul, July 24, 1947—Yun.).

IIA. Junggar: *Alt. region* (near Qinhe, 1800–1900 m, Nos. 904 and 1230, Aug. 2, 1956—Ching; Tsagan-Obo hill to nor.-west of Khobsair, 2000 m, No. 10530, June 22, 1959—Lee et al.); *Tarb.* (nor. of Dachen town, No. 2916, Aug. 13, 1957—Huang); *Jung. Alt.* (Toli district, No. 1060, Aug. 6; between Syat and Ven'tsyuan, 2000 m, No. 3417, Aug. 13; 20 km south of Ven'tsyuan, 2810 m, No. 1497, Aug. 14; 15 km nor.-west of Ven'tsyuan, No. 1623, Aug. 29—1957, Huang); *Tien Shan*.

General distribution: Balkh. region (Karkaralinsk hills), Jung.-Tarb., Nor. and Cent. Tien Shan; Europe, Mediterr., Balk.-Asia Minor (hilly), Near East (hilly), Caucasus (hilly), Middle Asia (West. Tien Shan), West. Sib. (south), East. Sib. (south), Nor. Mong. (Hang., Mong.-Daur.); introduced in USA and some other countries.

Section 3. Avenula (Dum.) Tzvel.[5]

7. **H. asiaticum** (Roshev.) Grossh. Fl. Kavk. 1 (1939) 215, quoad nom.; Fl. Tadzh. 1 (1957) 349; Krylov, Fl. Zap. Sib. 12, 1 (1961) 3099.—

---

[4]*Helictotrichon* sect. Pubescentes (Rouy) Tzvel. in Spiske rast. Gerb. Fl. SSSR [List of Plants in the Herbarium of Russian Flora], 17 (1967) 42.

[5]*Helictotrichon* sect. Avenula (Dum.) Tzvel. comb. nova = *Trisetum* sect. Avenula Dum. Observ. Gram. Belg. (1823) 122, p.p. [lectotype: *Helictotrichon pratense* (L.) Bess.].

Avenastrum asiaticum Roshev. in Izv. Bot. sada AN SSSR, 30 (1932) 770 and in Fl. SSSR, 2 (1934) 275, p.p.; Fl. Kirgiz. 2 (1950) 100; Fl. Kazakhst. 1 (1956) 197.—*Avenochloa asiatica* (Roshev.) Holub in Acta Horti Bot. Prag. (1962) 83.—*Avena versicolor* auct. non Vill.: Krylov, l.c. 2 (1928) 245.

Described from Cent. Tien Shan. Lectotype (near Nikolaevsk settlement on Dzhirgalan river, no. 995, June 13, 1910, A. Mikhel'son) in Leningrad.

In grasslands and on rocky slopes; in middle and upper mountain belts.

IA. **Mongolia:** *Mong. Alt.* (2 km west of Bulugun somon along road to Kharagaitu-Khutul, July 24; upper Indertiin-Gol 4–5 km below summer camp at Bulugun somon, July 25—1947, Yun.).

IIA. **Junggar:** *Alt. region* (near Qinhe, 1800–1900 m, Nos. 1188 and 1328, Aug. 2 and 3; between Qinhe and Chzhunkhaitsza village, 2440 m, No. 1427, Aug. 6—1956, Ching); *Jung. Alt., Tien Shan.*

General distribution: Jung.-Tarb., Nor. and Cent. Tien Shan; Middle Asia (West. Tien Shan, Alay, Giss.-Darv.), West. Sib. (Altay).

Note. This species, together with European high-mountain *H. versicolor* (Vill.) Pilg., Eurasian steppe *H. schellianum* (Hack.) Kitag. and North American hill steppe *H. hookeri* (Scribn.) Henr., forms a very closely related groups. Recently, it became clear that the Caucasian specimens regarded as *H. asiaticum* by R. Roshevitz and other investigators actually occupy an intermediate position between *H. asiaticum* and *H. versicolor* proles *caucasicum* Holub [in Preslia, 31 (1959) 51].

8. **H. dahuricum** (Kom.) Kitag. Lin. Fl. Mansh. (1939) 77.—*Avena dahurica* Kom. Fl. Kamch. 1 (1927) 159; Norlindh, Fl. mong. steppe, 1 (1949) 78.—*Avenastrum dahuricum* (Kom.) Roshev. in Fl. SSSR, 2 (1934) 275.—*Avenochloa dahurica* (Kom.) Holub in Acta Horti Bot. Prag. (1962) 84. —Ic.: Fl. SSSR, 2, Plate 21, fig. 56.

Described from Transbaikal region (Shilka river). Type in Leningrad.

In forest glades, meadows, on rocky slopes and rocks; up to lower mountain belt.

IA. **Mongolia:** *Cis-Hing.* (near Irekte railway station, June 15, 1902—Litv.). **General distribution:** East. Sib., Far East, China (Dunbei), Korea.

Note. Notwithstanding the vast separation between distribution ranges, this species is extremely close to the European high-mountain species *H. planiculme* (Schrad.) Pilg. and is commonly combined with it.

9. **H. schellianum** (Hack.) Kitag. Lin. Fl. Mansh. (1939) 78; Keng, Fl. ill. sin., Gram. (1959) 493; Krylov, Fl. Zap. Sib. 12, 1 (1961) 3099.—*Avena schelliana* Hack. in Acta Horti Petrop. 12 (1892) 419; Krylov, l.c. 2 (1928) 244; Pavlov in Byull. Mosk. obshch. ispyt. prir. 38 (1929) 18; Norlindh, Fl. mong. steppe, 1 (1949) 78.—*Avenastrum schellianum* (Hack.) Roshev. in Fl. SSSR, 2 (1934) 274; Fl. Kirgiz. 2 (1950) 100; Grubov, Consp. fl. MNR (1955) 67; Kazakhst. 1 (1956) 198.—*Avenochloa schelliana* (Hack.) Holub in Acta Horti Bot. Prag. (1962) 85.—*Avena pratensis* auct. non L.: Franch. Pl. David. 1 (1884) 333; Forbes and Hemsley, Index Fl. Sin. 3 (1904) 401; Pavlov, l.c.

18.—*Avenastrum dahuricum* auct. non Roshev.:? Pavlov, l.c. 18; Grubov, l.c. 67, **p.p.—Ic.**: Fl. SSSR, 2, Plate 21, figs. 2 and 4; Keng, l.c. fig. 425.

Described from Far East. Lectotype (Blagoveshchensk town, July 5, 1891, S. Korshinsky) in Leningrad.

In meadow steppes, forest glades and on rocky slopes; up to midmountain belt.

IA. **Mongolia:** *Mong. Alt.* (nor. trail of Khan-Taishiri mountain range 10 km south-east of Yusun-Bulak, July 14, 1947—Yun.); *Cis-Hing., Cent. Khalkha, East. Mong.* (Khuntu somon 5 km west of Toge-Gol, Aug. 7, 1949—Yun.; near Shilin-Khoto, 1959—Ivan.; Numutuin-Daba pass 2.5 km south of Erdeni-Tsagan somon, July 27, 1962—Lavr. et al.); *Gobi-Alt.*

General distribution: Aral-Casp., Balkh. region, Jung.-Tarb., Nor. Tien Shan; Europe (south-east.), Middle Asia (east.), West. Sib. (south.), East. Sib., Far East (south.) Nor. Mong., China (Dunbei, Nor., Nor.-West.), Korea.

Note. This species cannot be clearly distinguished from the north American hill steppe species *H. hookeri* (Scribn.) Henr. and, through *H. asiaticum* (Roshev.) Grossh., is intimately associated with the European high-mountain species *H. versicolor* (Vill.) Pilag.

38. Avena L.
Sp. pl. (1753) 79.

1. Rachilla pilose, with distinct joints under all florets or only under lower floret, easily shedding along joints in fruit and on desiccation; lemma with hairy callus, glabrous or rather pilose on back, with long awn twisted in lower part and later geniculately curved .... 2.

+ Rachilla glabrous but sometimes scabrous, without joints, breaking irregularly in fruit and on desiccation; lemma with glabrous or subglabrous callus, almost invariably glabrous on back; without awn or with short erect or geniculately curved awn .... 3.

2. Joints present under all florets of spikelet, as a result of which rachilla easily disintegrates into individual segments; less often, fully developed joint only under lowest floret but then lower segment of rachilla distinctly flattened from outside or even has a longitudinal groove and is set off from callus of lemma of next flower by distinctly manifest constriction with underdeveloped joint; scar on callus of lemma suborbicular .... 2. **A. fatua L.**

+ Joint present only under lowest floret of spikelet, as a result of which all florets of spikelet shed together; lower segment of rachilla convex throughout length from outside and directly passing into callus of lemma of next (second from below) flower; scar on callus of lemma of lowest flower in spikelet oval .... 4. **A. trichophylla C. Koch.**

3. Spikelets with 2–5 florets and elongated, usually scabrous, rachilla segments; glumes shorter than spikelets; lemma coriaceous-membranous, with distinct ribs throughout length; tip of lowest flower with 2 short and broad lobes .... 1. **A. chinensis** (Fish. ex Roem. et Schult.) Metzg.

+   Spikelets with 2–3 florets and short smooth rachilla segments; glumes as long as spikelets; lemma coriaceous with distinct ribs only in upper half; tip of lowest flower unevenly dentate or with 2 very short and narrow lobes .... 3. **A. sativa** L.

1. **A. chinensis** (Fisch. ex Roem. et Schult.) Metzg. Europ. Cereal. (1824) 53; Franch. Pl. David. 1 (1884) 339; Nevski in Tr. Sredneaz. gos. univ., ser. 8b, 17 (1934) 6; Norlindh, Fl. mong. steppe, 1 (1949) 79.—*A. nuda* var. *chinensis* Fisch. ex Roem. et Schult. Syst. Veg. 2 (1817) 669.—*A. nuda* auct. non L.: Forbes and Hemsley, Index Fl. Sin. 3 (1904) 401; Keng, Fl. ill. sin., Gram. (1959) 487. — Ic.: Keng, l.c. fig. 420.

Described from China. Type probably in Berlin. Isotype in Leningrad.

Cultivated in oases and found as introduced or wild plant along roadsides, around irrigation ditches and in wastelands; up to mid-mountain belt.

IA. **Mongolia:** *East. Mong.* (near Khukh-Khoto town, near debris in Tuchen town, July 21, 1884—Pot.); *Ordos* (Ortous,No. 2934—David).

General distribution: China (Dunbei, Nor., Nor.-West., Cent., East., South-West.), Korea; occasionally cultivated in some other countries.

2. **A. fatua** L. Sp. pl. (1753) 80; Forbes and Hemsley, Index Fl. Sin. 3 (1904) 401; Diels in Filchner, Wissensch. Ergebn. 10, 2 (1908) 248; Krylov, Fl. Zap. Sib. 2 (1928) 240; Pavlov in Byull. Mosk. obshch. ispyt. prir. 38 (1929) 18; Pampanini, Fl. Carac. (1930) 74; Roshev. in Fl. SSSR, 2 (1934) 267; Persson in Bot. notiser (1938) 274; Fl. Kirgiz. 2 (1950) 98; Grubov, Consp. fl. MNR (1955) 67; Fl. Kazakhst. 1 (1956) 195; Fl. Tadzh. 1 (1957) 346; Keng. Fl. ill. sin., Gram. (1959) 487; Bor, Grasses Burma, Ceyl., Ind. and Pakist. (1960) 434.—*A. fatua* var. *glabrata* Peterm. Fl. Bienitz (1841) 13.—*A. fatua* var. *basifixa* Malz. in Tr. Byuro po prikl. bot. 7,5 (1914) 329.— *A. aemulans* Neski in Tr. Sredneaz. gos. univ., ser. 8b, 17 (1934) 5. —Ic.: Fl. SSSR, 2, Plate 20, figs. 2–5; Fl. Tadzh. 1, Plate 40, fig. 4.

Described from Europe. Type in London (Linn.).

As weed in oases, along roadsides and in wastelands; up to upper mountain belt.

IA. **Mongolia:** *Val. Lakes, Gobi-Alt.* (in barley crops near Bain-Leg somon camp, July 25, 1948—Grub.); *Khesi.*

IB. **Kashgar:** *Nor., West.*

IC. **Qaidam:** *plains* (Nomokhun-Gol river, Aug. 3, 1884—Przew.).

IIA. **Junggar:** *Alt. region, Tarb., Tien Shan.*

IIIA. **Qinghai:** *Nanshan* ("Si-ning-fu"-Diels, l.c.).

IIIB. Tibet: *Weitzan* (Nruchu area on right bank of Yangtze river, 3900 m, July 25, 1900—Lad.); *South.* (Gyantze, No. 30, July to Sept., 1904—Walton).

IIIC. Pamir (Pas-Rabat village, Aug. 3, 1909—Divn.).

General distribution: Aral-Casp., Balkh. region, Jung.-Tarb., Nor. and Cent. Tien Shan; Europe, Mediterr., Balk.-Asia Minor, Near East, Caucasus, Middle Asia, West. Sib. (south), East. Sib. (south), Far East (south), Nor. Mong., China, Himalayas, Korea, Japan, Nor. Africa; introduced in Nor. and South Amer.

Note. Along with the typical variant having lemma rather pilose dorsally and well-developed joints under each flower of spikelet, the varieties mentioned above are also often encountered in Cent. Asia: *A. fatua* var. *glabrata* Peterm. with lemma glabrous dorsally and *A. fatua* var. *basifixa* Malz. (=*A. aemulans* Nevski) with underdeveloped upper joints of rachilla, probably of hybrid origin: *A. fatua* X *A. sativa*. Specimens of the latter variety are occasionally treated erroneously as *A. trichophylla* C. Koch.

3. **A. sativa** L. Sp. pl. (1753) 79; Franch. Pl. David. 1 (1884) 333; Forbes and Hemsley, Index Fl. Sin. 3 (1904) 402; Krylov, Fl. Zap. Sib. 2 (1928) 239; Roshev. in Fl. SSSR, 2 (1934) 267; Fl. Kirgiz. 2 (1950) 98; Grubov, Consp. fl. MNR (1955) 67; Fl. Kazakhst. 1 (1956) 194; Fl. Tadzh. 1 (1957) 347; Keng, Fl. ill. sin., Gram. (1959) 487; Bor, Grasses Burma, Ceyl., Ind. and Pakist. (1960) 434. —Ic.: Fl. SSSR, 2, Plate 20, fig. 6; Fl. Tadzh. 1, Plate 40, fig. 3; Keng, l.c. fig. 419.

Described from Europe. Type in London (Linn.).

Cultivated and often found as introduced or wild plant along roadsides and around irrigation ditches and in waste lands; up to upper mountain belt.

IA. Mongolia: *East. Gobi.*
IB. Kashgar: *Nor., West., South.,*
IIA. Junggar: *Tien Shan, Dzhark.*
IIIA. Qinghai: *Nanshan.*

General distribution: cultivated in almost all extratropical countries of both hemispheres; apparently native of the Mediterranean.

4. **A. trichophylla** C. Koch in Linnaea, 21 (1848) 393; Roshev. in Fl. SSSR, 2 (1934) 269; Fl. Kirgiz. 2 (1950) 96; Fl. Tadzh. 1 (1957) 347.—*A. ludoviciana* Dur. in Bull. Soc. Linn. Bordeaux, 20 (1855) 41; Roshev., l.c. 269; Persson in Bot. notiser (1938) 274; Fl. Kazakhst. 1 (1956) 196; Bor, Grasses Burma, Ceyl., Ind. and Pakist. (1960) 434. —Ic.: Fl. SSSR, 2, Plate 20, figs. 10–16; Fl. Tadzh. 1, Plate 40, fig. 1.

Described from Caucasus. Type in Berlin. Isotype in Leningrad.

On rocky and melkozem slopes, in riverine pebble beds, often as weed in field crops, along roadsides and around irrigation ditches; up to midmountain belt.

IB. Kashgar: *West.* ("Kentalek, 2700 m, No. 204, July 10, 1931"—Persson, l.c.).

General distribution: Aral-Casp., Balkh. region (south.), Nor. Tien Shan; Europe (south.), Mediterr., Balk.-Asia Minor, Near East, Caucasus, Middle Asia, Himalayas; introduced in other countries.

Note. It is not unlikely that the solitary reference of this species for Kashgar actually pertains to *A. fatua* var. *basifixa* Malz. However, *A. trichophylla* is quite well distributed in Middle Asia (collections available even from Alay valley) and the Himalayas and thus there is every good possibility of its introduction into Kashgar.

39. Cynodon Rich.
in Pers. Synops. pl. 1 (1805) 85, nom. conserv.

1. **C. dactylon** (L.) Pers. Synops. pl. 1 (1805) 85; Franch. Pl. David. 1 (1884) 334; Forbes and Hemsley, Index Fl. Sin. 3 (1904) 402; Roshev. in Fl. SSSR, 2 (1934) 285; Fl. Kirgiz. 2 (1950) 104; Fl. Kazakhst. 1 (1956) 200; Fl. Tadzh. 1 (1957) 454; Keng. Fl. ill. sin., Gram. (1959) 472; Bor, Grasses Burma, Ceyl., Ind. and Pakist. (1960) 469; Krylov, Fl. Zap. Sib. 12, 1 (1961) 3100.—*Panicum dactylon* L. Sp. pl. (1753) 58. —Ic.: Fl. SSSR, 2, Plate 22, fig. 1; Keng, l.c. fig. 403.

Described from South. Europe. Type in London (Linn.).

On rocky and melkozem slopes, in coastal sand and pebble beds; often as weed along roadsides, in field crops and wastelands; up to midmountain belt.

IB. **Kashgar:** *Nor.* (south. slope of Tien Shan, Sept. 16, 1895—Rob.); *West.* (along Yarkand-Darya river, June 4, 1889—Rob.; around Yangigisar, May 24, 1909—Divn.); *East.* (Chitkal area near Turfan, Sept. 14, 1898—Klem; Toksun oasis, No. 7219, June 7, 1958—Lee and Chu).

IIA **Junggar:** *Tien Shan, Jung. Gobi* (in region of Savan settlement, No. 1314, July 9, 1957—Huang); *Dzhark.* (floodplain of Ili river near crossing from Kul'dzha to Ketmen', Aug. 21, 1957—Yun.); *Balkh.-Alak.* (4 km east of Dachen, Aug. 10; between Durbul'dzhin and Dachen, Aug. 10—1957, Huang).

General distribution: Aral-Casp., Balkh. region, Jung.-Tarb., Nor. and Cent. Tien Shan; Europe (Cent. and South.), Mediterr., Balk.-Asia Minor, Near East, Caucasus, Middle Asia, West. Sib. (south-west. Altay), China (Nor., Nor.-West., Cent., East., South-West., South., Hainan, Taiwan), Himalayas, Korea (south.), Japan (south), Indo-Malay., Nor. Amer. (south.), South Amer., Afr., Austral., New Zealand.

40. Chloris Sw.
Prodr. Veg. Ind. Occid. (1788) 25.

1. **Ch. virgata** Sw. Fl. Ind. Occid. 1 (1797) 203; Forbes and Hemsley, Index Fl. Sin. 3 (1904) 404; Roshev. in Fl. SSSR, 2 (1934) 286; Hao in Engler's Bot. Jahrb. 68 (1938) 579; Kitag. Lin. Fl. Mansh. (1939) 67; Walker in Contribs. U.S. Nat. Herb. 28 (1941) 597; Norlindh, Fl. mong. steppe, 1 (1949) 79; Grubov, Consp. fl. MNR (1955) 67; Fl. Kazakhst. 1 (1956) 200; Keng, Fl. ill. sin., Gram. (1959) 469; Bor, Grasses Burma, Ceyl., Ind. and Pakist. (1960) 468.— *Ch. caudata* Trin. in Mém. Ac. Sci. St.-Pétersb. Sav. Etrang. 2 (1831) 144; Franch. Pl. David. 1 (1884) 334; Danguy in Bull. Mus. nat. hist. natur. 3 (1914) 11; Pavlov in Byull. Mosk. obshch. ispyt. prir. 38

(1929) 19.—*Ch. alberti* Regel in Acta Horti Petrop. 7, 2 (1881) 650.—*Ch. barbata* auct. non Sw.:? Henderson and Hume, Lahore to Yarkand (1873) 341; ? Deasy, In Tibet and Chin. Turk. (1901) 405. —Ic.: Fl. SSSR, 2, Plate 22, fig. 2; Keng, l.c. fig. 401.

Described from Antilles archipelago. Type in Stockholm. Plate IV, fig. 3.

On rocky and rubble slopes, in pebble beds of rivers and lakes and solonetz; often as weed along roadsides and around irrigation ditches, in plantations and various other crops; up to midmountain belt.

IA. **Mongolia:** *East. Mong.* (left bank of Huang He river below Hekou town, July 21, 1884—Pot.; near Shilin-Khoto, 1959—Ivan.); *Bas. lakes* (valley of Buyantu river near Kobdo town, Aug. 5; Tuguryuk river, Aug. 16—1930, Bar.; near crossing on Kobdo river along Kobdo-Ulangom road, July 23, 1944—Yun.); *Val. lakes, Gobi-alt., East. Gobi, Alash. Gobi, Ordos* (Huang He valley, July 25, 1871—Przew.; second terrace of Huang He river near Dalat town, Aug. 10, 1957—Petr.).

IB. **Kashgar:** *Nor., West., South.* (Chira oasis near Khotan, 1400 m, Aug. 2, 1885—Przew.); *East.* (Turfan vicinity, Sept. 22, 1879—A. Reg.; 10 km from Utuan' state farm, No. 6871, July 29, 1958—Lee and Chu).

IIA. **Junggar:** *Tien Shan, Jung. Gobi* (south-east of Ulyungur lake,Aug. 3, 1876—Pot.; Darbata river valley at its intersection with Karamai-Altay road, June 20, 1957—Yun.).

IIIA. **Qinghai:** *Nanshan* ("Kokonor, Scha-Chu-Yi, 2900 m, 1930"—Hao, l.c.).

General distribution : Balkh. region, Jung.-Tarb., Nor. Tien Shan; Near East, Caucasus (Dagestan), Middle Asia (Karakums and Kyzylkums), Nor. Mong. (Hang., Mong.-Daur.), China, Himalayas, Korea, Japan, Indo-Malay., Nor. Amer. (south.), South Amer., Afr., Austral.

Note. The above two references for West. Kashgar of the related species *Ch. barbata* Sw. probably pertain to this species; *Ch. barbata* is also a widely distributed tropical weed.

41. Tripogon Roem. et Schult.
Syst. Veg. 2 (1817) 34.

1. Tip of lemma with 0.9–1.7 mm long awn, emerging between two very short lobes; lobes terminating in short 0.1–0.8 mm long awn or cusp; anthers 0.8–1.3 mm long .... 1. T. chinensis (Franch.) Hack.

+ Tip of lemma with up to 0.5 mm long cusp or awn, emerging between two lobes; lobes usually without awn or cusp; anthers 1.5–2.2 mm long .... 2. T. purpurascens Duthie.

1. **T. chinensis** (Franch.) Hack. in Bull. Herb. Boiss., ser. 2, 3 (1903) 503; Forbes and Hemsley, Index Fl. Sin. 3 (1904) 404; Roshev. in Fl. SSSR, 2 (1934) 287; Kitag. Lin. Fl. Mansh. (1939) 95; Norlindh, Fl. mong. steppe, 1 (1949) 80; Grubov, Consp. fl. MNR (1955) 67; Keng, Fl. ill. Sin., Gram. (1959) 462.—*Nardurus filiformis* var. *chinensis* Franch. Pl. David. 1 (1884) 339. —Ic.: Fl. SSSR, 2, Plate 22, fig. 4; Keng, l.c. fig. 395.

Described from China. Type in Paris. Plate IV, fig. 1; map 9.
On rocky and rubble slopes and rocks; up to midmountain belt.

IA. **Mongolia:** *Cent. Khalkha* (Ikhe-Dzara hills, Aug. 25, 1926—Lis.; Sorgol-Khairkhan hill 180 km south-south-west of Ulan Bator along old road to Dalan-Dzadagad, Aug. 15, 1943—Yun.; Bain-Baratuin somon, rocks on south. slope of Bain-Ul hill, July 12, 1948—Grub.); *East. Mong.* (Cent. Kerulen, gorge in Mergen-Khamar hills, 1899—Pal.; "Khongkhor-Obo, 14VIII, 1926, leg. Eriksson"—Norlindh, l.c.; Dariganga somon 5 km nor. of Sain-Khudugin-Khid, Sept. 18, 1949—Yun.; near Shilin—Khoto, 1959—Ivan.; Numuttsin-Daba crossing 25 km south of Erdeni-Tsagan somon, July 27, 1962—Lavr. et al.); *East. Gobi* (Tumur-Khada, Roerich Exped., No. 566, July 30, 1935—Keng; Jichi-Ola, steeps [sic] in canyon bottom at 1200 m, 1925—Chaney).
General distribution: East. Sib. (Arguni river basin), China (Dunbei, Nor., Nor.-West., Cent., East.), Korea.

2. **T. purpurascens** Duthie in Ann. Roy. Bot. Gard. Calc. 9 (1901) 75; Bor, Grasses Burma, Ceyl., Ind. and Pakist. (1960) 522.
Described from West. Himalayas. Type in Calcutta. Plate IV, fig. 2; map 9.

On rocky slopes and rocks; up to midmountain belt.

IA. **Mongolia:** *Gobi-Alt.* (1.5–2 km nor.-east of Noyan somon, July 24; Ser'-Ula hilly area 12–15 km south-south-west of Bain-Gobi somon, Sept. 4—1943, Yun.); *East. Gobi* (Ail'-Bain somon 1 km west of Ulegei-Khid monastery, Sept. 20 and 21, 1940—Yun.); *Alash. Gobi* (Chintunaya gorge 30 km from In'chuan'town, June 18, 1958—Petr.).
IB. **Kashgar:** *East.* (Khotun-Sumbule district, near Lotogou, 1700 m, No. 6252, Aug. 1, 1958—Lee and Chu; Ulyasutai-Tsagan river valley 4–5 km below Balinte settlement on Karashar-Urumchi road, July 31, 1958—Yun.).
General distribution: Himalayas (west.)

Note. The authentic specimen of this species avaialble in the herbarium of Bot. Inst. AN SSSR ("Fl. of N.W. Himalaya, distr. Tehri Garhwal, Kulni Valley, 2000 m, No. 23543, May 21, 1900, J. Duthie"), in spite of vast discontinuity in the distribution range, is wholly similar to specimens mentioned above.

42. Beckmannia Host
Gram. Austr. 3 (1805) 5.

1. **B. syzigachne** (Steud.) Fern. in Rhodora, 30 (1928) 27; Roshev. in Fl. SSSR, 2 (1934) 288; Kitag. Lin. Fl. Mansh. (1939) 63; Norlindh, Fl. mong. steppe, 1 (1949) 80; Grubov, Consp. fl. MNR (1955) 68; Fl. Kazakhst. 1 (1956) 201; Bor, Grasses Burma, Ceyl., Ind. and Pakist. (1960) 527; Krylov, Fl. Zap. Sib. 12,1 (1961) 3101.—*Panicum syzigachne* Steud. in Flora, 29 (1846) 19.—*Beckmannia eruciformis* var. *baicalensis* V. Kusn. in Tr. Byuro po prikl. bot. 6, 9 (1913) 584; Krylov, l.c. 2 (1928) 249.—*B. baicalensis* (V. Kusn.) Hult. Fl. Kamtch. 1 (1927) 119.—*B. eruciformis* auct. non Host: Franch. Pl. David. 1 (1884) 322; Forbes and Hemsley, Index Fl. Sin. 3 (1904) 405;

Pavlov in Byull. Mosk. obshch. ispyt. prir. 38 (1829) 19; Keng, Fl. ill. sin., Gram. (1959) 474. —Ic.: Fl. SSSR, 2, Plate 22, fig. 6; Keng, l.c. fig. 406.

Described from Japan. Type in Leiden.

In marshy, usually rather saline coastal meadows, pebble beds on banks and along banks of water reservoirs; up to lower mountain belt.

IA. **Mongolia:** *Mong. Alt.* (Bulugun river floodplain at site of discharge of Ulyaste-Gol river into it, July 20, 1947—Yun.); *Cis-Hing.* (near Yaksha railway station, No. 2085, 1954—Wang); *Cent. Khalkha, East. Mong., Bas. lakes, Val. lakes* (Ongiin-Gol floodplain below Khushu-Khida, 1948—Yun.); *Gobi-alt.* (Bain-Tukhum area, Aug. 4; south. slope of Bain-Tsagan hills, Sept. 16—1931, Ik.-Gal.); *East. Gobi* (Urten-Gol river valley, July 22, 1909—Czet.).

IIA. **Junggar:** *Alt. region* (south of Shara-Sume, No. 2839, Oct. 8, 1956—Ching); *Jung. Gobi.*

General distribution: Balkh. region, Jung.-Tarb., Nor. Tien Shan (near Issyk-Kul' lake); Europe (Kama river basin), West. and East. Sib., Far East, Nor. Mong., China (Dunbei, Nor., Nor.-West., East., South-West., South.), Himalayas (east.), Korea, Japan, Nor. Amer.

43. Enneapogon Desv. Ex Beauv.
Ess. Agrost. (1812) 81.

1. **E. borealis** (Griseb.) Honda in Rep. First Sci. Exped. Manch., sect. IV, 4 (1936) 101; Fl. Kazakhst. 1 (1956) 202.—*Pappohorum boreale* Griseb. in Leded. Fl. Ross. 4 (1852) 404; Pavlov in Byull. Mosk. obshch. ispyt. prir. 38 (1939) 19; Roshev. in Fl. SSSR, 2 (1934) 297; Kitag. Lin. Fl. Mansh. (1939) 84; Norlindh, Fl. mong. steppe, 1 (1949) 81; Grubov, Consp. fl. MNR (1955) 68. —Ic.: Fl. SSSR, 2, Plate 23, fig. 2; Fl. Kazakhst. 1, Plate 16, fig. 2.

Described from Transbaikal region (Selenga river). Type in Leningrad. Plate V, fig. 1; map 5.

IA. **Mongolia:** *Khobd.* (hills along left bank of Kobdo river, Sept. 22, 1895—Klem.; Bayan-Ul hill slopes toward Kobdo river, Aug. 30, 1950—N. Kuznetsov); *Cent. Khalkha, East. Mong., Bas. lakes, Val. lakes, Gobi-Alt., East. Gobi, West. Gobi, Alash. Gobi* (south. part of Alashan hills, Aug. 2, 1880—Przew.; "Khara-Dzagh, 9, IX 1927, leg. Hummel"—Norlindh, l.c.); *Ordos* (15 km east of Dzhasak town, Aug. 16, 1957—Petr.).

IB. **Kashgar:** *East.*

IIA. **Junggar:** *Jung. Gobi* (Altay somon, west. trail of Tszolen-Bogdo mountain range, Aug. 4, 1947—Yun.; 40 km south of Beidashan', No 196, Sept. 27, 1957—Huang).

General distribution: Jung.-Tarb., Nor. Tien Shan (Issyk-Kul' lake basin); East. Sib. (Minusinsk basin, upper Yenisei and Selenga river basin), China (Nor.).

Note. An interesting biological feature of *E. borealis* is its ability to form fairly numerous cleistogamic spikelets within squamiform sheaths of highly contracted (reniform) shoots on a mat base. As a result of the presence of such spikelets, this species retains the ability to reproduce even after continuous grazing in pastures. This feature is also characteristic of the closely related species *E. desvauxii* Beauv. widely distributed in the steppes of Nor. and South America (from the USA to Argentina). *E.*

desvauxii differs from *E. borealis* only in slightly more profuse pubescence of rachilla segments. The discontinuity between the distribution ranges of these species suggests the existence of fairly close contacts between the steppes of Asia and America in the relatively recent past.

44. Phragmites Adans.
Fam. pl. 2 (1763) 34 and 559.

1. **Ph. communis** Trin. Fund. Agrost. (1820) 134; Franch. Pl. David. 1 (1884) 334; Hemsley, Fl. Tibet (1902) 203; Forbes and Hemsley, Index Fl. Sin. 3 (1904) 409; Simpson in J. Linn. Soc. London (Bot.) 41 (1913) 452; Danguy in Bull. Mus. nat. hist. natur. 20 (1914) 146; Krylov, Fl. Zap. Sib. 2 (1928) 250; Pavlov in Byull. Mosk. obshch. ispyt. prir. 38 (1929) 19; Rehder in J. Arnold Arb. 14 (1933) 4; Komarov in Fl. SSSR, 2 (1934) 304; Kitag. Lin. Fl. Mansh. (1939) 86; Walker in Contribs, U.S. Nat. Herb. 28 (1941) 598; Norlindh, Fl. mong. steppe, 1 (1949) 82; Fl. Kirgiz. 2 (1950) 104; Grubov, consp. fl. MNR (1955) 68; Fl. Kazakhst. 1 (1956) 205; Fl. Tadzh. 1 (1957) 388; Keng, Fl. ill. sin., Gram. (1959) 337; Bor, Grasses Burma, Ceyl., Ind. and Pakist. (1960) 416. —*Ph. isiacus* (Delile) Kunth, Rév. Gram. (1829) 80; Komarov, l.c. 305; Fl. Kazakhst. 1(1956) 205.— *Ph. humilis* De Not. Cat. Horti Genuen. (1846) 27.—*Ph. vulgaris* (Lam.) Crép. Man. Fl. Belg., ed. 2 (1866) 345; Pilger in Hedin, S. Tibet, 6, 3 (1922) 93; Pampanini, Fl. Carac. (1930) 74.—*Arundo phragmites* L. Sp. pl. (1753) 81.—*A. vulgaris* Lam. Fl. Franç. (1778) 615, nom. illeg.—*A. isiaca* Delile, Descr. Egypt. Hist. Nat. 2 (1813) 52, nom. illeg. —**Ic.** : Fl. SSSR, 2, Plate 23, fig. 9; Keng, l.c. fig. 277.

Described from Europe. Type in London (Linn.).

Along banks of lakes and rivers, around springs and wells, in marshes and marshy meadows; often forms pure thickets; on sand, solonchak and rocky slopes with high groundwater; up to upper mountain belt.

IA. **Mongolia**: all regions.

IB. **Kashgar**: all regions.

IC. **Qaidam**: *plains* (East. Qaidam, Aug. 11, 1884—Przew.; Qaidam, Aug. 1, 1901—Lad.); *hilly* (near Kurlyk-Nor lake, 2800 m, June 27, 1901—Lad.).

IIA. **Junggar**: all regions.

IIIA. **Qinghai**: *Nanshan* (Nanshan hills, June-July, 1879—Przew.; "Kokonor region, leg. Rock"—Rehder, l.c.).

IIIB. **Tibet**: *Chang Tang* ("Jugdi, 17 X 1896, leg. Hedin"—Hemsley, l.c.; "Mandarlik, 3430 m, 16 VII, Kash-Otak, 2916 m, 20 VIII 1900, leg. Hedin"—Pilger, l.c.).

IIIC. **Pamir**: (Tagdumbash-Pamir, in Kara-Chukur river valley and Beik tributary, 4300 m, July 18, 1905—Alekseenko).

General distribution: cosmopolitan (but not found in most regions of the Arctic and Antarctic).

Note. Apart from the type variety *Ph. communis* var. *communis*, two more varieties of this highly polymorphic species are found in Cent. Asia. One of them, *Ph. communis* var. *pseudodonax* Rabenh. [in Bot. Centralbl. 1 (1846) 242] with very large plant dimensions on the whole and leaf blades 2.5–5 cm broad, is so far known only from one location (oasis among spring brooks, Ekhin-Gol area, Bain-undur somon, West. Gobi, Aug. 21, 1948—Grub.) but is probably more widely distributed. Judging from its distribution range (found predominantly in the deltas and lowlands of larger rivers in the USSR territory), it is entirely possible that this variety, which has lost the capacity for sexual reproduction, represents the oldest ecogeographical race of the genetic group related to *Ph. communis* Trin. s.l. It is not without justification that V.L. Komarov treated it as an independent species (Fl. SSSR, 2, p. 305), "*Ph. isiaca* (Delile) Kunth" (its correct name in the rank of species *Ph. altissima* Benth.). The other variety, *Ph. communis* var. *humilis* (De Not.) Parl. [Fl. ital. (1848) 767; = *Ph. humilis* De Not. l.c.], with long creeping shoots, low stems and narrow (0.5–1.5 cm broad) leaf blades often longitudinally folded, is much less isolated and widely distributed in Cent. Asia (especially in solonchak).

45. Orinus Hitchc.
in J. Washington Ac. Sci. 23 (1933) 136.

1. Glumes glabrous; lemma only dorsally and along sides interrupted-pilose; those of lower floret in spikelet 5–6 mm long...
    .... 1. **O. kokonorica** (Hao) Keng.
+   Glumes sparsely pilose; lemma profusely pilose almost throughout surface; those of lower floret in spikelet 4–5 mm long ....
    .... 2. **O. thoroldii** (Stapf) Bor.

1. **O. kokonorica** (Hao) Keng, Fl. ill. sin., Gram. (1959) 284.— *Cleistogenes kokonorica* Hao in Engler's Bot. Jahrb. 68 (1938) 582.—*Kengia kokonorica* (Hao) Packer in Bot. notiser, 113, 3 (1960) 293. —Ic.: Keng, l.c. fig. 230.

Described from Qinghai. Type in Berlin.

On rubble and rocky slopes and in pebble beds of river and lake valleys; in upper and middle mountain belts.

IIIA. **Qinghai:** *Nanshan* ("Kokonor, 3340 m, No. 998, 24 VIII 1930, leg. Hopkinson"—Hao, l.c., type!); *Amdo* (Xining, near Gunkho town 2 km from Satszygoi, on top of knoll, Aug. 6, 1959—Petr.).

General distribution: endemic.

Note. The third known species of this small genus, *O. anomala* Keng [in Acta Bot. Sin. 9, 1 (1960) 68], described from South-West. China ("Sikang, VII 1940, leg. K.L. Chü") and distinguished by even more sparse pubescence of spikelets, is very close to *O. kokonorica*.

2. **O. thoroldii** (Stapf) Bor in Kew Bull. (1952) 454; Keng, Fl. ill, sin., Gram. (1959) 284; Bor, Grasses Burma, Ceyl., Ind. and Pakist. (1960) 519.— *Diplachne thoroldii* Stapf in Hemsley, J. Linn. Soc. London (Bot.) 30 (1894) 121. —*Orinus arenicola* Hitchc. in J. Washington Ac. Sci. 23 (1933) 136.—

Cleistogenes thoroldii (Stapf) Roshev. in Fl. Kirgiz. 2 (1950) 107, quoad nom.
—Ic.: Hitchc. l.c. fig. 2; Keng, l.c. fig. 229.

Described from Tibet. Type in London (K).

In sand and pebble beds of river and lake valleys; in high mountain belt.

IIIB. **Tibet: *South.*** ("Tibet, sandy soil in valleys, 5200 m, No. 120, leg. Thorold"—Stapf, l.c., type!; Khambajong, No. 147, July 26, 1903—Younghusband).

General distribution: Himalayas (West., Kashmir).

46. Cleistogenes Keng
in Sinensia, 5 (1934) 149.

1. Lemma of chasmogamic spikelets (on terminal panicle) 3–4.5 mm long, acute or with cusp at tip, without lateral lobes; cleistogamic spikelets (on short axillary branches in sheaths of middle and upper cauline leaves) with 2–6 florets. Greyish-green plant with erect or suberect stems .... 1. **C. songorica** (Roshev.) Ohwi.

+   Lemma of chasmogamic spikelets 4.5–6.5 mm long, with two small lobes at tip; cusp or erect awn emerging between lobes; cleistogamic spikelets with one, less often two florets .... 2.

2. Uppermost internode (ignoring panicles) erect, highly elongated, usually almost as long as rest of stem, coiled in dry condition; leaf blades 0.5–2 mm broad. Awn of lemma of chasmogamic spikelets 2.5–6 mm long and, in cleistogamic spikelets, 6–9 mm long; palea with two 0.2–0.6 mm long cusps at tip ....
.... 4. **C. squarrosa** (Trin.) Keng .

+   Uppermost internode invariably less than one-third as long as rest of stem. Awn of lemma of chasmogamic spikelets up to 2, less often 2.5 mm long and, in cleistogamic spikelets, up to 4 mm long; palea without cusp or with up to 0.2 mm long cusps .... 3.

3. Stems with numerous close-spaced nodes in upper half, coiled in dry condition; three upper internodes (ignoring panicles) half or less than half as long as its 2 lowest internodes (ignoring highly shortened internodes at base of stem); leaf blades 1–3 mm broad, usually flat, easily breaking in dry condition ....
.... 3. **C. kitagawai** Honda.

+   Stems with generally uniformly spaced nodes, suberect in dry condition; three upper internodes usually longer than two lower internodes, less frequently almost as long; leaf blades 0.6–2 mm broad, usually longitudinally convoluted, persistent on dry plant .... 2. **C. festucacea** Honda.

Section 1. Pseudorinus Tzvel.[6]

1. C. songorica (Roshev.) Ohwi in J. Jap. Bot. 18 (1942) 540; Conert in Engler's Bot. Jahrb. 78, 2 (1959) 228.—C. chinensis (Maxim.) Keng in Sinensia, 5 (1934) 152, quoad pl.—C. mutica Keng in J. Washington Ac. Sci. 28 (1938) 299; Grubov, Consp. fl. MNR (1955) 68; Keng, Fl. ill. sin., Gram. (1959) 298.—C. thoroldii (Stapf) Roshev. in Fl. Kirgiz. 2 (1950) 107, quoad pl.; Fl. Kazakhst. 1 (1956) 208.—Diplachne songorica Roshev. in Fl. SSSR, 2 (1934) 311.—Kengia mutica (Keng) Packer in Bot. notiser, 113, 3 (1960) 292.—K. songorica (Roshev.) Packer, l.c. 293. —Ic.: Fl. SSSR, 2, Plate 23, fig. 15; Keng, l.c. (1959) fig. 242.

Described from East. Kazakhstan. Type in Leningrad. Plate V. fig. 3; map 7.

In desert steppes, on rocky and rubble slopes, in flat pebble beds; represents one of the characteristic species of snakeweed-stipa barren steppes; up to lower mountain belt.

IA. **Mongolia:** *Mong. Alt.* (along Buyantu river, Aug. 28, 1930—Bar.; Bain somon, near Khalyun, Aug. 24, 1943—Yun.; south. trail of Khan-Taishiri mountain range 5 km south-west of Dzun-Bulak, Sept. 3, 1948—Grub.); *Cent. Khalkha* (on way from Orle-Sume to Baishintin-Sume 43 km from Khashatyn-Khuduk, Aug. 16, 1927—Zam.; south of Del'ger-Tsogtu somon on road to Dalan-Dzadagad, Aug. 14, 1940—Yun.; 8 km nor.-east of Dzun-Tsagan-Tokhoi, Aug. 21, 1952—Davazhamts); *East. Mong.* (around Dariganga, near Nyudungin-Ula, Aug. 25, 1927—Zam.); *Val. lakes, Gobi-alt., East. Gobi, Alash. Gobi* (30 km south-east of Min'tsin' town, July 3; Tengeri sand near Bayan-Khoto, Aug. 4—1958, Petr.); *Khesi* (80 km west-nor. west of Yunchan on road to Chzhan"e, Oct. 8; 30 km south-east of Chzhan"e town, 2000 m, July 12—1958, Petr.).

IB. **Kashgar:** *East.* (Ulyasutai-Chagan valley 3-5 km below Balinte settlement on Karashar-Urumchi road, July 31, 1958—Yun.; Khotun-Sumbule district, near Lotogou, 1700 m, No. 6250, Aug. 1; near Khomote, No. 6919, Aug. 1; 2 km south of Bortu village, 1700 m, No. 6933; Aug. 2; near Tsinshuikhe river, No. 7643, Aug. 28—1958, Lee and Chu).

IC. **Qaidam:** *hilly* (Erdeni-Obo area, 3300 m, Aug. 7, 1901—Lad.).

IIA. **Junggar:** *Jung. Alt.* (Borotala river valley 28 km west of Borotala settlement, Aug. 17, 1957—Yun.; 15 km east of Ven'tsyuan', 1480 m, Aug. 29, 1958—Huang; 30 km nor. of Sairam-Nur lake, Aug. 31, 1959—Petr.); *Jung. Gobi Zaisan* (nor. trails of Saur mountain range 13 km west-nor.-west of Kheisangou settlement on Burchum-Karamai road, July 10, 1959—Yun.).

General distribution: Balkh. region (south-east.), Jung.-Tarb. (south. foothills of Jung. Alatau), Nor. Tien Shan; West. Sib. (Altai: along Chulyshman river), East. Sib. (south.: Yenisei upper course).

Note. This and western Himalayan species C. gatacrei (Stapf) Bor occupy a distinct position in the genus and have much in common with species of the relatively more primitive genus Orinus Hitchc. We place them in a distinct section, Pseudorinus Tzvel.

---

[6]*Cleistogenes* sect. Pseudorinus Tzvel. sect. nova.—A sectione typica praesertim lemmatis lanceolato-ovatis apice integris muticis et spiculis cleistogamis 2-5 floris differt. Typus; C. songorica (Roshev.) Ohwi.

Section 2. Cleistogenes

o **C. caespitosa** Keng, in Sinensia, 5 (1934) 154; Gubanov, Consp. Fl. Outer Mong. (1996) 17.

IA. **Mongolia:** *East. Gobi.*

2. **C. festucacea** Honda in Rep. First. Sci. Exped. Manch., sect. IV, 4 (1936) 98; Conert in Engler's Bot. Jahrb. 78, 2 (1959) 218.—*C. foliosa* Keng in J. Washington Ac. Sci. 28 (1938) 290; ej. Fl. ill. sin., Gram. (1959) 290.— *Kengia festucacea* (Honda) Packer in Bot. notiser, 113, 3 (1960) 292.—*K. foliosa* (Keng) Packer, l.c. 292.—?*Cleistogenes hancei* auct. non Keng: Norlindh, Fl. mong. steppe, 1 (1949) 85.—Ic.: Keng, l.c. (1959) fig. 236.

Described from China (Dunbei). Type in Tokyo.

In desert steppes, on rubble and rocky slopes and rocks, sometimes on fixed sand; up to lower mountain belt.

IA. **Mongolia:** *East. Gobi* ("Peiyan-Obo, Roerich Exped., No. 755, 9 VIII; Madenii-Amok, Roerich Exped., No. 778, 10 VIII; near Tumur-Hada, Roerich Exped., No. 825, 13 VIII 1935"—Keng, l.c. [1938]; near Naran-Obo, Roerich Exped., No. 886, Aug. 21, 1935—Keng, isotype *C. foliosae* Keng).

General distribution: China (Dunbei, Nor.).

Note. Following the monograph on the genus (Conert, l.c.) based on studies of types *C. festucacea* Honda and *C. foliosa* Keng, we combine these species under *C. festucacea* Honda. Another northern Chinese species, *C. caespitosa* Keng [in Sinensia, 5 (1934) 154] with less xeromorphic habit but totally similar in all other respects to *C. festucacea*, is very close to this species. *C. festucacea* as well as *C. caespitosa* occupy a somewhat intermediate position between highly specialised desert-steppe species *C. squarrosa* (Trin.) Keng and one of the more primitive species of section *C. hancei* Keng extensively found in East. China.

3. **C. kitagawai** Honda in Rep. First Sci. Exped. Manch., sect. IV, 4 (1936) 99; Conert in Engler's Bot. Jahrb. 78, 2 (1959) 218.—*C. striata* Honda, l.c. 100.—*Kengia kitagawai* (Honda) Packer in Bot. notiser, 113, 3 (1960) 292.—*K. striata* (Honda) Packer, l.c. 292.—*Diplachne serotina* auct. non Link: Krylov, Fl. Zap. Sib. 2 (1928) 253.—*D. sinensis* auct. non Hance: Roshev in Fl. SSSR, 2 (1934) 311.—*Cleistogenes serotina* auct. non Keng:? Norlindh, Fl. mong. steppe, 1 (1949) 84; Grubov, Consp. fl. MNR (1955) 68.—*C. chinensis* auct. non Keng: Krylov, Fl. Zap. Sib. 12, 1 (1961) 3101.

Described from China (Dunbei). Type in Tokyo. Plate VI, fig. 1; map 7.

On rocky slopes and rocks and in upland steppes; up to lower mountain belt.

IA. **Mongolia:** *Cent. Khalkha* (Sorgol-Khairkhan hill 180 km south-west of Ulan Bator on road to Dalan-Dzadagad, Aug. 15, 1943—Yun.); *East. Mong.* (Tamtsag-

Bulak somon 20 km south-west of Lag-Nur along road to Yugodzyr, May 13, 1944—Yun.; 22 km west of Tamtsag-Bulak along road to Matad somon, July 25, 1962—Lavr. et al.).

General distribution: West. Sib. (Altay), East. Sib. (south.), Far East (south.), Nor. Mong. (Hent., Hang., Mong.-Daur.), China (Dunbei, Nor.).

Note. Morphologically, this species is totally isolated and easily differentiated from Palaeo-Mediterranean *C. serotina* (L.) Keng as well as from the eastern Asian species *C. hancei* Keng (=*Diplachne sinensis* Hance) and *C. hackelii* (Honda) Honda [=*C. chinensis* (Maxim.) Keng] among which specimens of *C. kitagawai* are usually placed. Unlike in the case of *C. squarrosa* (Trin.) Keng, fruiting stems of *C. kitagawai* are coiled not in the lower but in the upper half, this feature representing a characteristic adaptation to anemochory.

4. **C. squarrosa** (Trin.) Keng in Sinensia, 5 (1934) 156; Roshev. in Fl. Kirgiz. 2 (1950) 107; Grubov, Consp. fl. MNR (1955) 68; Fl. Kazakhst. 1 (1956) 206; Conert in Engler's Bot. Jahrb. 78, 2 (1959) 226; Keng, Fl. ill. sin., Gram. (1959) 293.—*C. squarrosa* var. *longe-aristata* (Rendle) Keng, l.c. (1934) 156.—*C. andropogonoides* Honda in Rep. First Sci. Exped. Manch., sect. IV, 4 (1936) 98.—*Molinia squarrosa* Trin. in Ledeb. Fl. alt. 1 (1829) 105.—*Diplachne squarrosa* (Trin.) Maxim. in Bull. Soc. Natur. Moscou, 54, 1 (1879) 71, in adnot.; Forbes and Hemsley, Index Fl. Sin. 3 (1904) 411; Krylov, Fl. Zap. Sib. 2 (1928) 252; Roshev. in Fl. SSSR, 2 (1934) 310.—*D. squarrosa* var. *longe-aristata* Rendle in Forbes and Hemsley, l.c. 411.—*Kengia squarrosa* (Trin.) Packer in Bot. notiser, 113, 3 (1960) 292.—*K. andropogonoides* (Honda) Packer, l.c. 292.—Ic. : Ledeb. Ic. pl. fl. ross. 3 (1831) Table 227; Fl. SSSR, 2, Plate 23, fig. 12; Keng, l.c. (1959) fig. 239.

Described from Altay. Type in Leningrad. Plate VI, fig. 2; map 8.

Represents a characteristic species of snakeweed and snakeweed-stipa desert steppes, especially in sandy and loamy soils; also found scattered on rubble and rocky slopes, among rocks and in flat pebble beds; up to midmountain belt.

IA. Mongolia: *Mong. Alt., Cis-Hing.* (near Yaksha railway station, Aug. 19, 1902—Litw.); *Cent. Khalkha, East. Mong., Bas. lakes, Val. lakes, Gobi-Alt., East. Gobi.*

IB. Kashgar: *East.* (in Bagrashkul' lake region, 1600 m, No. 1817, Sept. 1, 1957—Huang; 8 km beyond Balinte settlement along Karashar-Yuldus road, Aug. 1, 1958—Yun.).

IIA. Junggar: *Jung. Alt.* .(ascent to Kuzyun' pass, July 2, 1908—B. Fedczenko; Dzhair mountain range 25 km west of Aktal settlement along road to Chuguchak, Aug. 3, 1957—Yun.; 15 km east of Ven'tsyuan', 1920 m, Aug. 26, 1957—Huang); *Jung. Gobi* (on way from Tien Shan-Laoba to Myaoergou settlement, Aug. 3, 1957—Huang; east. Trails of Saur mountain range 60 km nor. of Kosh-Tologoi along road to Shara-Sume, July 4, 1959—Yun.); *Zaisan.*

General distribution: Aral-Casp., Balkh. region, Jung.-Tarb., Nor. and Cent. Tien Shan; Europe (south-east), Caucasus (Dagestan), West. Sib. (south.), East. Sib. (south., Cent., Yakutsk), Nor. Mong., China (Dunbei, Nor.).

Note. The typical manner of propagation of diaspores characteristic of this genus (in this case, diaspores of axillary cleistogamic spikelets), similar to 'tumbleweed', is greatly manifest in this species. On desiccation, its stems coil even in specimens collected before fruiting, which generally does not occur in other species of the genus. Specimens from China outside Cent. Asia and sometimes even from East. Mongolia, have usually slightly longer (5.5–7 mm) awns of Chasmogamic spikelets and are classified as a distinct variety, C. *squarrosa* var. *longe-aristata* (Rendle) Keng, l.c. (=C. *andropogonoides* Honda, l.c.).

47. Eragrostis Wolf
Gen. Pl. Vocab. Char. Def. (1776) 23.

1. Perennial, 30–100 cm tall with greatly thickened of funiform roots and hairy sheaths of radical leaves; leaf blades very long and narrow, 0.6–3 mm broad. Lemma 2.2–3 mm long; anthers 0.8–1.4 mm long. .... E. curvula (Schrad.) Nees.
+ Annual, (6) 10–40 (60) cm tall with much thinner filiform roots and glabrous, less often sparsely hairy sheaths of radical leaves. Lemma (1.3) 1.5–2.3 (2.6) mm long; anthers 0.2–0.4 mm long .... 2.
2. Stalk spikelets with 1–3(5) distinctly perceptible 'crater-like' glands (in the form of processes with broad but shallow pit at centre); spikelets 1.4–4 mm broad; lemma ovate; panicles with branches emerging singly, less often in pairs; invariably without long hairs in lower nodes. Leaf blade along margin and sheath along more prominent ribs with distinct but occasionally several few 'crater-like' glands .... 3.
+ Stalk of spikelets with 'crater-like' glands; spikelets 1.2–1.6 mm broad; lemma broadly lanceolate; panicles with branches emerging singly and up to five; usually with, less often without, long hairs in lower nodes. Leaf blade and sheath without 'crater-like' glands .... 3. E. pilosa (L.) Beauv.
3. Spikelets 2.5–4 mm broad, with (5) 10–30 (40) florets; lemma broadly ovate, 1.9–2.6 mm long; anthers 0.3–0.4 mm long; caryopsis subspherical .... 1. E. cilianensis (All.) Vign.-Lut.
+ Spikelets 1.4–2.5 mm broad, with (3) 5–16 (20) florets; lemma 1.5–2 mm long; anthers 0.2–0.3 mm long; caryopsis broadly ellipsoidal ....4.
4. Lemma broadly ovate, obtuse at tip; stalk of lateral spikelets 0.6–4 mm long .... 2. E. minor Host.
+ Lemma ovate, somewhat attenuated and usually rather pointed at tip; stalk of lateral spikelets 2.5–10 mm long; panicles larger on average. .... 4. E. suaveolens A. Beck. ex Claus.

1. E. cilianensis (All.) Vign.-Lut. in Malpighia, 18 (1904) 386; Kitag. Lin. Fl. Mansh. (1939) 75; Keng, Fl. ill. sin., Gram. (1959) 317; Bor, Grasses

Burma, Ceyl., Ind. and Pakist. (1960) 503.—*E. major* Host, Gram. Austr. 4 (1809) 14; Forbes and Hemsley, Index Fl. Sin. 3 (1904) 416.—*E. megastachya* (Koel.) Link, Hort. Bot. Berol. 1 (1827) 187; Franch. Pl. David. 1 (1884) 335; Danguy in Bull. Mus. nat. hist. natur. 20 (1914) 146; Krylov, Fl. Zap. Sib. 2 (1928) 256; Roshev. in Fl. SSSR, 2 (1934) 316; Fl. Kirgiz. 2 (1950) 111; Grubov, Consp. fl. MNR (1955) 69; Fl. Kazakhst. 1 (1956) 210; Fl. Tadzh. 1 (1957) 451.—*E. poaeoides β. compacta* Regel in Bull. Soc. natur. Moscou, 41, 4 (1868) 295.—*E. megastachya* var. *compacta* (Regel) Kryl. Fl. Zap. Sib. 2 (1928) 257.—*E. starosselskyi* Grossh. in Bot. mater. Gerb. Glavn. bot. sada RSFSR, 4 (1923) 18; Roshev., l.c. 319; Fl. Kirgiz. 2 (1950) 111.—*Briza eragrostis* L. Sp. pl. (1753) 70.—*Poa cilianensis* All. Fl. Pedem. 2 (1785) 246.—*P. megastachya* Koel. Descr. Gram. (1802) 181. —Ic. Fl. SSSR, 2, Plate 24, figs. 8–9; Keng, l.c. fig. 257.

Described from Nor. Italy. Type probably in Turin.

On rocky and melkozem slopes, in sand and pebble beds of rivers; up to midmountain belt.

IA. **Mongolia:** *Bas. lakes* ("Environs de Kobdo, sables, 1500 m, 22 IX 1895, leg. Chaffanjon"—Danguy, l.c.).

IB. **Kashgar:** *East.* (near Chukur settlement, Aug. 28, 1929—Pop.).

IIA. **Junggar:** *Tien Shan* (Ili valley 42 km east of bridge on Kash, Aug. 29, 1937—Yun.).

General distribution: Aral-Casp. (south-east.), Balkh. region, Jung.-Tarb., Nor. and Cent. Tien Shan; Europe (cent. and south.), Mediterr., Balk.-Asia Minor, Near East. Caucasus, Middle Asia, China (Dunbei, Nor., Nor.-West., East.), Korea, Japan, Indo-Malay.; introduced in many other countries of both hemispheres.

Note. Specimens of this species from Cent. Asia as well as most specimens from Middle Asia and East. Transbaikal have greatly contracted panicle branches and appear to belong to a distinct ecogeographical race described before as variety *E. megastachya* var. *compacta* (Regel) Kryl. and independent species *E. starosselskyi* Grossh. (l.c.). It is more likely, however, that shortened branches of panicles in this case represent only a consequence of fungal disease with a wholly definitive geography.

E. curvula (Schrad.) Nees, Fl. Afr. Austr. (1841) 397; Keng, Fl. ill. sin., Gram. (1959) 305; Bor, Grasses Burma, Ceyl., Ind. and Pakist. (1960) 507.—*Poa curvula* Schrad. in Anz. Gesellsch. Wissensch. Goettingen, 3 (1821) 2073.—Ic.: Keng, l.c. fig. 306.

Described from South Africa. Type in Goettingen.

Cultivated as soil-binding, drought-resistant edible plant; sometimes wild.

IIA. **Junggar:** *Jung. Gobi* (between Manas and Kuitun rivers, No. 3842, Oct. 11, 1956—Ching).

General distribution: South Africa; cultivated in many other countries of the world.

2. **E. minor** Host, Gram. Austr. 4 (1809) 15; Forbes and Hemsley, Index Fl. Sin. 3 (1904) 416; Krylov, Fl. Zap. Sib. 2 (1928) 256; Pavlov in Byull.

Mosk. obshch. ispyt. prir. 38 (1929) 20; Pampanini, Fl. Carac. (1930) 74;
Roshev. in Fl. SSSR, 2 (1934) 315; Fl. Kirgiz. 2 (1950) 108; Grubov, Consp.
fl. MNR (1955) 69; Fl. Kazakhst. 1 (1956) 209.—*E. poaeoides* Beauv. Ess.
Agrost. (1812) 162; Kitag. Lin. Fl. Mansh. (1939) 76; Norlindh, Fl. mong.
steppe, 1 (1949) 87; Fl. Tadzh. 1 (1957) 450; Keng, Fl. ill. sin., Gram. (1959)
317; Bor, Grasses Burma, Ceyl., Ind. and Pakist. (1960) 512.—*Poa eragrostis*
L. Sp. pl. (1753) 68.—Ic.: Fl. SSSR, 2, Plate 24, fig. 6; Fl. Kazakhst. 1, Plate
16, fig. 9; Keng, l.c. fig. 258.

Described from Italy. Type in London (Linn.).

In sand and pebble beds of rivers, on rubble and rocky slopes, talus,
often as weed in oases, along roadsides; up to midmountain belt.

IA. **Mongolia:** *Khobd.* (hills on left bank of Kobdo river, Sept. 22, 1895—Klem.);
Mong. Alt. (valley of Buyantu river near Kobdo town, Aug. 8; Tuguryuk river, Aug.
15—1930, Bar.); *Cent Khalkha* (Del'girkh hill on way from Ulan Bator to Del'ger-
Hangay, July 29, 1931—Ik.-Gal.; south of Del'ger-Tsogtu somon along road to Dalan-
Dzadagad, Aug. 14, 1940—Yun.); *East. Mong., Bas. lakes, Gobi-Alt., East. Gobi,
West. Gobi* (Tsagan-Bogdo mountain range, Aug. 4; Bain-Gobi somon, Shara-
Khulusun area, Aug. 8—1943, Yun.); *Alash. Gobi* (around Chzhunwei, Sabotou, July
26, 1957; Tengeri sand near Bayan-Khoto, Aug. 9, 1958—Petr.); *Ordos, Khesi* (nor.—
Nanshan foothill near Dadzhin town, Aug. 11, 1880—Przew.)

IB. **Kashgar:** *Nor., South.* (nor. Kunlun foothills, 1600 m, June 13; same site, Zava-
Kurgan oasis, Sept. 17—1889, Rob.; nor. slope of Russky mountain range, 2500 m, on
wet sand of Mal'dzha river, June 15, 1890—Rob.; Niya district, No. 9576, June 26,
1959—Lee et al.); *East.*

IIA. **Junggar:** *Jung. Alt., Tien Shan, Jung. Gobi* (nor.-West.: nor. bank of last lake of
Manas river, June 24, 1957—N. Kuznetsov; east.: nor. slope of Baitak-Bogdo mountain
range, Ulyastu-Gol gorge 3–4 km from estuary, Sept. 18, 1948—Grub.); *Zaisan*
(between Ch. Irtysh and Saur mountain range, Aug. 21, 1906—Sap.; right bank of Ch.
Irtysh below Burchum river, Sary-Dzhasyk gorge, June 15, 1914—Schisch.).

General distribution: Aral-Casp., Balkh. region, Jung.-Tarb., Nor. and Cent. Tien
Shan; Europe (cent. and south.), Mediterr., Balk.-Asia Minor, Near East, Caucasus,
Middle Asia, West. Sib. (Altay), East. Sib. (south), Far East (south), Nor. Mong., China
(Dunbei, Nor., Nor.-West., Cent., East.), Himalayas (west.), Korea, Japan, Indo-
Malay.; introduced in many other countries.

Note: The above mentioned specimen of V.I. Roborowsky from the nor. slope of
Russky mountain range (isolated and highest altitude report of this species!) differs
from all other specimens of *E. minor* Host in total absence of 'crater-like' glands on
stalks of spikelets as well as on leaves and fully merits recognition at least as a variety,
E. minor var. *roborovskii* Tzvel. var. nova.—A varietate typica pedicellis et foliis
eglandulosis differt. Typus: "Kunjlunj, declivitas borealis jugi Rossici, in arena humidi
in valle fl. Maljdzha, 15 VI 1890, V. Roborovsky" in Herb. Inst. Bot. Ac. Sci. URSS—LE
conservatur.

3. **E. pilosa** (L.) Beauv. Ess. Agrost. (1812) 162; Franch. Pl. David. 1
(1884) 335; Forbes and Hemsley, Index Fl. Sin. 3 (1904) 417; Danguy in
Bull. Mus. nat. hist. natur. 20 (1914) 147; Krylov, Fl. Zap. Sib. 2 (1928) 257;
Pavlov in Byull. Mosk. Obshch. ispyt. prir. 38 (1929) 20; Roshev. in Fl.
SSSR, 2 (1934) 315; Kitag. Lin. Fl. Mansh. (1939) 76; Norlindh, Fl. mong.

steppe, 1 (1949) 86; Fl. Kirgiz. 2 (1950) 108; Grubov, Consp. fl. MNR (1955) 69; Fl. Kazakhst. 1 (1956) 209; Fl. Tadzh. 1 (1957) 44; Keng, Fl. ill. sin., Gram. (1959) 313; Bor, Grasses Burma, Ceyl., Ind. and Pakist. (1960) 512.— *Poa pilosa* L. Sp. pl. (1753) 68.—Ic.: Fl. SSSR, 2, Plate 24, fig. 4; Keng, l.c. fig. 255.

Described from Italy. Type in London (Linn.).

In sand and pebble beds of rivers, on rocky slopes, often as weed along roadsides, in wastelands and in farms; up to midmountain belt.

IA. **Mongolia:** *Mong. Alt.* (Kobdo town, Aug. 8, 1879—Pot.; Buyantu river valley, Aug. 27, 1930—Bar.; Khasagtu-Khairkhan hills near discharge of Dundutseren-Gol river onto rocky trail, Sept. 15, 1930—Pob.); *Cent. Khalkha, East. Mong., East. Gobi, Ordos* (Tsaidamin-Chzhao monastery, July 28; Ushkyut-Tokum area, Aug. 17—1884, Pot.).

IIA. **Junggar:** *Jung. Alt.* (Urtaksary river, 2000 m, Aug. 7, 1878—A. Reg.), *Jung. Gobi* (west.: Takiansi, Aug. 24, 1878—A. Reg.; nor.: "Bords de l'Irtich, 29 VIII, 1895, leg. Chaffanjon"—Danguy, l.c.; left bank of Urungu river 85 km beyond Din'syan' settlement, July 13, 1959—Yun.); *Dzhark.* (near Kul'dzha, June 1877—A. Reg.); *Balkh.-Alak.* (near Chuguchak town—Skachkov).

General distribution: Aral-Casp., Balkh. region, Jung.-Tarb., Nor. and Cent. Tien Shan; Europe (cent. and south.), Mediterr., Balk.-Asia Minor, Near East, Caucasus, Middle Asia, West. Sib. (south.), East. Sib. (south., Cent. Yakutsk), Far East (south.), Nor. Mong., China, Korea, Japan, Indo-Malay.; introduced in many other countries.

Note. Two specimens from East. Mong. ("22 km west of Baishintu-Sume, 10 IX 1931—Pob." and "near Shilin-khoto, 1959—Ivan") represent variety E. pilosa var. *imberbis* Franch. (l.c. 335) described from Inner Mongolia, without long hairs in axils of lemma, which is characteristic of typical *E. pilosa*. Species E. *jeholensis* Honda described from Inner Mongolia outside Cent. Asia [in Rep. First Sci. Exped. Manch., Sect. 4, 2 (1935) 6] and E. *multispicula* Kitag. [in J. Jap. Bot. 39, 8 (1964) 250], judging from their first diagnosis, do not differ significantly from *E. pilosa*.

4. **E. suaveolens** A. Beck. ex Claus in Beitr. Pflanzenk. Russ. Reich, 8 (1851) 266; Roshev. in Fl. SSSR, 2 (1934) 316; Fl. Kazakhst. 1 (1956) 210.— *E. poaeoides* β. *suaveolens* (A. Beck. ex Claus) Schmalh. Fl. Sredn. i Yuzhn. Rossii, 2 (1897) 631.—Ic.: Fl. SSSR, 2 Plate 24, fig. 7.

Described from land along Volga. Type in Leningrad.

In sand and pebble beds of river and lake valleys.

IIA. **Junggar:** *Jung. Gobi* (crossing of Dyurbel'dzhin on Ch. Irtysh, Aug. 12, 1876—Pot.; west. part of Junggar basin between Orkha and Baikhoutsza villages, June 25, 1957—Yun.).

General distribution: Aral-Casp., Balkh. region; Europe (south-east.).

48. Koeleria Pers.
Syn. pl. 1 (1805) 97.

1. Vegetative shoots with 5–8 developed leaves; shoot bases surrounded by extremely numerous sheaths of dead leaves and

appear bulbous, thickened; leaf blades very stiff, glaucous-green, with both sides very densely covered in setiform spinules. Lemma rather obtuse at tip .... K. glauca (Schrad.) DC.

+ Vegetative shoots with 2–4 developed leaves; bases of shoots with less numerous sheaths of dead leaves; not bulbous, thickened; leaf blades less stiff, green or greyish-green, covered in rather diffuse (particularly on under surface) spinules or hairs. Lemma acute at tip .... 2.

2. Stems puberulent for almost entire length. Spicate panicles fairly dense, rather greyish-violet; spikelets 3.2–4.2 mm long, glabrous .... 2. K. atroviolacea Domin.

+ Stems puberulent only close to base of panicles (less than 1 cm), with few hairs occasionally even under nodes .... 3.

3. Glumes and lemma covered with rather long hairs; spikelets rather greyish-violet. Leaf blades relatively less rigid, light greyish usually somewhat pilose ....
.... 1. K. altaica (Domin) Kryl.

+ Glumes usually glabrous, less often setulose along keel; lemma glabrous or covered with very short semiappressed hairs .... 4.

4. Sheaths of dead leaves surrounding shoot bases usually not disintegrating into longitudinal fibres; leaf blades rather pilose or glabrous; panicles greenish or pinkish-violet, relatively broader than long; spikelets 3.5–5.5 mm long; lemma glabrous or covered with very short hairs; anthers 1.5–2.3 mm long ....
.... 3. K. cristata (L.) Pers.

+ Sheaths of dead leaves surrounding shoot bases largely disintegrating into longitudinal fibres; blades of all leaves (as also their sheaths) glabrous; panicles light-green, usually narrower than long (especially after anthesis); spikelets 3.2–4.5 mm long; glumes and lemmas invariably glabrous; anthers 1.2 to 1.8 mm long .... 4. K. mukdenensis Domin.

1. K. altaica (Domin) Kryl. Fl. Zap. Sib. 2 (1928) 261; Grubov, Consp. fl. MNR (1955) 69; Fl. Kazakhst. 1 (1956) 213.—K. eriostachya subsp. caucasica var. altaica Domin in Bibl. Bot. 65 (1907) 163.—K. caucasica (Domin) B. Fedtsch. Rast. Turkest. [Plants of Turkestan] (1915) 120, quoad pl.; Pavlov in Byull. Mosk. Obshch. ispyt. prir. 38 (1929) 20; Gontscharov in Fl. SSSR, 2 (1934) 327, p.p.; Fl. Kazakhst. 1 (1956) 213.—Ic.: Fl. Kazakhst. 1, Plate 16, fig. 11.

Described from Altay. Type in Berlin.

In meadows, on rocky slopes and in riverine pebble beds; in upper mountain belt.

IA. **Mongolia:** *Khobd.* (upper course of Kharkira river, July 10 and 12, 1879—Pot.; Kharkira hills, Mosten' area south-west of Ulangom, July 25, 1903—Gr.-Grzh.); *Mong. Alt.* (between Dain-Gol and Ak-Korum lake, June 29, 1903—Gr.-Grzh.; 10 km south-east of Yusun-Bulak, nor. Trails of Khan-Taishiri mountain range, July 14; south. slope of Tamchi-Daba pass, July 16; Bus-Khairkhan mountain range, about 3000 m, July 17—1947, Yun.); *Gobi-Alt.* (Barun-Saikhan mountain range, Nomgon hill, Aug. 23, 1953—Dashnyam).

General distribution: Jung.-Tarb., Cent. Tien Shan (along Sarydzhas river); West. Sib. (Altay), Nor. Mong. (Fore Hubs., Hang., Mong.-Daur.).

2. **K. atroviolacea** Domin in Bibl. Bot. 65 (1907) 252; Krylov, Fl. Zap. sib. 2 (1928) 263; Gontscharov in Fl. SSSR, 2 (1934) 336.—*K. geniculata* Domin, l.c. 253; Gontscharov, l.c. 336.—Ic.: Fl. SSSR, 2, Plate 25, fig. 16.

Described from East. Siberia (West. Sayans). Type in Leningrad.

In grasslands, on rocky slopes and in riverine pebble beds; in upper mountain belt.

IIIA. **Qinghai:** *Nanshan* (pass 6 km west of Xining, cereal grass-herb steppe with alpine features, Aug. 5, 1959—Petr.).

General distribution: West. Sib. (Altay), East. Sib. (Sayans).

Note. Specimens from Qinghai, unlike south. Siberian typical specimens of *K. atroviolacea* Domin, have very dense mats and merit recognition as a variety: *K. atroviolacea* var. *tsinghaica* Tzvel. var. nova.—A varietate typica caespitibus densis differt.

Typus: "Trajectus 6 km ad occidentem versus opp. Sinin, 5 VIII 1959, M. Petrov" in Herb. Inst. Bot. Ac. Sci. URSS—LE conservatur.

3. **K. cristata** (L.) Pers. Syn. pl. 1 (1805) 97, quoad nom.; Franch. Pl. David. 1 (1884) 335; Forbes and Hemsley, Index Fl. Sin. 3 (1904) 410; Keng, Fl. ill. sin., Gram. (1959) 477; Bor, Grasses Burma, Ceyl., Ind. and Pakist. (1960) 444.—*K. gracilis* Pers. l.c. 97, nom. Illeg.; Krylov, Fl. Zap. Sib. 2 (1928) 264; Pavlov in Byull. Mosk. obshch. ispyt. prir. 38 (1929) 20; Gontscharov in Fl. SSSR, 2 (1934) 230; Kitag. Lin. Fl. Mansh. (1939) 81, p.p.; Norlindh, Fl. mong. steppe, 1 (1949) 88; Fl. Kirgiz. 2 (1950) 112; Grubov, Consp. fl. MNR (1955) 69; Fl. Kazakhst. 1 (1956) 214; Fl. Tadzh. 1 (1957) 353.—*K. macrantha* (Ledeb.) Schult. et Schult. f. Mant. 2 (1824) 345.—*Aira cristata* L. Sp. pl. (1753) 63, p.p.—*A. macrantha* Ledeb. in Mém. Ac. Sci. St.-Pétersb. 5 (1812) 515.—Ic.: Fl. SSSR, 2, Plate 25, fig. 9; Fl. Kazakhst. 1, Plate 16, fig. 12; Keng, l.c. fig. 408.

Described from Europe. Type in London (Linn.).

In steppes, on rocky slopes, sand and pebble beds of river and lake valleys, sometimes in thin hill forests; up to upper mountain belt.

IA. **Mongolia:** *Khobd., Mong. Alt., Cis-Hing.* (near Trekhrech'e, 750–800 m, No. 1217, July 10, 1951—Li et al.); *Cent. Khalkha, East. Mong.* (nor.); *Val. lakes* (Mailakhtsin-Ama area between Tsagan-Olom and Dzag-Baidarik, Aug. 27, 1943; water divide between Tuin-Gol and Tatsain-Gol, 2212 m, July 9, 1947—Yun.); *Gobi-Alt., East. Gobi* (In'shan' hills, Muni-Ul mountain range, June 2, 1871—Przew.; "Khujirtu-Gol, No. 1085, 2 VI 1927, leg. Hummel; Khongkhor-Obo, No. 403, 18 VI,

1927, leg. Eriksson; in montibus Wu-Kung-Pa, No. 6523, 4, VIII 1927, leg. Söderbom; Khonin-chagan-Cholo-Gol, No. 1308, 1 VIII 1927, leg. Hummel"—Norlindh, l.c.).

IB. **Kashgar:** *Nor., West.* (Bostan-Terek village, July 10, 1929—Pop.; Sulu-Sakal valley 25 km from Irkeshtam, July 26, 1935—Olsuf'ev; nor. Slope of Kingtau mountain range 1 km nor. of Kosh-Kulak settlement, June 10, 1959—Yun.).

IIA. **Junggar:** *Alt. region, Jung, Alt.* (45 km south of Toli settlement, No. 1401, Aug. 8, 1957—Huang); *Tien Shan, Jung. Gobi* (nor.: near Burchum settlement, No. 3186, Sept. 16, 1956—Ching; east. trail of Saur mountain range 60 km nor. of Kosh-Tologoi on road to Shara-Sume, July 4, 1959—Yun.).

IIIA. **Qinghai:** *Nanshan* (25 km east of Gunhe, 3100 m, Aug. 7, 1959—Petr.).

IIIB. **Tibet:** *Weitzan* (Nruchu area on Yangtze river, 3900 m, July 25, 1900; nor. Slope of Burkhan-Budda mountain range, Khatu gorge, 4500–4700 m, July 18, 1901—Lad.).

General distribution: Aral-Casp., Balkh. region, Jung.-Tarb., Nor. and Cent. Tien Shan; Europe, Mediterr., Balk.-Asia Minor, Near East, Caucasus, Middle Asia (hilly), West. Sib. (south.), East Sib. (south.), Far East (cent. and south.), Nor. Mong., China (Dunbei, Nor., Nor.-West., Cent., East., South-West.), Korea, Japan, Himalayas (west.), Nor. Amer., Afr. (hilly).

Note. Unusually polymorphic variable species. Typical specimens of the species have hairy leaves but variety *K. cristata* var. *glabra* Regel [in Acta Horti Petrop. 7, 2 (1881) 630] with totally glabrous leaves is found in Tien Shan and Altay hills. Other varieties include *K. cristata* var. *pilifera* (Domin) Tzvel. comb. nova (=*K. gracilis* var. *pilifera* Domin, l.c. 217), occasionally seen in Mongolia, with puberulent lemma and *K. cristata* var. *sabulosa* (Reverd.) Tzvel. comb. nova [=*K. gracilis* var. *sabulosa* Reverd. in Fl. Krasnoyarskogo Kraya (Flora of Krasnoyarsk Region) 2 (1964) 59], confined to river sand, with broad greyish-green leaf blades and shoot bases wrapped in numerous sheaths of dead leaves. Specimens of the latter variety (*K. thonii* Domin described from Yenisei sand probably belong to it) are outwardly very similar to *K. glauca* (Schrad.) DC. and collectors often identify them as such. Variety "*K. gracilis* var. *hummelii* Norlindh" (l.c. 90) was evidently described from very small specimens collected from extreme conditions of existence of the species and hardly merits a distinct name.

K. glauca (Schrad.) DC. Catal. Pl. Horti Monspel. (1813) 116; Krylov, Fl. Zap. Sib. 2 (1928) 259; Gontscharov in Fl. SSSR, 2 (1934) 324; Grubov, Consp. fl. MNR (1955) 69, p.p.; Fl. Kazakhst. 1(1956) 212.—*Aira glauca* Schrad. Fl. Germ. 1 (1806) 256.—*Poa glauca* Schkuhr, Catal. Pl. Hort. Wittenb. (1799) 49, non Vahl (1790).—Ic.: Fl. SSSR, 2, Plate 25, fig. 2.

Described from Cent. Europe. Type in Berlin (?).

On spreading sand, in sandy steppes.

Outside Cent. Asia, known only from one site in Nor. Mongolia (midcourse of Selenga river, Orkhon-Selenga water divide, Aug. 20, 1931, No. 322—N. Desyatkin).

General distribution: Aral-Casp. (nor.), Balkh. region (nor.); Europe, West. Sib., East. Sib. (south.). Nor. Mong., (Mong.-Daur.).

4. **K. mukdenensis** Domin in Bibl. Bot. 65 (1907) 171.—*K. tokiensis* subsp. *mongolica* Domin, l.c. 130; Norlindh, Fl. mong. steppe, 1 (1949) 88.—*K. gracilis* var. *gracillima* Honda in Rep. First Sci. Exped. Manch., sect. IV,

2 (1935) 8.—*K. gracilis* var. *mukdenensis* (Domin) Kitag. Lin. Fl. Mansh. (1939) 81.—Ic.: Domin, l.c. Table 11, fig. 1.

Described from China (Dunbei). Type in Leningrad.

In steppes, on rocky slopes and in fixed sand; up to lower mountain belt.

IA. Mongolia: *Cis-Hing.* (Khuntu somon 5 km west of Toge-Gol, Aug. 7, 1949—Yun.); *Cent. Khalkha, East. Mong., Bas. lakes* (east of Ulei-Daban in Ubsu-Nur lake region, June 11, 1879—Pot.; Borig-Del' sand south-east of Bain-Nur lake, July 25, 1945—Yun.); *Val. lakes* (5 km east of Tatsain-Gol along road from Arbai-Khere to Bayan-Khongor, July 9, 1947—Yun.); *East. Gobi* (Khar-Sair, Roerich Exped., No. 470, July 22, 1935—Keng; Gurban-Saikhan somon, Sumbur-Ula hill east of Tabiin-Chzhis, June 23, 1949—Yun.).

General distribution: East. Sib. (south), Nor. Mong., China (Dunbei).

Note. Unlike the closely related species *K. cristata* (L.) Pers. distributed almost exclusively in the hilly regions of Cent. Asia, *K. mukdenensis* is found mainly in flat steppes. Types *K. mukdenensis* and *K. tokiensis* subsp. *mongolica* Domin preserved in Leningrad (LE) are similar in structure of sheaths of dead leaves at shoot bases as well as in other rather significant characteristics, but the specimen of type *K. mukdenensis* collected during fruiting is relatively larger and longer-leaved and hence its panicles are markedly narrower than long. Concomitantly, *K. tokiensis* Domin, described from Japan and to which Domin added subspecies *K. tokiensis* subsp. *mongolica*, differs quite distinctly from *K. mukdenensis* as well as from *K. cristata* in largely puberulent stems; it thus evidently deserves to be retained as an independent species placed between *K. cristata* and *K. asiatica* Domin. Species *K. ledebourii* Domin, *K. seminuda* (Trautv.) Gontsch., *K. sibirica* (Domin) Gontsch. and *K. ascoldensis* Roshev., adopted in Flore SSSR, should probably be treated as much later synonyms of *K. tokiensis*.

49. Catabrosa Beauv.
Ess. Agrost. (1812) 97.

1. **C. aquatica** (L.) Beauv. Ess. Agrost. (1812) 97; Krylov, Fl. Zap. Sib. 2 (1928) 270; Pavlov in Byull. Mosk. Obshch. ispyt. prir. 38 (1929) 20; Pampanini, Fl. Carac. (1930) 75; Nevski in Fl. SSSR, 2 (1934) 445; Fl. Kirgiz. 2 (1950) 138; Grubov, Consp. fl. MNR (1955) 72; Fl. Kazakhst. 1 (1956) 253; Fl. Tadzh. 1(1957) 235; Keng, Fl. ill. sin., Gram. (1959) 321; Bor, Grasses Burma, Ceyl., Ind. and Pakist. (1960) 528; Ikonnikov, Opred. rast. Pamira [Key to Plants of Pamir] (1963) 62.—*Aira aquatica* L. Sp. pl. (1753) 64. —Ic.: Fl. Kazakhst. 1, Plate 19, fig. 4; Keng, l.c. fig. 263.

Described from Europe. Type in London (Linn.).

Along banks of water reservoirs, often in water, marshy meadows and riverine pebble beds; up to upper mountain belt.

IA. Mongolia: *Cent. Khalkha* (Dzhirgalante river basin near Botog, 47° N. Lat., 104°–105° E. Long., Sept. 11, 1925—Krasch. and Zam.; around Ikhe-Tukhum-Nur lake, 46.5° N. Lat. and 104–105° E. Long., June, 1926—Zam.; Tsagan-Nur lake lowland along Ulan-Bator—Tsetserleg road, June 26, 1948—Grub.; Buridu somon, midcourse of Mukhrom-Gol river, July 14, 1952—Davazhamts); *Bas. lakes* (Kharkhira river valley, Sept. 1; near Ulangom town, Sept. 1; sand near Khara-Bury, Oct. 4—1931, Bar.); *Gobi-Alt.* (Dundu-Saikhan mountain range foothill, Aug. 19, 1931—Ik.-Gal.; 35–40 km west-nor.-west of Dalan-Dzadagad, July 13, 1943—Yun.).

IC. Qaidam: *hilly* (Sarlyk-Ula hills, 3000 m, May 20; Ichegyn-Gol river, 3300 m, June 16—1895, Rob.); *plains* (Burkhan-Budda mountain range, along Khatu river, 3350 m, Aug. 12, 1884—Przew.; same site, Khatu gorge, 3350 m, June 25, 1905—Lad.).

IIA. Junggar: *Alt. region* (near Qinhe, No. 894, Aug. 2, 1957—Huang); Jung. Alt. (nor. part of Dzhair mountain range 24 km nor.-nor.-east of Toli settlement on road to Temirtam, Aug. 5, 1957—Yun.; near Toli settlement, No. 686, Aug. 6, 1957—Huang); *Tien Shan* (near Kul'dzha town, May 25, 1877—A. Reg.; B. Yuldus basin 30–35 km south-west of Bain Bulak, Aug. 10, 1958—Yun.); *Jung. Gobi* (Khubchigin-Nuru mountain range west of Adzhi-Bogdo, Aug., 1947—Yun.); *Zaisan* (on road from Burchum to Khobsair, No. 10480, June 21, 1959—Lee et al.).

IIIA. Qinghai: *Nanshan* (valley of Sharagol'dzhin river, Sungi-Nur area, 3300 m, June 8; same site, Yayakhu area, 3300 m, June 13—1894, Rob.).

General distribution: Aral-Casp., Balkh. region, Jung.-Tarb., Nor. and Cent. Tien Shan, East. Pam.; Europe, Mediterr., Balk.-Asia Minor, Near East, Caucasus, Middle Asia (hills), West. Sib. (south), East. Sib. (south.), Nor. Mong., China (Dunbei, Nor., Nor.-West.), Himalayas, Nor. Amer.

50. Melica L.

Sp. pl. (1753) 66.

1. Spikelets 9–15 mm long; glumes 7–11 mm long; lemma of lowest florets 7–11 mm long; anthers 2–3.5 mm long. Plant up to 150 cm tall; leaf blade 4–12 mm broad, flat; ligules 2–7.5 mm long .... 2.

+ Spikelets 4–8.5 mm long; glumes up to 8.2 mm long; lemma up to 8 mm long; anthers up to 2 mm long. Generally, very low plant; leaf blade up to 5 mm broad, flat or longitudinally convoluted 3.

2. Stems scabrous under nodes and panicle. Panicles 10–20 cm long, very dense, with many spikelets .... 1. M. altissima L.

+ Stems smooth. Panicles 12–30 cm long, usually spreading at anthesis, with relatively few spikelets ....
.... 9. M. turczaninowiana Ohwi.

3. Panicles with one fully developed lower floret and 2–3 rudimentary upper florets; lemma with numerous long (almost as long as lemma) hairs along sides; panicles 5–11 cm long, dense. Leaf blades 1.5–4 mm broad, often longitudinally convoluted, puberulent toward top; ligules 2–4 mm long ....
.... 8. M. transsilvanica Schur.

+ Spikelets with (1) 2–3 (4) fully developed lower florets and 1–3 rudimentary upper florets ; lemma largely scabrous but glabrous, less often covered with short (less than 1/3 lemma) hairs .... 4.

4. Stalk of spikelet in lower part with very short and in upper part with very long spinules; panicles 4–12 cm long, with few spikelets, racemose (only lowest branches sometimes bear 2–3

spikelets each); anthers 1.3–1.7 mm long. Leaf blades 2–4 mm broad, flat; ligules up to 0.3 mm long ....3. **M. nutans** L.

+ Stalk of spikelet puberulent in upper part; panicles 6–30 cm long, usually with many spikelets; branches often bearing more than 3 spikelets each. Leaf blades flat or longitudinally convoluted .... 5.

5. Stems smooth under panicle .... 6.

+ Stems highly scabrous under panicle due to deflexed spinules ...
   .... 7.

6. Ligules 2–6 mm long, not emarginate, truncate or mostly rounded; leaf blades 2–4 mm broad, usually flat. stalk of spikelet smooth or almost so in lower part; spikelets 5–7 mm long; glumes 4–5.5 mm long, only slightly shorter than spikelets; lemma of lowest flowers 4.5–6 mm long, glabrous .... 5. **M. scabrosa** Trin.

+ Ligules emarginate : 0.5–1.5 mm long in midportion and up to 2.5 mm long along sides; leaf blades 0.6–2.5 mm broad, usually longitudinally convoluted. Stalk of spikelet scabrous in lower portion; spikelets 4–5.6 mm long; glumes 2–3.5 mm long, considerably shorter than spikelets; lemma of lowest flowers 3–5.2 mm long, usually covered in diffuse stiff hairs, sometimes without hairs .... 10. **M. virgata** Turcz. ex Trin.

7. Ligules up to 0.3 mm long, coriaceous-membranous. Panicles 15–30 cm long; spikelets very narrow, 5–9 mm long and 1.5–3 mm broad; lower glumes 2–3 mm long, with solitary rib, upper glumes 3–4 mm long, with 3(5) ribs; lemma lanceolate, rather obtuse at tip, 3.8–5.2 mm long .... 4. **M. przewalskyi** Roshev.

+ Ligules over 0.5 mm long, membranous. Panicles 6–20 cm long; spikelets broader; lower glumes 3.5–7 mm long, with 3–5 ribs visible only in lower part; upper glumes with 3–9 ribs nearly as long as spikelet .... 8.

8. Ligules emarginate: 0.5–1.5 mm long in midportion and up to 3 mm long along sides; panicles with relatively few spikelets; spikelets 6.8–8.3 mm long, with 2–3 fully developed florets; lemma of lowest florets 6–8 mm long, obtuse or weakly emarginate at tip; anthers 1.2–1.8 mm long .... 2. **M. kozlovii** Tzvel.

+ Ligules not emarginate, truncate or rather rounded; panicles with many spikelets, fairly dense; lemma of lowest florets 3–6 mm long, emarginate at tip; anthers 0.6–1 mm long .... 9.

9. Spikelets 4–7 mm long, with 2–3 fully developed florets, usually silvery; lemma of lowest florets 3–4.5 mm long, oblong-ovate, membranous only at tip; ligules 2.5–6.5 mm long, glabrous on back side .... 6. **M. tangutorum** Tzvel.

+ Spikelets 6–8 mm long, with (1) 2 fully developed florets, usually pinkish-violet; lemma of lowest florets 4–6 mm long, oblong,

membranous and slightly enlarged in upper third; ligules 0.6–1.5 mm long, densely puberulent on back side ....
.... **7. M. tibetica** Roshev.

1. M. altissima L. Sp. pl. (1753) 66; Krylov, Fl. Zap. Sib. 2 (1928) 273; Lavrenko in Fl. SSSR, 2 (1934) 350; Fl. Kirgiz. 2 (1950) 115; Fl. Kazakhst. 1 (1956) 217; Fl. Tadzh. 1 (1957) 220. —Ic.: Fl. Kazakhst. 1, Plate 16, fig. 14.
Described from Siberia (?). Type in London (Linn.).
In thin forests and among shrubs, in forest glades; up to midmountain belt.

IIA. Junggar: *Jung. Alt.* (Toli district near Arba-kegen, village, No. 1326, Aug. 7, 1957—Huang); *Tien Shan* (Borgaty brook on nor. flank of Kash river valley, 1600–2000 m, July 5, 1879—A. Reg.).

General distribution: Aral-Casp., Jung.-Tarb., Nor. and Cent. Tien Shan; Europe (cent. and south-east.), Balk.-Asia Minor, Caucasus, Middle Asia (West. Tien Shan; Giss.-Darv.), West. Sib. (south.), East. Sib. (south-west.).

2. M. kozlovii Tzvel. sp. nova.—Planta perennis, laxe caespitosa, 30–60 cm alta; culmi erecti, sub paniculis scabri; vaginae scabrae; ligulae membranaceae, glabrae, emarginatae: in parte media 0.5–1.5 mm lg., lateribus ad 3 mm lg.; laminae 0.8–2.7 mm lt., laxe convolutae vel partim planae, supra pilis brevissimis vel aculeolis elongatis, subtus aculeolis brevissimis tectae. Paniculae 6–16 cm lg., secundae, paucispiculae, ramis valde abbreviatis, scabriusculis, spiculas 1–5 gerentibus, pedunculis 1–6 mm lg., flexuosis, in parte inferiore laevibus vel scabriusculis, in parte superiore breviter pilosis; spiculae 6.8–8.3 mm lg., partim griseo-violacea tinctae; flores evoluti in numero 2–3, rudimentales in numero 2–3, appendicem clavatam formantes; glumae late ellipticae, inferior 5.5–7 mm lg., submembranacea, 3–5 nervis, superior 7–8.2 mm lg., in parte superiore et marginibus membranacea, 5–9-nervis; lemmata florium infimorum 6–8 mm lg., late oblonga, subcoriacea, in parte superiore (1/5–1/4) membranacea, apice obtusa vel leviter emarginata, 7–9-nervia, scabra; paleae ellipticae, pilis brevissimis vel aculeolis elongatis tectae; antherae 1.2–1.8 mm lg.

Typus: "Tzaidam, monasterium Dulan-Chit, 3700 m, No. 367, 9 VIII 1901, V. Ladygin". In Herb. Inst. Bot. Acad. Sci. URSS—LE conservatur.

Affinitas. A speciebus proximis—*M. tangutorum* Tzvel. et *M. tibetica* Roshev. paniculis paucispiculatis, spiculis majoribus, antheris majoribus et ligulis emarginatis bene differt. In habitu haec species *M. nutantem* L. simulans, sed pedunculis apice breviter pilosis et ligulis multo longioribus facile differt.

Described from Qinghai. Type in Leningrad. Plate VII, fig. 2; map 8.
On rocks and rocky slopes; in middle and upper mountain belts.

IA. Mongolia: *Khesi* (15 km nor. of Yunchan town, rocky slopes of Beidashan' hills, June 28, 1958—Petr.).

IIIA. Qinghai: *Amdo* (Dulan-Khit mon., 3700 m, No. 367, Aug. 9, 1901—Lad., type!).

General distribution: endem.

3. M. nutans L. Sp. pl. (1753) 66; Forbes and Hemsley, Index Fl. Sin. 3 (1904) 418; Danguy in Bull. Mus. nat. hist. natur. 20 (1914) 147; Krylov, Fl. Zap. Sib. 2 (1928) 274; Lavrenko in Fl. SSSR, 2 (1934) 351; Kitag. Lin. Fl. Mansh. (1939) 82; Fl. Kazakhst. 1(1956) 218; Bor, Grasses Burma, Ceyl., Ind. and Pakist. (1960) 590. —Ic.: Fl. SSSR, 2, Plate 24, fig. 4; Fl. Kazakhst. 1, Plate 16, fig. 18.

Described from Nor. Europe. Type in London (Linn.).

In forests, among shrubs; in midmountain belt.

IIA. Junggar: *Tien Shan* (Nilki brook in Kash river basin, 2300 m, June 8, 1879—A. Reg.; "Montagnes entre Gorgosse et le Sairam-Nor, 1720 m, 17 VII 1895, leg. Chaffanjon"—Danguy, l.c.).

General distribution: Jung.-Tarb., Nor. Tien Shan; Europe, Balk.-Asia Minor, Caucasus, West. Sib., East. Sib. (south.), Far East (south.), China (Dunbei), Himalayas, Korea, Japan.

4. M. Przewalskyi Roshev. in Bot. mater. Gerb. Glavn. bot. sada RSFSR, 2 (1921) 25; Keng, Fl. ill. sin., Gram. (1959) 240.—*M. polyantha* Keng in Sunyatsenia, 6 (1941) 77. —Ic.: Keng, l.c. (1941) fig. 1; Keng, l.c. (1959) fig. 192.

Described from Qinghai. Type in Leningrad. Map 8.

In grasslands and on rocky slopes; in upper mountain belt.

IIIA. Qinghai: *Nanshan* (along Yusun-Khatyma river, 3000–3300 m, No. 582, July 12, 1880—Przew., lectotype!).

IIIB. Tibet: *Weitzan* (Yangtze river basin, in Kabchzha-Kamba village on Khichu river, 4000 m, July 20, 1900—Lad.).

General distribution: China (Nor.-West.).

5. M. scabrosa Trin. in Mém. Ac. Sci. St.-Pétersb. Sav. Etrang. 2 (1835) 146; Franch. Pl. David. 1 (1884) 338; Kanitz in Széchenyi, Wissensch. Ergebn. 2 (1898) 737; Forbes and Hemsley, Index Fl. Sin. 3 (1940) 419; Kitag. Lin. Fl. Mansh. (1939) 82; Walker in Contribs U.S. Nat. Herb. 28 (1941) 597; Norlindh, Fl. mong. steppe, 1 (1949) 92; Keng, Fl. ill. sin., Gram. (1959) 244. —Ic.: Keng, l.c. fig. 197.

Described from Nor. China (Beijing district). Type in Leningrad.

On rocky slopes, rocks and in riverine pebble beds; up to midmountain belt.

IA. Mongolia: *East. Mong.* ("Khongkor-Obo, 7 VIII 1926; Naiman-Ul, 26 VII 1928; 7.5 km ad bor.-orient. Khadain-Sume, 26 VI 1926, leg. Eriksson"—Norlindh, l.c.); *East. Gobi* (Madenii-Amok, Roerich Exped., No. 525, July 21, 1935; Tumur-Hada, Roerich Exped., No. 565, July 30, 1935—Keng); *Alash. Gobi* (Alashan mountain range, Tszosto gorge, May 16, 1908—Czet.; "Mouth of Hsi-yeh-kou, leg. Ching"—Walker, l.c.).

IIIA. Qinghai: *Nanshan* ("Ku-lang-hsien, 14 VI 1879, leg. Széchenyi"—Kanitz, l.c.).

General distribution: China (Dunbei, Nor., Nor.-West., Cent., South-West.), Korea.

Note. Specimens of this species collected by G.N. Potanin from the southernmost part of the distribution range—Kamsk upland (between Pan'shamyr and Sin'chontszan villages in Syaochzhin'kho river valley, July 27, 1893; between Shin'gaitszy and Chin'-chzhevan' villages, July 31, 1893—Pot.)—differ from typical specimens of *M. scabrosa* Trin. in very short leaf blades profusely pilose on both sides (but not scabrous) and slightly shorter (4–5.5 mm long) spikelets and merit recognition at least as a variety, *M. scabrosa* var. *potaninii* Tzvel. var. nova.—A varietate typica foliorum superiorum laminis ab utroque latere brevissime, sed copiose pubescentibus et spiculis paulo minoribus differt. Typus: "Kam, inter pag. Schinggajtzy et Tzinjtzshevanj, 31 VII 1893, G. Potanin" in Herb. Inst. Bot. Ac. Sci. URSS—LE conservatur.

6. **M. tangutorum** Tzvel. sp. nova.—Planta perennis laxe caespitosa; 30–80 cm alta; culmi erecti, sub paniculis scabri; vaginae scaberrimae; ligulae membranaceae, glabrae, 2.5–6.5 mm lg.; laminae 1–3.5 mm lt., laxe convolutae vel partim planae, utrinque scaberrimae. Paniculae 10–20 cm lg., subsecundae, sat densae, sed in parte inferiore interruptae, ramis valde abbreviatis, scabriusculis, pedunculis 1–7 mm lg., flexuosis, in parte inferiore sublaevibus, in parte superiore longiuscule pilosis; spiculae 4–7 mm lg., argyraceae; flores evoluti in numero 2–3, rudimentales in numero 2–3, appendicem clavatam formantes; glumae ellipticae, inferior spicula paulo brevior, submembranacea, 3–5-nervis, superior spicula aequans, in dimidio superiore membranacea, 5–7-nervis; lemmata florium infimorum 3–4.5 mm oblongo-ovata, subcoriacea, in parte superiore (1/7–1/6) membranacea, apice emarginata, 7–9 nervia, scaberrima; paleae ellipticae, aculeolis elongatis et gracilibus tectae; antherae 0.7–1 mm lg.

Typus: "china, prov. Kansu, in deserto graminoso jugi Boreali-Tetungensi (prope opp. Daigu), No. 123, 21 VI 1872, N. Przewalski". In Herb. Inst. Bot. Ac. Sci. URSS—LE conservatur.

Affinitas. A species proxima—*M. tibetica* Roshev. Praesertim ligulis mul to longioribus glabris et lemmatis solum in apice (non in tertia superiore) membranaceis bene differt.

Described from Qinghai. Type in Leningrad. Plate VII, fig. 1; map 8.

On rocky and rubble slopes and in riverine pebble beds; up to upper mountain belt.

IA. **Mongolia:** *Khesi* (between Yunchen-Syan' and Shan'dan'-Syan' near Syaku village, July 21, 1875—Pias.).

IIIA. **Qinghai:** *Nanshan* (herbaceous barren land in Severo-Tetungsk mountain range, No. 123, June 21, 1872—Przew., type!; Loukhu-Shan' mountain range, July 17, 1908—Czet.); *Amdo* (along Churmyn river, 3000–3200 m, April 29, 1880—Przew.). **General distribution:** endemic.

7. **M. tibetica** Roshev. in Bot. mater. Gerb. Glavn. bot. sada RSFSR, 2 (1921) 27; Keng, fl. ill. sin., Gram. (1959) 241. —Ic.: Keng l.c., fig. 195.

Described from Tibet. Type in Leningrad. On rocky slopes and rocks, in upper mountain belt.

IIIB. **Tibet:** Weitzan (Yangtze river, around Kabchzha-Kamba village on Khichu river, 4000 m, No. 338, July 20, 1900—Lad., type!). **General distribution:** Endemic.

8. **M. transsilvanica** Schur, Enum. Fl. Transsilv. (1886) 764; Lavrenko in Fl. SSSR, 2 (1934) 345; Fl. Kirgiz. 2 (1950) 115; Fl. Kazakhst. 1 (1956) 216; Keng, Fl. ill. sin., Gram. (1959) 245.—*M. ciliata* auct. non L.: Krylov in Fl. Zap. Sib. 2 (1928) 272. —Ic.: Fl. SSSR, 2, Plate 26, fig. 3; Keng, l.c. fig. 199.

Described from Europe (Transylvania). Type in Vienna (?).

On rocky and rubble slopes, riverine pebble beds, talus and among shrubs; up to midmountain belt.

IIA. Junggar: *Alt. region* (near Shara-Sume, 1100 m, No. 2427, Aug. 26; same site, No. 2550, Aug. 27—1956, Ching); *Jung. Alt.* (Kepel' river valley on west. Offshoots of Barlyk, June 25, 1905—Obruchev); *Tien Shan* (Sairam-Nur lake, July 1877—A. Reg.; around Urumchi town, Toutun'khe river near Litszyagou, No. 375, July 14, 1956—Ching; south. slope of Borokhoro mountain range 4 km beyond Nizhn. Ortai settlement on Sairam-Nur-Kul'dzha road, Aug. 19; nor. slope of Ketmen' mountain range 1 km nor. of Sarbushin village on Kzyl-Kure road, Aug. 21; same site, 3–4 km beyond Sarbushin village, Aug. 23, 1957—Yun.).

General distribution: Aral-Casp. (Mugodzhary), Balkh. region, Jung.-Tarb., Nor. and Cent. Tien Shan; Europe (cent. and south.), Balk.-Asia Minor, Caucasus, Middle Asia (West. Tien Shan, Alay), West. Sib. (south), East. Sib. (Minusinsk trough).

9. **M. turczaninowiana** Ohwi in Acta Phytotax, et Geobot. 1, 2 (1932) 142; Lavrenko in Fl. SSSR, 2 (1934) 349; Kitag. Lin. Fl. Mansh. (1939) 82; Grubov, Consp. fl. MNR (1955) 69; Keng, Fl. ill. sin., Gram. (1959) 241.—*M. gmelinii* Turcz. ex Trin. in Mém. Ac. Sci. St.-Pétersb., sér. VI, 1 (1831) 368, non Roth (1789); Franch. pl. David. 1 (1884) 336; Forbes and Hemsley, Index Fl. Sin. 3 (1904) 418; Danguy in Bull. Mus. nat. hist. natur. 20 (1914) 147. —Ic.: Fl. SSSR, 2, Plate 26, fig. 8; Keng, l.c. fig. 193.

Described from Transbaikal region. Type in Leningrad.

On rocky slopes, rocks, talus and among shrubs; up to midmountain belt.

IA. Mongolia: *Cis-Hing.* (near Yaksha railway station, June 13, 1902—Litw.); *East. Mong.* (near Hailar town, 1960—Ivan.); *East. Gobi* (Muni-Ula hills, June 16, 1871—Przew.).

General distribution: East. Sib. (south), Far East (south.), Nor. Mong. (Hent., Hang.), China (Dunbei, Nor.), Korea.

10. **M. virgata** Turcz. ex Trin. in Mém. Ac. Sci. St.-Pétersb., sér. VI, 1 (1831) 369; Franch. Pl. David. 1 (1884) 337; Pavlov in Byull. Mosk. obshch. ispyt. prir. 28 (1929) 21; Lavrenko in Fl. SSSR, 2 (1934) 349; Norlindh, Fl. mong. steppe, 1 (1949) 92; Grubov, Consp. Fl. MNR (1955) 69; Keng, Fl. ill. sin., Gram. (1959) 242. —Ic.: Keng, l.c. fig. 196.

Described from Transbaikal region. Type in Leningrad.

On rocky slopes and rocks; up to midmountain belt.

IA. Mongolia: *Cent. Khalkha, East. Mong.* ("Naiman-Ul, 26 VII; Khadain-Sume, 22 VIII, 1928, leg. Eriksson"—Norlindh, l.c.; near Shilin-Khoto, 1959—Ivan.); *Gobi-Alt.* IIIA. Qinghai: *Nanshan* (66 km west of Xining town, 2800 m, Aug. 5, 1959—Petr.). General distribution: East. Sib. (south.), Nor. Mong., China (Dunbei, Nor.).

51. Aeluropus Trin.
Fund. Agrost. (1820) 143.

1. Lemma with long hairs along sides near margin, abruptly acuminate at tip; branches of panicles interrupted, spikelets in two distinct rows, aggregated and declinate from rachis .... 3. **A. pungens** (M.B.), C. Koch.

+ Lemma glabrous along sides, gradually acuminate towards tip; branches of panicle approximate and spikelets often not in two distinct rows .... 2.

2. Anthers 1.2–1.6 mm long. Plant 15–40 cm tall; leaf blade flat, 3–6 mm broad .... 1. **A. sinensis** (Debeaux) Tzvel.

+ Anthers 0.6–0.8 mm long. Plant 6–30 mm tall; leaf blade flat or longitudinally convoluted, 1–3 mm broad .... 2. **A. micrantherus** Tzvel.

1. **A. sinensis** (Debeaux) Tzvel. comb. nova. —*A. litoralis* var. *sinensis* Debeaux in Acta Soc. Linn. Bordeaux, 33 (1879) 50; Forbes and Hemsley, Index Fl. Sin.3 (1904) 421; Kitag. Lin. Fl. Mansh. (1939) 58; Keng, Fl. ill. sin., Gram. (1959) 329. —Ic.: Keng, l.c. fig. 269.

Described from Nor. China. Type in Paris.

In solonchak and saline meadows.

IA. **Mongolia:** *Alash. Gobi* (45 km nor.-west of In'chuan' town, Aug. 4, 1957—Petr.).

General distribution: China (Dunbei: south-west.; Nor.).

Note. This species belongs to a group of very closely related species grouped formerly under the name *A. litoralis* (Gouan) Parl. s.l., replacing one another from west to east. This group includes: *A. litoralis* (Gouan) Parl. s.s. (Mediterranean, southern European USSR and partly Caucasus), *A. intermedius* Regel (almost the whole of Kazakhstan from Lower Volga to Altay and Junggar Alatau), *A. korshinskyi* Tzvel. (Syr Darya and Amu Darya basins), *A. micrantherus* Tzvel. described below and *A. sinensis* (Debeaux) Tzvel. *A. sinensis* should be regarded as the most primitive species of this series, followed by *A. litoralis*, most closely related to it, i.e., relatively mesophilic species whose distribution ranges are farthest removed from one another. However, another Central Asian species, *A. micrantherus*, on the contrary, can be regarded as much 'younger' in the evolutionary context of species of this series.

2. **A. micrantherus** Tzvel. sp. nova.—*A. litoralis* auct. non Parl.: Norlindh, Fl. mong. steppe, 1 (1949) 93; Grubov, Consp. fl. MNR (1955) 70.—Planta perennis, 6–30 cm alta; culmi procumbentes vel ascendentes, in nodis inferioribus ramosi; vaginae glabrae, rarius subglabrae; laminae vulgo 1–3 mm lt., planae vel laxe convolutae, scabrae, griseo-virides; ligulae membranaceae, brevissimae (ad 0.2 mm lg.), margine pilis ad 0.3–0.5 mm lg. dense tectae. Paniculae spiciformes 2–7 cm lg. et 3–8 mm lt.; spiculae subsessiles, (2) 3–6 (8)-florae; lemmata vulgo 2.5–3.2 mm lg., ovata vel late ovata, apice breviter acutata vel mucronulata, glabra; antherae 0.6–0.8 mm lg; caryopsis circa 1 mm lg.

Typus: "Mongolia, Gobi australis, 15–20 km ad orientes versus Tzagan-Bogdo, Bilgechu-Bulak, in salsis, No. 14121, 31 VII 1943, A. Junatov". In Herb. Inst. Bot. Acad. Sci. URSS—LE conservatur.

Affinitas. A speciebus proximis: *A. litoralis* (Gouan) Parl., *A. intermedius* Regel et *A. pungens* (Bieb.) C. Koch praesertim antheris minoribus (0.6–0.8, non 1.2–1.6 mm lg.) bene differt.

Described from Mongolia. Type in Leningrad. Plate V, fig. 2; map 6. In solonchak and saline meadows.

IA. Mongolia: *Bas. lakes* (valley of Dzergen river, Oct. 7, 1930—Bar.); *West. Gobi* (Bil'gekhu-Bulak region 15–20 km east of Tsagan-Bogdo, July 31—Yun., type!; Dzakhoi-Dzaram region in foothill plain south of Mong. Altay, Aug. 18–1943, Yun.); *Alash. Gobi* (South. Gobi, Yanchi village, June 17; Suvan'no area, June 24; between Koko-Buryuk and Gantsy-Dzak, July 8, 1886—Pot.; "Edsen-Gol, Tsondol, 24 VI; Baishing-tei, 3 VII; Gashun-nor, prope Princes Encampment, 15 VII 1928; Bayan-Bogdo Camp, Wen-tsun-hai-tze, 17 VI, 1929, leg. Söderbom"—Norlindh, l.c.); *Hesi* (near Chzhan"e village, solonchak on meander scar of Beitakhe river, July 18, 1958—Petr.).

IB. Kashgar: *Nor.* (Maralbashi oasis, Chuderlik village, May 6, 1909—Divn.; 6 km south of Shamal village, No. 1514, Sept. 13, 1958—Lee and Chu); *West.* (Yangigisar vicinity, May 27, 1909—Divn.); *Takla Makan* (8–10 km south-east of Karasai village along road to Karakash, May 25, 1959—Yun.; same site, No. 146, May 25, 1959—Lee et al.).

IIA. Junggar: *Jung. Gobi* (east.: near Nom-Tologoi settlement, June 8, 1877—Pot.). General distribution: endemic.

Note. This endemic Central Asian species is easily distinguished from all other species of the genus in much smaller anthers. In Kashgar, it is sometimes found together with *A. pungens* (M.B.) C. Koch but does not hybridise with it, being probably a cleistogamous or apomictic species.

3. A. pungens (M.B.) C. Koch in Linnaea, 21 (1848) 408; Tzvelev in Novosti sist. vyssh. rast. (1966) 28.—*Poa pungens* M.B. Beschreib. Länd. zw. Flüss. Terek u.Kur (1800) 130.—*A. litoralis* auct. non Parl.: ? Danguy in Bull. Mus. nat. hist. natur. 6 (1911) 7; Danguy, l.c. 20 (1914) 147; Pilger in Hedin, S. Tibet, 6, 3 (1922) 93; Roshev. in Fl. SSSR, 2 (1934) 357, p.p.; Fl. Kirgiz. 2 (1950) 119; Fl. Kazakhst. 1 (1956) 219, p.p.; Fl. Tadzh. 1 (1957) 134; Keng, Fl. ill. sin., Gram. (1959) 328, excl. var. —Ic.: Keng, l.c. fig. 268; Tzvelev, l.c. 29, figs. 8–10.

Described from Transcaucasus region (Kura river valley). Type in Leningrad.

In solonchak and saline meadows, in particular sandy and loamy soils; up to midmountain belt.

IA. Mongolia: *Khesi* ("Route de Soutcheou a Kantcheau, Yen-che, 1500 m, 26 VI 1908, leg. Vaillant"—Danguy, l.c. [1911]; Sangun sand 12 km south-west of An'si town, July 25, 1958—Petr.).

IB. Kashgar: *Nor.* (near Uchturfan, May 16, 1908—Divn.; near Chonza village between Kashgar and Maralbash, Aug. 1, 1929—Pop.; 7 km south of Bachu village, No. 1422, Sept. 6, 1958—Lee and Chu); *West., South.* (Keriya oasis, May 9, 1959—Yun.; near Khotan town, May 27, 1959—Lee et al.). *East., Lob-Nor* ("Lop-nor, Just-chapgham, 817 m, 24 VI 1900, leg. Hedin"—Pilger, l.c.).

IIA. Junggar: *Tien Shan* (Kash river between Ulastai and Nilki villages, June 30, 1879—A. Reg.); *Jung. Gobi* (nor.-west.; near Karamai village, June 19, 1957—Huang; west.: "Ebi-Nor, 30 VII 1895, leg. Chaffanjon"—Danguy, l.c. [1914]; meteorological station at Kzyl-Tuz near Junggar outlet, Aug. 16, 1957—Yun.; south. Manas river between Savan and Paotai, June 11; near Syaedi village, June 13; 21 km nor.-west of Paotai, June 17—1957, Huang; right bank of Manas river 6–8 km nor. of Paotai, June 11, 1957—Yun.); *Zaisan, Dzhark.*

General distribution: Aral-Casp., Balkh. region, Jung.-Tarb., Nor. Tien Shan (Issyk-Kul' lake); Near East (nor.-east.), Caucasus (Casp. region), Middle Asia.

Note. Among those closer to *A. litoralis* (Gouan) Parl., this species is morphologically most isolated and, in our view, is of hybrid origin from the common ancestor of group *A. litoralis* s.l., whose distribution range lay north of Tethys and another common ancestor of the group, *A. lagopoides* (L.) Thwaites s.l. [including *A. repens* (Desf.) Parl.] with its distribution range falling south of Tethys.

52. Dactylis L.

Sp. pl. (1753) 71.

1. **D. glomerata** L. Sp. pl. (1753) 71; Forbes and Hemsley, Index Fl. Sin. 3 (1904) 421; Simpson in J. Linn. Soc. London (Bot.) 41 (1913) 453; Krylov, Fl. Zap. Sib. 2 (1928) 277; Pampanini, Fl. Carac. (1930) 75; Ovczinnikov in Fl. SSSR, 2 (1934) 361; Fl. Kirgiz. 1 (1950) 120; Fl. Kazakhst. 1 (1956) 220; Fl. Tadzh. 1 (1957) 133; Keng, Fl. ill. sin., Gram. (1959) 327; Bor, Grasses Burma, Ceyl., Ind. and Pakist. (1960) 530. —Ic.: Fl. SSSR 2, Plate 27, fig. 15; Keng, l.c. fig. 267.

Described from Europe. Type in London (Linn.).

In meadows, forest glades, thin forests, among shrubs, on rocky slopes and rocks; up to upper mountain belt.

IIA. Junggar: *Alt. region, Tarb.* (nor. of Chuguchak town, No. 2917, Aug. 13, 1957—Huang); *Jung. Alt.* (Toli dist., Aug. 6; 20 km nor. of Ulasutai, 2100 m, Aug. 27—1957, Huang); *Tien Shan.*

General distribution: Jung.-Tarb., Nor. and Cent. Tien Shan; Europe, Mediterr., Balk.-Asia Minor, Near East, Caucasus, Middle Asia, West. Sib., East. Sib. (south), China (Nor.-West., Cent., South-West.), Himalayas; introduced in many other countries of both hemispheres.

53. Schismus Beauv.

Ess. Agrost. (1812) 73.

1. **S. arabicus** Nees, Fl. Afr.Austr. (1841) 422; Krylov, Fl. Zap. Sib. 2 (1928) 278; Roshev. in Fl. SSSR, 2 (1934) 365; Fl. Kirgiz. 2 (1950) 121; Fl. Kazakhst. 1 (1956) 221; Fl. Tadzh. 1 (1957) 360; Bor, Grasses Burma, Ceyl., Ind. and Pakist. (1960) 481.

Described from Arabia. Type in Berlin.

On sand and in pebble beds of river and lake valleys, on talus; up to midmountain belt.

IB. **Kashgar:** *Nor.* (from Guilyu village up to Shakh'yar village, No. 3681, Aug. 20, 1957—Huang); *South.* (nor. slope of Russky mountain range, Karasai village, 3000 m, June 13, 1890—Rob.).

IIA. **Junggar:** *Tarb.* (Khobuk river near Shatszychai village, No. 1029, May 18, 1959—Lee et al.); *Tien Shan* (Sev. Borborogus-sun river, 1000–1300 m, April 28, 1879—A. Reg.); *Jung. Gobi* (south.: Savan district, between Paotai and Sykeshu, June 10; 5 km from Paotai toward San'daokhetszy, June 11; Manas district, between Paotai and Syaedi, june 12; in Syaedi state farm, June 14; near "Tridtsatyi Polk" state farm, July 9—1957, Huang; east.: Bulugun river floodplain near confluence of Ulyaste-Gol river, July 20, 1947—Yun.); *Zaisan* (left bank of Ch. Irtysh, Mai-Kain area, June 7, 1914—Schischk.); *Dzhark.* (Suidun town, May 8, 1878—A. Reg.).

General distribution: Aral-Casp., Balkh. region, Jung.-Tarb., Nor. and Cent. Tien Shan; Asia Minor, Near East, Caucasus (east. and south. Transcaucasus), Middle Asia, Himalayas (west., Kashmir), Nor. Africa; introduced in other countries.

54. Poa L.
Sp. pl. (1753) 67.

1. Plant forming small dense mats with vegetative shoots bulbous, thickened at base. Spikelets usually modified into greatly shortened shoots (vivipary) .... 28. **P. bulbosa** L.
+ Shortened vegetative shoots, if present, not bulbous, thickened at base 2

2. Panicles 10–25 cm long, constitute 1/3 or 1/4 length of stem; their branches quite thick and long, lower ones usually only slightly shorter than panicle, widely spreading; lemma 4–6 mm long, glabrous or subglabrous. Greyish-green plant with long and fairly thick rootstock .... 1. **P. subfastigiata** Trin.
+ Panicles of different habit; their branches less than 2/3 length of panicle .... 3.

3. Sheath of cauline leaves closed for 2/3–4/5 length from base, flattened laterally, with highly prominent, slightly winged keel. Panicles 10–30 cm long, widely spreading. .....12. **P. remota** Forsell.
+ Sheath of cauline leaves closed for less than 2/3 length from base, keel weak, not winged .... 4.

4. Lemma together with callus glabrous .... 5.
+ Lemma rather pilose in lower part .... 7.

5. Plant 20–100 cm tall with short creeping subsurface shoots. Anthers 1.5–2.5 mm long .... 13. **P. sibirica** Roshev.
+ Plants 5–25 cm tall, without creeping subsurface shoots .... 6.

6. Panicles 2–5 cm long, rather compressed and dense; anthers 1.2–1.6 mm long .... 24. **P. poiphagorum** Bor.

+ Panicles 4–10 cm long, rather spreading; anthers 0.3–0.6 mm long .... 33. **P. tibeticola** Bor.

7(4). Plant with creeping subsurface shoots; shortened vegetative shoots usually present .... 8.

+ Plant without creeping subsurface shoots; shortened vegetative shoots usually absent .... 18.

8. Callus of lemma without tuft of long undulating hairs, sometimes with stray hairs .... 9.

+ Callus of lemma with well-developed tuft of long undulating hairs .... 12.

9. Panicles compressed and dense, with very short branches. Greyish-green plant with long and fairly thick rootstock .... 2. **P. tibetica** Munro ex Stapf.

+ Panicles spreading, with long slender, branches. Green plant with short creeping subsurface shoots .... 10.

10. Plant 10–50 cm tall, usually forming mats; shoots at base surrounded by cap of several sheaths of dead leaves .... 6. **P. lipskyi** Roshev.

+ Plant 40–80 cm tall, not forming mat; shoots at base without distinctly manifest cap of sheaths of dead leaves .... 11.

11. Sheath of cauline leaves closed up to half length from base; ligules 2–3.5 mm long; leaf blade 1–2.5 mm broad .... 7. **P. megalothyrsa** Keng.

+ Sheath of cauline leaves closed only near base; ligules 3–7 mm long; leaf blades 2.5–5 mm broad .... 4. **P. asperifolia** Bor.

12 (8). Blade of cauline leaves 2–4 mm broad and up to 3 (4) cm long, shortly acuminate at tip, with both sides smooth. Panicles 3–6 cm long, spreading, each branch with 1–2 deviating and 1–3 supporting lanceolate spikelets; lemma 4–5.5 mm long .... 11. **P. tangii** Hitchc.

+ Blade of cauline leaves very long and narrow, gradually acuminate towards tip, often with scattered spinules towards tip. Panicles usually with several spikelets, at least on part of each branch .... 13.

13. Palea with spinules along keels that gradually elongate towards base then transform into short hairs and, between keels, with scattered short hairs; lemma puberulent in lower part between ribs, less often glabrous. Plant with short creeping subsurface shoots, usually forming loose mats .... 14.

+ Palea with spinules only along keels, glabrous between keels; lemma glabrous between ribs. Plant with long creeping subsurface shoots, usually not forming mats .... 15.

14. Spikelets 4–5.5 mm long; lemma 3.2–4 mm long..........................
.... **P. kenteica** Ivanova.
+ Spikelets 5.5–8 mm long; lemma 4.2–5.5 mm long.....................
.... 10. **P. smirnovii** Roshev.
15. Plant forming tufts of surface shoots surrounded by common cap of sheaths of dead leaves; tufts with single reproductive shoot and one or more vegetative shoots with long leaf blade up to 1.5 mm broad, longitudinally folded and appearing setaceous Lemma 2.5–3.7 mm long... 3. **P. angustifolia** L.
+ Shoots single, less often approximate but not surrounded by common cap of sheaths of dead leaves; leaf blade of vegetative shoots usually not aceouset. Lemma 3.3–4.7 mm long .... 16.
16. Plant 5–18 cm tall. Panicles 2–3.5 cm long; branches usually horizontally divergent, with closely packed spikelets only at tip; lemma pinkish-violet, often golden at tip ...... 5. **P. calliopsis** Litv.
+ Plant 15–100 cm tall. Panicles 3.5–15 cm long; branches usually with interrupted spikelets; lemma usually lightly coloured, often pale green .... 17.
17. Plant green, usually 25–100 cm tall; leaf blade relatively slender, strongly diverging from stem................................ 8. **P. pratensis** L.
+ Plant greyish or glaucous green, usually 15–35 cm tall. Leaf blade very thick, weakly diverging from stem ....
.... 9. **P. pruinosa** Korotky.
18 (7). Anthers 0.3–0.8 mm long .... 19.
+ Anthers 1.1–2.5 mm long .... 21.
19. Branches of panicle smooth; anthers 0.5–0.8 mm long ....
.... 30. **P. annua** L.
+ Branches of panicle rough; anthers 0.3–0.6 mm long................ 20.
20. Plant 10–40 cm tall, forming loose mats. Callus of lemma with tuft of long undulating hairs .... 29. **P. acroleuca** Steud.
+ Plant 5–10 cm tall, forming dense mats. Callus of lemma glabrous .... 31. **P. rossbergiana** Hao.
21. Branches of panicle smooth or almost so 22
+ Branches of panicle scabrous throughout length .... 23.
22. Greyish-green plant forming very dense mats with shoots surrounded at base by cap of sheaths of dead leaves. Panicles dense with ovate spikelets .... 27. **P. alpina** L.
+ Green plant, forming loose mats, with shoots not surrounded by cap of sheaths of dead leaves. Panicles usually widely spreading with narrowly ovate spikelets .... 32. **P. supina** Schrad.
23. Panicles somewhat spreading, with fairly long branches; longest branches only 1/2 to 2/3 as long as panicle. Plant green or with slight greyish tinge, forming relatively loose mats; stems almost

invariably covered in sheaths of leaves up to 1/2 or more their length........24.tr

+ Panicles very dense, usually spicate, with greatly shortened branches; longest branches 1/8 to 1/3 as long as panicle. Greyish or glaucous green plant forming very dense mats; stems covered with sheaths for less than half their length .... 29.

24. Ligules 0.2–0.8 mm long.................................... 21. P. nemoralis L.

+ Ligules 1–6 mm long..25.

25. Callus of lemma glabrous..26.

+ Callus of lemma with tuft of long undulating hairs.................27.

26. Ligules 3–6 mm long...............................18. P. elanata Keng.

+ Ligules 1–2.7 mm long....................................20. P. krylovii Reverd.

27. High-altitude, densely caespitose plant 10–35 cm tall; shoots covered at base with numerous sheaths of dead leaves. Spikelets almost invariably pinkish-violet, slightly lustrous ....
....15. P. altaica Trin.

+ Plant usually of very low hills, 20–100 cm tall, forming very loose mats; shoots only with few sheaths of dead leaves at base. Spikelets pale green or with slight pinkish-violet tinge, not lustrous...28.

28. Predominantly meadow plant, up to 100 cm tall; stems usually glabrous, occasionally slightly scabrous, with 3–5 interrupted nodes. Panicles usually widely spreading, with long branches ..
....23. P. palustris L.

+ Predominantly steppe plant, up to 50 cm tall; stems almost invariably rather scabrous, with 2–3 interrupted nodes. Panicles slightly spreading, with very short branches ....
....25. P. relaxa Ovcz.

29 (23). Lemma covered with short hairs in lower part between ribs........30.

+ Lemma pilose only along ribs in lower part ....32.

30. High-altitude plant 5–20 cm tall; at least some vegetative shoots present in mats .... 19. P. koelzii Bor.

+ Plants of plains and low hills, 20–50 cm tall; mats only with reproductive shoots ....31.

31. Stems and leaf blades highly scabrous on both sides. Glumes almost as long as lemma .... 26. P. reverdattoi Roshev.

+ Stems and leaf blades smooth, or almost so, towards bottom. Glumes almost 2/3 length of lower lemma ....
.... 16. P. argunensis Roshev.

32. Callus of lemma glabrous; lemma (2.5) 3–4 (4.5) mm long, usually with dull pinkish-violet tinge........................ 14. P. albertii Regel.

+ Callus of lemma with tuft of long undulating hairs, sometimes only with stray ones ....33.

33. Lemma 2.5–3.5 mm long, almost invariably with dull pinkish-violet tinge. Stems usually rather scabrous under panicle, occasionally smooth .... 17. P. attenuata Trin.

+ Lemma 3–4.3 mm long, usually pale green, occasionally with faint pinkish-violet tinge. Stems smooth or slightly scabrous under panicle .... 22. P. ochotensis Trin.

Section 1. Arctopoa
(Griseb.) Tzvel.

1. P. subfastigiata Trin. in Ledeb. Fl. alt. 1 (1829) 96; Krylov, Fl. Zap. Sib. 2 (1928) 300; Roshev. in Fl. SSSR, 2 (1934) 425; Kitag. Lin. Fl. Mansh. (1939) 88; Norlindh, Fl. mong. steppe, 1 (1949) 101; Grubov, Consp. fl. MNR (1955) 71; Keng, Fl.ill. sin., Gram. (1959) 151. —Ic.: Ledeb. Ic. pl. fl. ross. 3 (1831) Table 224; Fl SSSR, 2, Plate 32, fig. 5; Keng, l.c. fig. 100.

Described from Altay. Type in Leningrad.

In solonetz meadows, along banks of water reservoirs, in riverine sand and pebble beds; up to lower mountain belt.

IA. Mongolia: *Cent. Khalkha, East. Mong., Gobi-Alt.* (Dzun-Saikhan mountain range, along Tsagan-Gol river, Aug. 20, 1931—Ik.-Gal.).

General distribution: West. Sib. (Altay), east. Sib. (south, Cent. Yakutia), Far East (south.), Nor. Mong., China (Dunbei).

2. P. tibetica Munro ex Stapf in Hook. f. Fl. Brit. Ind. 7 (1896) 339; Hemsley, Fl. Tibet (1902) 204; Krylov, Fl. Zap. Sib. 2 (1928) 289; Pavlov in Byull. Mosk. Obshch. ispyt. prir. 38 (1929) 23; Pampanini, Fl. Carac. (1930) 76; Roshev. in Fl. SSSR, 2 (1934) 425; Norlindh, Fl. mong. steppe, 1 (1949) 100; Fl. Kirgiz. 2 (1950) 136; Grubov, Consp. fl. MNR (1955) 72; Fl. Kazakhst. 1 (1956) 238; Fl. Tadzh. 1 (1957) 161; Keng, Fl. ill. sin., Gram. (1959) 157; Bor, Grasses Burma, Ceyl., Ind. and Pakist. (1960) 561; Pazii in Bot. mater. Gerb. Inst. bot. AN Uzbek SSR, 17 (1962) 32; Ikonnikov, Opred. rast. Pamira [Key to Plants of Pamir] (1963) 57.—*P. macrocalyx β. tianschanica* Regel in acta Horti Petrop. 7 (1880) 619.—*P. tianschanica* (Regel) Hack. ex O. Fedtsch in Acta Horti Petrop. 21 (1903) 441.—*P. pricei* Simpson in J. Linn. Soc. London (Bot.) 41 (1913) 452; Grubov, l.c. 71.—*P. ciliatiflora* Roshev in: Sev. Monoliya (Nor. Mogolia), 1 (1926) 163.—*P. fedtschenkoi* Roshev. in Izv. Bot. sada AN SSSR, 30 (1932) 297; Roshev., l.c. (1934) 421; Fl. Kirgiz. 2 (1950) 135. Ic.: Simpson, l.c. Tab. 23, figs. 4–12; Fl. SSSR, 2, Plate 32, fig. 4; Fl. Tadzh. 1, Plate 20; Keng, l.c. fig. 108.

Described from West. Himalayas. Type in London (K).

In saline meadows and solonchak, in rather saline sand and pebble beds of river and lake valleys; up to upper mountain belt.

IA. Mongolia: *Khobd.* ("In the desert-plateau valley of the Saklya River, No. 74, 1910, leg. Price"—Simpson, l.c.); *Mong. Alt., Bas. lakes, Val. lakes, Gobi-Alt., East.*

Gobi, West. Gobi (south. slope of Tagan-Bogdo mountan range, Suchzhi-Bulak collective, Aug. 4, 1943—Yun.).

IC. Qaidam: *plains* (Syrtyn valley, Ikhin-Shirik area, about 3300 m, June 20, 1895—Rob.); *hilly* (Ichegyn-Gol river, about 3300 m, June 19, 1895—Rob.).

IIA. Junggar: *Tien Shan* (near Nan'shan'kou village, May 25, 1877—Pot.; near Nilki village, 3300 m, June 17; near Zagastai-Gol river, 3000 m, Sept. 5—1879, A. Reg.; B. Yuldus basin 30-35 km south-west of Bain-Bulak settlement, Aug. 10, 1958—Yun.).

IIIA. Qinghai : *Nanshan* (sharagol'dzhin river valley, Khuitun area, 4000 m, June 12; same site, Yayakha area, 3300 m, June 13—1894, Rob.).

IIIB. Tibet: *Chang Tang* (nor. slope of Przewalsky mountain range, about 4600 m, Aug. 24, 1890—Rob.); *Weitzan* (south. bank of Orin-Nur lake, 4500 m, July 21; nor. slope of Burkhan-Budda mountain range, 3800–5000 m, Aug. 2—1884, Przew.; valley of Alyk-Norin river, 4000 m, June 7; nor. slope of Burkhan-Budda mountain range, Khatu gorge, 3500 m, June 19—1901, Lad.).

IIIC. Pamir (Tagarma valley, July 23, 1913—Knorring).

General distribution: Cent. Tien Shan, East. Pam.; West. Sib. (Altay), East. Sib. (south-west.), Nor. Mong. (Hang.), Himalayas (west.).

Section 2. Poa

3. **P. angustifolia** L. Sp. pl. (1753) 67; Roshev. in Fl. SSSR, 2 (1934) 388; Persson in Bot. notiser (1938) 274; Fl. Kirgiz. 2 (1950) 130; Grubov, Consp. fl. MNR (1955) 70; Fl. Tadzh. 1 (1957) 157; Keng, Fl. ill. sin., Gram. (1959) 154; Bor, Grasses Burma, Ceyl., Ind. and Pakist. (1960) 555; Pazii in Bot. mater. Gerb. Inst. bot. AN Uzbek SSR, 17 (1962) 30; Ikonnikov, Opred. rast. Pamira [Key to Plants of Pamir] (1963) 58.—*P. pratensis* var. *angustifolia* (L.) Smith, Fl. Brit. (1800) 105; ? Franch. Pl. David. 1(1884) 337; ? Forbes and Hemsley, Index Fl. Sin. 3 (1904) 426, Krylov, Fl. Zap. Sib. 2 (1928) 298. —Ic.: Fl. SSSR, 2 , Plate 20, figs. 2 and 3; Keng, l.c. fig. 102.

Described from Europe. Type in London (Linn.).

In steppes, arid meadows, on rocky slopes, in riverine sand and pebble beds; up to upper mountain belt.

IA. Mongolia: *East. Mong.* (near Kharkhonte railway station, June 7, 1902—Litw.; "Doyen, June 29, 1934, leg. Eriksson"-Norlindh, l.c.; near Hailar town, June 11, 1951—Li et al.).

IB. Kashgar: *West.* ("Jerzil, 2800 m, July 16, 1930"—Persson, l.c.; near Upal village, 1800 m, No. 205, June 10, 1959—Lee et al.).

IIA. Junggar: *Alt. region, Jung. Alt.* (near Ven'tsyuan', 2400 m, No. 1436, Aug. 14, 1957—Huang); *Tien Shan, Jung. Gobi* (south: near Shikhedzy settlement, No. 664, June 6, 1957—Huang; near Staryi Kuitun settlement 2–3 km east of Kuitun state farm, June 30, 1957—Yun.); *Dzhark.*

IIIC. Pamir (Ulug-Tuz gorge in Charlym river basin, June 27, 1909—Divn.; "Aj-Bolong, 3400 m, 20 VII, 1930"—Persson, l.c.).

General distribution: Aral-Casp., Balkh. region, Jung.-Tarb., Nor. and Cent. Tien Shan, East. Pam.; Europe, Mediterr., Balk.-Asia Minor, Near East, Caucasus, Middle Asia, West. Sib., East. Sib. (south.), Nor. Mong. (Hang., Mong.-Daur.), China (Dunbei, Nor., Nor.-West.), Himalayas (west., Kashmir), Korea, Japan, Nor. Amer.

4. P. asperifolia Bor in Kew Bull. (1952) 130; ej. Grasses Burma, Ceyl., Ind. and Pakist. (1960) 556.

Described from Tibet. Type in London (K).

IIIB. Tibet: *South* ("Pembu La, 10–15 miles north of Lhasa, IX 1904, leg. Walton"—Bor, l.c., type!).

IIIC. Pamir: (Ulug-Tuz gorge in Charlym river basin, June 28, 1909—Divn.).

General distribution: Himalayas (west.).

5. P. calliopsis Litv. in Fl. SSSR, 2 (1934) 414 and 755; Persson in Bot. notiser (1938) 275; Fl. Kirgiz. 2 (1950) 132; Fl. Tadzh. 1 (1957) 160; Bor, Grasses Burma, Ceyl., Ind. and Pakist. (1960) 556; Pazii in Bot. mater. Gerb. Inst. bot. AN UzbekSSR, 17 (1962) 30; Ikonnikov, Opred. rast. Pamira [Key to Plants of Pamir] (1963) 58.— *?P. phariana* Bor in Kew Bull. (1948) 141; id. l.c. (1960) 559. —Ic. : Fl. SSSR, 2, Plate 21, fig. 11; Fl. Tadzh. 1, Plate 19.

Described from East. Pamir. Type in Leningrad.

In grasslands and riverine pebble beds; in upper mountain belt.

IIIB. Tibet: *Chang Tang* (south of Kerii along Kyuk-Egil' river, 4300 m, June 28, 1885—Przew.; nor. slope of Przewalsky mountain range, 4600–5000 m, Aug. 20, 1890—Rob.; upper course of Tiznaf river 3–4 km east of Saryk-Daban settlement, 4800 m, June 4, 1959—Yun.); *Weitzan* (left bank of Dychu river in Huang He basin, 4300–5000 m, June, 1884—Przew.; nor. slope of Amnen-Kor mountain range, 4150 m, June 4; water divide between Yangtze and Mekong, Gur-La pass, 5000 m, Aug. 25–1909, Lad.; Yantgze basin, Enitak area on left bank of Yalun'tszyan, 4100 m, May 5; Alyk-Norin river valley, 4030 m, June 7—1901, Lad.).

IIIC. Pamir.

General distribution: Cent. Tien Shan, East. Pam.; Middle Asia (Pam. Alay), China (south-west.), Himalayas.

Note. Part of the Pamir and almost all the Tibetan specimens of this species differ from typical specimens in glumes with slightly narrower membranous margin, and possibly belong to a distinct, although very weakly differentiated, ecogeographical race for which the name *P. phariana* Bor (l.c.) probably is correct. One of the specimens cited above (from Gur-La pass) represents a viviparous variety of the species *P. calliopsis* var. *vivipara* Tzvel. var. nova.—A varietate typica spiculis viviparis differt. Typus : "Inter systemata fl. Jantze et Mekong, in trajecto Gur-La, 5000 m, No. 493, 25 VIII 1900, V. Ladygin" in Herb. Inst. Bot. Ac. Sci. URSS—LE conservatur.

O **P. glauca** Vahl. Fl. Dan, 6, 17 (1790) 3; Gubanov, Consp. Fl. Outer Mong. (1996) 22.

IA. Mongolia. *Mong. Alt.*

P. kenteica Ivanova in Bot. mater. Gerb. Bot. inst. AN SSSR, 7 (1937) 278; Grubov, Consp. fl. MNR (1955) 70.—*P. turczaninvoii* Serg. in Sist. zam. Gerb. Tomsk. gos. univ. 83 (1965) 11.

Described from Nor. Mongolia (Hentey mountain range). Type in Leningrad.

In grasslands and riverine pebble beds; in bald peak zone of hills. Found in border region of Nor. Mongolia (Hentey).

General distribution: East. Sib. (south.: Sokhondo bald peak), Nor. Mong. (Hent.).

Note. This narrowly endemic species is evidently of hybrid origin from *P. arctica* R. Br. x *P. altaica* Trin.

6. **P. lipskyi** Roshev. in Izv. Bot. sada AN SSSR, 30 (1932) 303; Roshev. in Fl. SSSR, 2 (1934) 421; Fl. Kirgiz. 2 (1950) 135; Fl. Kazakhst. 1 (1956) 237; Fl. Tadzh. 1 (1957) 171; Pazii in Bot. mater. Gerb. Inst. bot. AN Uzbek SSR, 17 (1962) 33; Ikonnikov, Opred. rast. Pamira [Key to Plants of Pamir] (1963) 59.—*P. dschungarica* Roshev. l.c. (1932) 778; Roshev, l.c. (1934) 415; Fl. Kirgiz. 2 (1950) 133; Fl. Kazakhst. 1 (1956) 235; Fl. Tadzh. 1 (1957) 174; Ikonnikov, l.c. 59.—*P. taldyksuensis* Roshev. in Fl. Kirgiz. 2 (1950) 129, diagn. ross.—*P. kungeica* Golosk. in Vestn. AN Kaz. SSR, 1 (1955) 73; Fl. Kazakhst. 1 (1956) 237. —Ic.: Fl. SSSR, 2, Plate 31, fig. 12.

Described from Junggar Alatau. Type in Leningrad.

On talus, rocky and rubble slopes, grasslands, riverine pebble beds; in upper mountain belt.

IIA. Junggar: *Jung. Alt.* (below Koketau pass, July 21, 1909—Lipsky; Toli district, 2400 m, No. 1231, Aug. 6; west of Ven'tsyuan, 2640 m, No. 2031, Aug. 25—1957, Huang); *Tien Shan.*

IIIB. Tibet: *Chang Tang* (south of Cherchen settlement, 4190 m, No. 9423, June 4, 1959—Lee et al.).

IIIC. Pamir (Kenkol gorge, near Togai-Bashi village, July 30, 1913—Knorring; Kara-Dzhilga river basin, Gon-Arek tributary, 4000–4500 m, July 22, 1942—Serp.).

General distribution: Jung.-Tarb., Nor. and Cent. Tien Shan, East. Pam.; Middle Asia (Pamir-Alay).

Note: Aside from typical specimens of this species having lemma pilose between ribs, those with lemma glabrous between ribs are found within China as well as in the USSR. The latter specimens are sometimes treated as independent species *P. dschungarica* Roshev. (l.c.). It is more likely, however, that this characteristic in the present case, as in that of the closely related speices *P. smirnovii* Roshev. and *P. malacantha* Kom., is not of much taxonomic significance. Himalayan species *P. polycolea* Stapf is quite closely related to *P. lipskyi*.

7. **P. megalothyrsa** Keng, Fl. ill. sin., Gram. (1959) 209, diagn. sin. —Ic.: Keng, l.c. fig. 164 (type!). Planta perennis, 40–80 cm alta, innovationes breviter repentes emittens; culmi et vaginae glabrae et laeves; ligulae 2–3.5 mm lg.; laminae 1–2.5 mm lt., planae, supra scaberulae, subtus sublaeves. Paniculae 10–18 cm lg., late effusae, ramis paulo scaberulis; spiculae 5–7 mm lg., 3–6 florae, roseo-violaceo tinctae; glumae late lanceolatae, trinerviae, quam lemmata subsesqui breviores; lemmata 4–5.2 mm lg., in parte inferiore secus nervos breviter pilosa, callo glabro; paleae secus

carinas in dimidio superiore scabrae, infra glabrae et laeves; antherae 2-3 mm lg.

Described from South-West. China. Type ("Sikang, No. 5186"—keng, l.c.) in Nanking.

In thin forests, forest glades; in middle and upper mountain belts.

IIIB. Tibet: *Weitzan* (Yangtze basin, along Khichu river near Kabchzha-Kamba village, among pines, 4000 m, July 20, 1900—Lad.).
General distribution: China (South-West.).

Note. This species falls in a large group of relatively primitive East Asian and North American species of the genus (*P. stenantha* Trin., *P. platyantha* Kom. and others). Among them, those of high altitudes are *P. lipskyi* Roshev., *P. smirnovii* Roshev. and *P. malacantha* Kom.

8. **P. pratensis** L. Sp. pl. (1753) 67; Franch. Pl. David. 1 (1884) 337, p.p.; ?Hemsley, Fl. Tibet (1902) 204; Forbes and Hemsley, Index Fl. Sin. 3 (1904) 426; Krylov, Fl. Zap. Sib. 2 (1928) 297, p.p.; Pavlov in Byull. Mosk. Obshch. ispyt. prir. 38 (1929) 23; Pampanini, Fl. Carac. (1930) 76; Roshev. in Fl. SSSR (1934) 388; Fl. Kirgiz. 2 (1950) 129; Grubov, Consp. fl. MNR (1955) 71; Fl. Kazakhst. 1 (1956) 229, p.p.; Fl. Tadzh. 1 (1957) 155; Keng, Fl. ill. sin., Gram. (1959) 152; Bor, Grasses Burma, Ceyl., Ind. and Pakist. (1960) 559; Pazii in Bot. mater, Gerb. Inst. bot. AN Uzbek SSR, 17 (1962) 30; Ikonnikov, Opred. rast. Pamira [Key to Plants of Pamir] (1963) 58.—?*P. stenachyra* Keng, l.c. 155, diagn. sin.—?*P. dolichachyra* Keng, l.c. 157, diagn. sin. —Ic.: Fl. SSSR, 2, Plate 29, fig. 1; Keng, l.c. fig. 101.

Described from Europe. Type in London (Linn.).

In meadows, riverine sand and pebble beds, along banks of water reservoirs, in thin forests, along roadsides and around irrigation ditches; up to upper mountain belt.

IA. Mongolia: *Khobd., Mong. Alt., Cis-Hing., Cent. Khalkha* (Ubur-Dzhirgalante river between Bogat and Agit hills, Sept. 15, 1925—Krasch. and Zam.; hill range intersecting Dalan-Dzadagad road 72 km south of Ulan Bator, Aug. 8, 1951—Kal.). *East. Mong.* (Ourato, July—David; along Kerulen river, June 9, 1899—Pot. and Sold.; near Hailar town—collections of several collectors); *Bas. Lakes* (near Ulangom, June 20, 1879—Pot.); *Gobi-Alt.* (Ikhe-Bogdo, Dzun-Saikhan and Dundu-Saikhan mountain ranges); *Alash. Gobi* (in central part of Alashan hill, June 13, 1873—Przew.; near Dynyuanin [Bayan-Khoto] oasis, June 3, 1908—Czet.).
IIA. Junggar: *Alt. region, Jung. Alt., Tien Shan, Jung. Gobi* (south.: Datszymyao village in Savan district, No. 1730, July 22; east.: in Beidashan'[Baitak-Bogdo] hills, 1700 m, No. 5224, Sept. 28—1957, Huang).
IIIA. Qinghai: *Nanshan* (in Yuzhno-Tetungsk mountain range forest belt, July 14, 1872—Przew.; Ushilin pass, 3325 m, July 14 1875—Pias.; Xining hills along Myndansha river, June 14, 1890—Gr.-Grzh.).
IIIC. Pamir (Kashkasu area in Ilyksu river valley, 4600 m, June 17, 1901—Alekseenko; on west. extremity of Tashkurgan, June 13, 1959—Yun.).
General distribution: Aral-Casp., Balkh. region, Jung.-Tarb., Nor. and Cent. Tien Shan, East. Pam.; Arct. (Europ.), Europe, Mediterr., Balk.-Asia Minor, Near East, Caucasus, Middle Asia (hilly), West. Sib., East. Sib., Far East, Nor. Mong., China

(Dunbei, Nor., Nor.-West., South-West.), Himalayas, Korea, Japan (introduced ?), Nor. Amer.; introduced in other countries.

Note. We append to this species *P. stenachyra* Keng (l.c.) and *P. dolichachyra* Keng (l.c.) described from Qinghai (so far only in Chinese). Their differences from *P. pratensis* are not clear to us.

9. **P. pruinosa** Korotky in Feddes repert. 13 (1915) 291; Roshev. in Fl. SSSR, 2 (1934) 394; Grubov, Consp. fl. MNR (1955) 71.—*P. sajanensis* Roshev. in Izv. Bot. Sada AN SSSR, 30 (1932) 774; Roshev., l.c. (1934) 393.—*P. pamirica* Roshev. l.c. (1934) 414; Fl. Kirgiz. 2 (1950) 132; Fl. Tadzh. 1 (1957) 158; Pazii in Bot. mater. Gerb. Inst. bot. AN Uzbek SSR,17 (1962) 33; Ikonnikov, Opred. rast. Pamira [Key to Plants of Pamir] (1963) 58.—*P. alpigena* auct. non Lindm.: Keng, Fl. ill. sin., Gram. (1959)155; ? Bor, Grasses Burma, Ceyl., Ind. and Pakist. (1960) 555. —**Ic.**: Fl. SSSR, 2, Plate 31, fig. 10; Keng, l.c. fig. 104.

Described from Transbaikal region (Eravin lake). Type in Leningrad.

In solonetz meadows, riverine sand and pebble beds, along banks of water reservoirs; up to upper mountain belt.

IA. Mongolia: *Khobd.* (Bukhu-Muren river valley 5–6 km beyond Bukhu-Muren somon, July 31, 1945—Yun.); *Mong. Alt., Bas. lakes* (bank of springs near Ubsu-Nur lake, July 10, 1880—Pot.); *Gobi-Alt.* (near peak of Ikhe-Bogdo mountain range, Sept. 11, 1943—Yun.).

IB. Kashgar: *West.* (nor. slope of Kingtau mountain range 3–4 km south-east of Kosh-Kulak settlement, June 10, 1959—Yun.).

IIA. Junggar: *Jung. Alt.* (Toli district, No. 1074, Aug. 6; between Ven'tsyuan and Taldy, 2300 m, No. 2091, Aug. 26—1957, Huang); *Tien Shan* (source of Khorgos river, 3000–3600 m, Aug. 10 and 11, 1878—A. Reg.; basin of B. Yuldus 30–35 km south-west of Bain-Bulak settlement, Aug. 10, 1958—Yun.).

IIIB. Tibet: *South.* (Khambajong, July 8–10, 1903—Young-husband).

IIIC. Pamir

General distribution: Jung.-Tarb., Nor. and Cent. Tien Shan, East. Pam.; Middle Asia (Pam.-Alay), East. Sib. (south.), Nor. Mong. (Fore Hubs., Hang., Mong.-Daur.), Himalayas (west.).

Note. Types *P. pruinosa* Korotkyi and *P. pamirica* Roshev are entirely similar while type *P. sajanensis* Roshev. (from Tunkin bald peaks) has a somewhat more mesophilic habit and brighter-coloured spikelets, taking it considerably closer to some forms of polymorphic arctic species *P. alpigena* (Fries) Lindm. It is quite possible, therefore, that *P. sajanensis* forms an independent, although poorly distinguished, ecogeographical race intermediate to *P. alpigena*. To this race can also be assigned the solitary specimen from Mong. Altay: "Tolbo-Kungei-Alatau mountain range, 3200 m, cobresia meadows, 5 VIII 1945—Yun."

o **P. raduliformis** Probat. in Novit. Syst. Pl. Vasc. 8 (1971) 25; Gubanov. Consp. Fl. Outer Mong. (1996) 22.

IA. Mongolia. *Mong. Alt.*

o **P. sabulosa** (Turcz. ex Roshev.) Roshev. in Fl. SSSR, 2 (1934) 394; Gubanov, Consp. Fl. Outer Mong (1996) 22.

1A. Mongolia. *Mong. Alt., Cen. Khalkha.*

o P. schischkinii Tzvel. in Novit. Syst. Pl. Vasc.11 (1974) 32; Gubanov, Consp. Fl. outer mong. (1996) 22.

1A. Mongolia. *Mong. Alt., Val. Lakes.*

o P. sergievskajae Probat. in Novit. Syst. Pl. Vasc. 8 (1971)28; Gubanov, Consp. Fl. Outer Mong. (1996) 22.

1A. Mongolia. *Cen. Khalkha.*

10. P. smirnovii Roshev. in Izv. Glavn. bot. sada SSSR, 28 (1929) 381; Roshev. In Fl. SSSR, 2 (1934) 424; Grubov, Consp. fl. MNR (1955) 71 .—*P. mariae* Reverd. in Sist. zam. Gerb. Tomsk. univ. 3–4 (1933) 2; Roshev, l.c. (1934) 396.—*P. arctica* subsp. *smirnovii* (Roshev.) Malysch. Vysokogorn. fl. Vost. Sayana (1965) 65.—*P. dschungarica* auct. non Roshev.: Grubov, l.c. 70.

Described from East. Siberia (East. Sayan—Tunkin bald peaks). Type in Leningrad.

In grasslands, riverine sand and pebble beds; in middle and upper (bald peak) mountain belts.

1A. Mongolia: *Khobd.* (near peak of Kharkira, July 12, 1879—Pot.); *Mong. Alt.* (along Aksu river [Belaya Kobdo], July 22, 1909—Sap.).

General distribution: West. Sib. (east. Altay), East. Sib. (Sayans), Nor. Mong. (Fore Hubs., Hang.).

Note. The specimens listed above lack pubescence between ribs of lemma, characteristic of typical *P. smirnovii.* Such specimens were described from West. Sayan as an independent species, *P. mariae* Reverd. (l.c.), nevertheless compelling recognition that even among close species *P. lipskyi* Roshev. and *P. malacantha* Kom., pubescence between ribs of lemma is not a regular feature.

11. P. tangii Hitchc. in Proc. Biol. Soc. Washington, 43 (1930) 94; Keng, Fl. ill. sin., Gram. (1959) 204. —Ic.: Keng, l.c. fig. 158.

Described from Nor. China (Shanxi province). Type in washington. Isotype in Leningrad.

In sand and pebble beds of river valleys, on rocky slopes and talus; in midmountain belt.

1A. Mongolia: *Alash. Gobi* (Oulachan, No. 2689, June 1866—David; Alashan mountain range, Tszosto gorge, May 12, 1908—Czet.).

General distribution: China (Nor., Nor.-West.).

oP. veresczaginii Tzvel. 1974, in Novit. Syst. Pl. Vasc. 11 (1974) 34; Gubanov, Consp. Fl. Outer Mong. (1996) 22.

1A. Mongolia. *Mong. Alt.*

Section 3. Homalopoa Dum.

12. **P. remota** Forsell. in Act. Inst. Linn. Upsal. 1 (1807) 1; Krylov, Fl. Zap. Sib. 2 (1928) 295; Roshev. in Fl. SSSR, 2 (1934) 385; Fl. Kazakhst. 1 (1956) 228; Pazii in Bot. mater. Gerb. Inst. bot. AN Uzbek, SSR, 17 (1962) 29.

Described from Sweden. Type in Uppsala.

In spruce and larch forests, marshes and along banks of brooks; in midmountain belt.

IIA. **Junggar:** *Tarb.* (larch forest in Khobuk river valley, July 20, 1914—Sap.); *Tien Shan* (Bogdo mountain range, 2300–2700 m, July 1878—A. Reg.).

General distribution: Jung.-Tarb., Nor. Tien Shan; Europe (cent. and east.), Caucasus (Greater Caucasus), West. Sib.

Section 4. Macropoa Hermann

13. **P. sibirica** Roshev. in Izv. Peterb. bot. sada, 12 (1912) 121; Roshev. in Fl. SSSR, 2 (1934) 380; Krylov, Fl. Zap. Sib. 2 (1928) 299; Pavlov in Byull. Mosk. obshch. ispyt. prir. 38 (1929) 23; Fl. Kirgiz. 2 (1950) 127; Grubov, consp. fl.MNR (1955) 71; Fl. Kazakhst. 1(1956) 228; Keng, Fl. ill. sin., Gram. (1959) 159; Pazii in Bot. mater. Inst. bot. AN Uzbek. SSR, 17 (1962) 31, excl. syn. —Ic.: Fl. SSSR, 2, Plate 28, fig. 11; Keng, l.c. fig. 110.

Described from East. Siberia. Lectotype in Leningrad.

In meadows, thin forests, among shrubs, sometimes in riverine pebble beds; up to midmountain belt.

IA. **Mongolia:** *Khobd.* (along Tszusylan river, July 3; upper course of Kharkira river, July 10—1879, Pot.); *Mong. Alt., Cis-Hing.* (near Irekte railway station, June 15, 1902—Litw.); *Gobi-Alt.* (Bityuten-Ama creek valley, Ikhe-Bogdo mountain range, Aug. 12, 1927—Simukova).

IIA. **Junggar:** *Tarb.* (south. Slope of Saur mountain range, valley of Karagaitu river, Bain-Tsagan creek valley, June 23, 1957—Yun.); *Jung. Alt.* (15 km nor.-west of Ven'tsyuan', 2000 m, No. 1621, Aug. 29, 1957—Huang); *Tien Shan* (B. Yuldus basin, Sept. 1878—Fetisov).

General distribution: Jung.-Tarb., Nor. and Cent. Tien Shan; Europe (nor.-east.), Middle Asia (West. Tien Shan), West. Sib., East. Sib., Far East., Nor. Mong. (Fore Hubs., Hent., Hang.), China (Dunbei), Korea.

Section 5. Stenopoa Dum.

14. **P. albertii** Regel in Acta Horti Petrop. 7 (1880) 611, quoad. pl. songor.; Roshev. in Fl. SSSR, 2 (1934) 412, p.p.; Fl. Kirgiz. 2 (1950) 132; Fl. Kazakhst. 1 (1956) 234.—*P. juldusicola* Regel, l.c. 612.—*P. litwinowiana* Ovcz. in Izv. Tadzh. bazy AN SSSR, 1, 1 (1932) 22; Roshev, l.c. 417; Fl. Kirgiz. 2 (1950) 134; Fl. Kazakhst. 1 (1956) 236; Fl. Tadzh. 1 (1957) 151;

Tzvel. in Bot. mater. Gerb. Bot. inst. AN SSSR, 20 (1960) 418; Bor, Grasses Burma, Ceyl., Ind. and Pakist. (1960) 558; Pazii in Bot. mater. Gerb. Inst. bot. AN Uzbek. SSR, 17 (1962) 27; Ikonnikov, Opred. rast. Pamira [Key to Plants or Pamir] (1963) 61.—*P. glauciculmis* Ovcz. l.c. 19; Roshev, l.c. 402; Fl. Tadzh. 1(1957) 154; Ikonnikov in Dokl. AN Tadzh, SSR, 20 (1957) 56; id. l.c. (1963) 61.—*P. marginata* Ovcz. l.c. 24; Roshev., l.c. 417; Fl. Kirgiz. 2 (1950) 134; Fl. Kazakhst.1 (1956) 236; Fl. Tadzh. 1 (1957) 149.—*P. densissima* Roshev. ex. Ovcz. l.c. 26; Roshev., l.c. 418; Fl. . Kirgiz. 2 (1950) 135.—*P. lapidosa* Drob. in Fl. Uzbek. 1 (1941) 538; Fl. Kirgiz. 2 (1950) 134.— *P. indattenuata* Keng, Fl. ill. sin., Gram. (1959) 198, diagn. sin.—? *P. crymophila* Keng, l.c. 200, diagn. sin.—*P. attenuata* auct. non Trin.: Deasy, In Tibet and Chin. Turk. (1901) 405; Hemsley, Fl. Tibet (1902) 204; Pilger in Hedin, S. Tibet, 6, 3 (1922) 94; Pampanini, Fl. Carac. (1930) 76; ? Hao in Engler's Bot. Jahrb. 68 (1938) 581; ? Walker in Contribs. U.S. Nat. Herb. 28 (1941) 598.—*P. poiphagorum* auct. non Bor: Keng, l.c. 200.—Ic.: Fl. SSSR, 2, Plate 31, fig. 3.

Described from Junggar Alatau. Lectotype ("Dschungarischer Alatau, 2300–2700 m, VIII, 1878, A. Regel") in Leningrad.

In grasslands, on rocky and rubble slopes, riverine pebble beds, on rocks; in upper mountain belt.

IA. **Mongolia:** *Mong. Alt.* (Adzhi-Bogdo mountain range, water divide between Ara-Tszuslan and Ikhe-Gol, rubble-rocky talus, Aug. 7, 1947—Yun.); *Gobi-Alt.* (Ikhe-Bogdo mountain range, midportion of Narin-Khurmit creek valley, June 28, 1941—Yun.

IB. **Kashgar:** *Nor.* (Sogdyn-Tau mountain range 10–12 km south-east of Akchii hydraulic power station on Kokshaal-Darya, 2900 m, Sept. 19, 1958—Yun); *West., South.* (upper course of Chira river 15–18 km south of Uku settlement, Sary-Bulyk area, May 19, 1959—Yun.).

IIA. **Junggar:** *Jung. Alt., Tien shan.*

IIIA. **Qinghai:** *Nanshan, Amdo* ("auf dem Höhenplateau Da-ho-ba, 4000 m, 28 VIII 1930"—Hao, l.c.; 20 km west of Gunkhe town, 2980 m, Aug. 6, 1959—Petr.).

IIIB. **Tibet:** *Chang Tang, Weitzan, South* (Khambajong, No. 141, July 26, 1903—Younghusband).

IIIC. *Pamir.*

General distribution: Jung.-Tarb., Nor. and Cent. Tien Shan, East. Pam.; Middle Asia (West. Tien Shan, Giss.-Darv., Pam.-Alay), China (Nor.-West., South.-West.), Himalayas.

Note. V.K. Pazii (l.c. 38–39) regards the specimen of *P. pratensis* L. from the vicinity of Sairam lake as the lectotype of *P. albertii*. It does not fully confirm with the original diagnosis of *P. albertii*, which makes no mention at all of the presence of a tuft of undulating hairs on the callus of the lemma. At the same time, there is every justification for regarding as lectotype of this species the specimen cited by us from Junggar Alatau (having isotype in addition), which evidently was overlooked by V.K. Pazii and which even R. Roshevitz actually treated as the lectotype of *P. albertii*. Small specimens of *P. attenuata* Trin. occasionally found without tuft of undulating hairs on callus of lemma, described from Transbaikal region under the name *P. dahurica* Trin., are very similar to *P. albertii*. However, the possible similarity of *P. albertii* with *P. dahurica*, remains an open question, in our opinion, until a more detailed treatment is carried out on abundant material of the group *P. ochotensis* Trin. s.l. from South.

Siberia. Specimens from Junggar (East. Tien Shan, near Danu pass, 3800 m, No. 558, July 23, 1957—Huang) and Weitzan (Yangtze river basin, valley of Rhombo-Mtso lake, Nkogun river, 4500 m, Aug. 6, 1900—Lad.) belong to the viviparous variety described from Shanxi province—*P. albertii* var. *vivipara* (Rendle) Tzvel. comb. nova [=*P. attenuata* var. *vivipara* Rendle in Forbes and Hemsley, Index Fl. Sin. 3 (1904) 423]. This variety was cited before for Qinghai as well (Walker, l.c.).

15. **P. altaica** Trin. in Ledeb. Fl. Alt. 1(1829) 97; Krylov, Fl. Zap. Sib. 2 (1928) 290; Pavlov in Byull. Mosk. Obshch. ispyt. prir. 38 (1929) 23; Roshev. in Fl. SSSR, 2 (1934) 290; Grubov, consp. fl. MNR (1955) 70; Fl. Kazakhst. 1 (1956) 230.—*P. tristis* Trin. in Mém. Ac. Sci. St. - Pétersb. Sav. Etrang. 2 (1835) 528; Krylov, l.c. 288; Roshev., l.c. 400; Fl. Kazakhst. 1 (1957) 231.—*P. caesia* auct. non Smith: Danguy, Bull. Mus. nat. hist. natur. 20 (1914) 147.—Ic.: Ledeb. Ic. pl. fl. ross. 3 (1831) Table 225; Fl. SSSR, 2, Plate 30, fig. 2.

Described from Altay. Type in Leningrad.

In grasslands, on rocky and rubble slopes, rocks; in upper (bald peak) belt.

IA. **Mongolia:** *Khobd.* (near Kharkira peak, July 24, 1879—Pot.); *Mong. Alt., Gobi-Alt.* (Ikhe-Bogdo mountain range, June 29–30, 1945—Yun.; same site, 3600–3700 m, July 29, 1948—Grub.).

IIA. **Junggar:** *Alt. region, Tarb.* (south. slope of Saur mountain range, valley of Karagaitu river, Bain-Tsagan creek valley, June 23, 1957—Yun.); *Tien Shan* (Bogdo mountain range, 3300 m, July 25, 1878; Kumbel' river, 3000–3300 m, May 31, 1879—A. Reg.; Manas river basin, July 18–23, 1957—Yun.; same site, Nos. 425 and 558, July 18 and 23, 1957—Huang; Narat hill in Kunges region, 3000 m, No. 9131, Aug. 9, 1958—Lee and Chu).

General distribution: West. Sib. (Altay), East. Sib. (Sayans), Nor. Mong. (Fore Hubs., Hent., Hang.).

Note. This species is very close to the Arctic species *P. glauca* Vahl, which replaces *P. altaica* on bald peaks of South. Siberia and Cent. Asia.

16. **P. argunensis** Roshev. in Fl. SSSR, 2 (1934) 404 and in Tr. Bot. inst. AN SSSR, ser. 1,2 (1936) 98; Keng, Fl. ill. sin., Gram. (1959) 197; Krylov, Fl. Zap. Sib. 12, 1 (1961) 3107.—*P. sphondylodes* auct. non Trin.: Danguy in Bull. Mus. nat. hist. natur. 20 (1914) 147. —Ic.: Fl. SSSR, 2, Plate 30, fig. 11; Keng, l.c. fig. 149.

Described from Transbaikal region. Lectotype (valley of Sukhoi Urulyungui near Orabuduk lake, July 9, 1930, Sharova et al.) in Leningrad.

In steppes, dry meadows, predominantly in sandy and loamy soils; up to lower mountain belt.

IA. **Mongolia:** *Cis-Hing.* (valley of Khalkha-Gol river 13 km south-east of Khamar-Daba, Aug. 11, 1949—Yun.); *Cent. Khalkha, East. Mong.* (nor.), *Bas. lakes* (Borig-Del' sand south-east of Bain-Nur lake, July 25, 1945—Yun.; 40 km south of Ulyasutai along road to Tsagan-Olom, sand in Shurygin-Gol area, July 17, 1947—Yun.); *Gobi-Alt.* (Noyan-Bogdo hills 4–5 km south of Noyan somon, July 25, 1943—Yun.); *West. Gobi* (Atas-Bogdo hills, Aug. 12, 1943—Yun.).

General distribution: West. Sib. (Altay), East. Sib. (south.), Nor. Mong. (Mong.-Daur.), China (Dunbei).

17. **P. attenuata** Trin. in Mém. Ac. Sci. St.-Pétersb. Sav. Etrang. 2 (1835) 527; Forbes and Hemsley, Index Fl. Sin. 3 (1904) 423, p.p.; Krylov, Fl. Zap. Sib. 2 (1928) 285, p.p.; Pavlov in Byull. Mosk. obshch. ispyt. prir. 38 (1929) 21; Roshev. in Fl. SSSR, 2 (1934) 403; Grubov, consp. fl. MNR (1955) 70; Fl. Kazakhst. 1(1956) 233; pazii in Bot. mater. Gerb. Inst. bot. AN Uzbek SSR, 17 (1962) 24.—*P. dahurica* Trin. in Mém. Ac. Sci. St. Pétersb., sér. VI, 4 (1836) 63; Roshev., l.c. 404; Kitag. Lin. Fl. Mansh. (1939) 87.—*P. attenuata* var. *dahurica* (Trin.) Griseb. in Ledeb. Fl. Ross. 4 (1852) 371.—*P. serotina* var. *botryoides* Trin. ex Griseb. l.c. 375.—*P. botryoides* (Griseb.) Roshev. in Fl. Zabaik. 1(1929) 83; id. l.c. (1934) 403; Pavlov, l.c. 22; Grubov, l.c. 70; Krylov, l.c. 12, 1(1961) 3106.—? *P. sinoglauca* Ohwi in J. Jap. Bot. 19 (1943) 169; Keng, Fl. ill. sin., Gram. (1959) 196.—*P. sphondylodes* var. *dahurica* (Trin.) Meld. in Norlindh, Fl. mong. steppe, 1 (1949) 99.—*P. sinattenuata* Keng, l.c. 189, diagn. sin.—*Sesleria pavlovii* Litv. in Byull. Mosk. obshch. ispyt. prir. 38 (1929) 24. —Ic: Fl. SSSR, 2, Plate 30, figs 8–9.

Described from Altay. Type in Leningrad

In steppes, arid, usually rather solonetz meadows, on rocky and rubble slopes and rocks; up to upper mountain belt.

IA. **Mongolia:** *Khobd., Mong. Alt., Cis-Hing.* (near Yaksha railway station, June 13, 1902—Litw.; same site, 1954—Wang); *Cent. Khalkha, East. Mong., Bas. lakes* (Margatsa hill 4–5 km south-east of Santu-Margats somon, Aug. 19, 1944—Yun.); *Gobi Alt., East. Gobi* (Muni-Ula hills, June, 1871—Przew.; "Khujirtu-gol, June 19; Khonin-Chaghan-Chölo-gol, 1 VIII 1927, leg. Hummel"—Meld. l.c.; near Dalan-Dzadagad, June 17, 1945—Yun.); *Alash. Gobi* (Alashan mountain range, Yamata gorge, May 7; same site, Tszosto gorge, May 12—1908, Czet.; Beisy monastery 30 km east of Bayan-Khoto, July 4, 1957—Petr.).

IB. **Kashgar:** *Nor.* (valley of Khanga river 25 km nor.-west of Balinte settlement on Karashar-Yuldus road, Aug. 1; valley of Muzart river 10–12 km above its discharge into Baisk basin, Chokarpa area, Sept. 7—1958, Yun.).

IIA. **Junggar:** *Alt. region, Tarb., Jung. Alt., Tien Shan, Jung. Gobi* (east.: nor. slope of Baga-Khabtak-Nuru mountain range, about 2000 m, Sept. 14, 1948—Grub.).

General distribution: Jung.-Tarb.; West. Sib. (Altay), East. Sib., Nor. Mong., China (Dunbei, Nor.).

Note. Typical specimens of this species are overall small in size with smooth, or almost smooth stems; they are found in relatively very high mountains and are quite rare in Cent. Asia (as in the rest of the distribution range of the species). The low-mountain variety *P. attenuata* var. *botryoides* (Griseb.) Tzvel. comb. nova (=*P. serotina* var. *botryoides* Trin. ex Griseb. l.c.) with strongly scabrous stems and leaf blades is far more widely distributed. Specimens with totally reduced tuft of undulating hairs on callus of lemma, *P. attenuata* var. *dahurica* (Trin.) Griseb. (l.c.), are found occasionally; they are considerably closer to *P. albertii* Regel.

18. **P. elanata** Keng, Fl. ill. sin., Gram. (1959) 181, diagn. sin. —Ic.: Keng, l.c. fig. 133 (type!)—Planta perennis, 30–70 cm alta, sat dense caespitosa;

culmi et vaginae laeves vel scabriusculae; ligulae 3–6.5 mm lg.; laminae 1–2 mm lt., utrinque scabrae. Paniculae 6–10 cm lg., subeffusae, ramis longis scaberrimis; spiculae 3–5 mm lg., 2–4-florae, rachilla scabriuscula; glumae lanceolatae, trinerviae, quam lemmata paulo breviores; lemmata 3–4 mm lg., lanceolata, secus carinam et nervos submarginales paulo et breviter pilosa, callo glabro; paleae secus carinas scabridulae; antherae 1.2–1.8 mm lg.

Described from Qinghai. Type in Nanking.

In sparse forests and among shrubs; in middle and upper mountain belts.

IIIA. Qinghai: *Nanshan* ("near Khuan"yuan' town, No. 5253"—Keng, l.c., type!).

IIIB. Tibet: *Weitzan* (Mekong river basin along Chokchu river, 4000 m, Aug. 31; same site, Dorchzhilin monastery, in spruce and juniper forests, 3700–4000 m, Sept. 6—1900, Lad.).

General distribution : endemic.

19. P. koelzii Bor in Kew Bull. (1948) 139; ej. Grasses Burma, Ceyl., Ind. and Pakist. (1960) 557.—*P. rangkulensis* Ovcz. et Czuk. in Izv. Otd. estestv. nauk AN Tadzh, SSR, 17 (1956) 40; Fl. Tadzh. 1(1957) 172; Ikonnikov, Opred. rast. Pamira [Key to Plants of Pamir] (1963) 61.

Described from West. Himalayas (Ladakh). Type in London (K).

On rocky and rubble slopes and in riverine pebble beds; in upper mountain belt.

IB. Kashgar: *Nor.* (Airi gorge in Uchturfan region, June 6, 1908—Divn.).

IIIA. Qinghai: *Nanshan* (pass through Altyntag mountain range 24 km south of Aksai settlement, 3460 m, Aug. 2, 1958—Petr.).

IIIB. Pamir (Ulug-Tuz gorge in Charlym river basin, June 28; foothill of Muztag-Ata hills, July 20; Ulug-Rabat pass, July 23—1909, Divn.).

General distribution: East. Pam.; Himalayas (west., Kashmir).

20. P. krylovii Reverd. in Sist. zam. Gerb. Tomsk. univ. 8 (1963) 3.

Described from East. Sib. (Minusinsk region). Type in Tomsk.

In upland steppes, on rocky slopes and rocks, in thin (predominantly larch) forests, among shrubs; in lower and middle mountain belts.

IA. Mongolia: *Mong. Alt.* (along Khartsiktei river, July 27, 1898—Klem.; Urtu-Gol river valley, Aug. 19; nor. slope of Khara-Dzarga mountain range along Khairkhan-Duru river, Aug. 25—1903, Pob.).

IIA. Junggar: *Tien Shan* (along Borborogussun river, 3000 m, June 15;—Mengute hill, July 2—1879, A. Reg.).

General distribution: East. Sib. (south-west.), Nor. Mong. (Hang.).

Note. This species is very close to *P. relaxa* Ovcz. but differs from it in the absence of tuft of undulating hairs on callus of lemma and in more mesophilic habit.

21. P. nemoralis L. Sp. pl. (1753) 69; ? Hemsley, Fl. Tibet (1902) 204; Forbes and Hemsley, Index Fl. Sin. 3 (1904) 425, excl. var.; Krylov, Fl. Zap. Sib. 2 (1928) 291; Pavlov in Byull. Mosk. obshch. ispyt. prir. 38 (1929) 22;

Roshev. in Fl. SSSR, 2 (1934) 400; Kitag. Lin. Fl. Mansh. (1939) 87;? Walker in Contribs, U.S. Nat. Herb. 28 (1941) 598; Fl. Kirgiz. 2 (1950) 130; Grubov, Consp. fl. MNR (1955) 70; Fl. Kazakhst. 1(1956) 231; Fl. Tadzh. 1(1957) 142; Keng, Fl. ill. sin., Gram. (1959) 164; Bor, Grasses Burma, Ceyl., Ind. and Pakist. (1960) 558; Pazii in Bot. mater. Gerb. Inst. bot. AN Uzbek, SSR, 17 (1962) 35. —Ic.: Keng, l.c. fig. 114.

Described from Europe. Type in London (Linn.).

In forests, among shrubs, in forest glades; up to upper mountain belt.

IA. **Mongolia**: *Cis-Hing.* (near Irekte railway station, June 16, 1902—Litw.); *East. Mong.* (near Trekhrech'e, 750 m, in scrub No. 1242, July 10, 1951—Wang).

IIA. **Junggar**: *Alt. region* (near Qinhe, 1900 m, No. 902, Aug. 2, 1956—Ching; right bank of Kairty river 20 km nor. of Koktogoi, Kuidyn river valley, July 15, 1959—Yun.); *Jung. Alt., Tien Shan.*

General distribution: Balkh. region (rare), Jung.-Terb., Nor. and Cent. Tien Shan; Europe, Mediterr., Balk.-Asia Minor, Near East, Caucasus, Middle Asia (hilly), West. Sib., East. Sib. (south-west), Far East, Nor. Mong. (Hent., Hang.), China (Dunbei, Nor., Nor.-West.), Himalayas (west., Kashmir), Korea, ? Japan, Nor. Amer.

Note. Specimens with totally reduced tuft of undulating hairs on callus of lemma (*P. nemoraliformis* Roshev. ?) are found in East. Tien Shan mountains.

22. **P. ochotensis** Trin. in Mem. Ac. Sci. St.-Pétersb., sér. VI, 1 (1831) 377; Roshev. in Fl. SSSR, 2 (1934) 395; Tzvel. in Arkt. fl. SSSR, 2 (1963) 152.—*P. sphondylodes* Trin. in Mém. Ac. Sci. St. Pétersb. Sav. Etrang. 2 (1835) 145; Franch. Pl. David. 1 (1884) 338; Forbes and Hemsley, Index Fl. Sin. 3 (1904) 427; Roshev., l.c. 407; Kitag. Lin. Fl. Mansh. (1939) 88;? Walker in Contribs, U.S. Nat. Herb. 28 (1941) 598; Norlindh, Fl. mong. steppe, 1 (1949) 95, p.p.; Grubov, consp. fl.MNR (1955) 71; Keng, Fl. ill. sin., Gram. (1959) 194.—*P. linearis* Trin. l.c. (1835) 145.—? *P. transbaicalica* Roshev. in Izv. Glavn. bot. sada SSSR, 28 (1929) 382; Roshev., l.c.(1934) 404.—? *P. subaphylla* Honda in Rep. First Sci. Exped. Manch., sect. 4, 4(1936) 102.—*P. sphondylodes* var. *erikssonii* Meld. in Norlindh, l.c. 99. —Ic.: Keng, l.c. fig. 146.

Described from Far East. Type in Leningrad.

On rocky slopes and rocks; up to midmountain belt.

IA. **Mongolia**: *Cis-Hing.* (near Yaksha station, June 11, 1902—Litw.; 5–7 km south-west of Dzhara-Ula hills, Aug. 8, 1949—Yun.); *East. Mong. , Gobi-alt.* (Bain-Tsagan mountain range, July-Aug., 1933—Khurlat and Simukova; central belt of Dundu-Saikhan mountain range, July 22; south. slope of Ikhe-Bogdo mountain range, Sept. 10; nor. slope of Ikhe-Bogdo, Nakhoitin-Khundei creek valley, Oct. 14—1943, Yun.); *East. Gobi* (Temur-Khada, Roerich Exped., No. 895, Aug. 23, 1935—Keng; near Bailinmyao, 1960—Ivan.); *Alash. Gobi* (rocky slopes of Alashan 50 km south-west of In'chuan' town, July 10, 1957; 50 km on In'chuan'—Bayan-Khoto highway, 1800 m, June 10, 1958—Petr.); *Khesi* (Gaolan', Kheishan' coal pits, June 29, 1958—Petr.).

General distribution: East. Sib. (south-east.), Far East (south.), China (Dunbei, Nor., Nor.-West.,) Cent., Korea, ? Japan.

Note : The type of *P. sphondylodes* Trin. is wholly similar to the type of *P. ochotensis* while the lectotype of *P. transbaicalica* Roshev. (valley of Nerchi river 200 km beyond Nerchinsk, 1908, V. Zyryanov) has strongly scabrous stems and very lax panicles, taking it considerably closer to *P. relaxa* Ovcz. *P. ochotensis* Trin., *P. attenuata* Trin., *P. argunensis* Roshev., *P. reverdattoi* Roshev., *P. albertii* Regel and *P. koelzii* Bor, adopted by us as independent species for Cent. Asia, are morphologically very poorly distinguished and related through transitional populations, and possibly warrant being combined into a single polytypic species, *P. ochotensis* Trin. s.l. However, this group of species is almost as closely associated with other species of section Stenopoa Dum., primarily with *P. glauca* Vahl and *P. altaica* Trin. and, through *P. relaxa* Ovcz. and *P. krylovii* Reverd., with *C. palustris* L. and *P. nemoralis* L.

23. **P. palustris** L. Syst. pl., ed. 10 (1759) 874, emend. Roth (1789); Krylov, Fl. Zap. Sib. 2 (1928) 293; Pavlov in Byull. Mosk. obshch. ispyt. prir. 38 (1929) 22; Roshev. in Fl. SSSR, 2 (1934) 397; Kitag. Lin. Fl. Mansh. (1939) 87; Fl. Kirgiz. 2 (1950) 130; Grubov, consp. fl. MNR (1955) 71; Fl. Tadzh. 1 (1957) 146; Bor, Grasses Burma, Ceyl., Ind. and Pakist. (1960) 559; Pazii in Bot. mater. Gerb. inst. bot. Uzbek SSR, 17 (1962) 37.—*P. serotina* Ehrh. Beitr. Naturk. 6 (1791) 83; Fl. Kazakhst. 1 (1956) 230.—*P. fertilis* Host, Gram. Austr. 3(1805) 10; Franch. Pl. David. 1(1884) 338. —Ic.: Fl. SSSR, 2, Plate 29, fig. 14.

Described from Europe. Type in London (Linn.).

In meadows, riverine sand and pebble beds, in sparse forests and among shrubs; up to upper mountain belt.

IA. **Mongolia:** *Mong. Alt.* (upper Kharagaitu-Gol, left bank of Bulgun tributary, Aug. 24, 1947—Yun.); *Cis-hing.* (near Yaksha railway station, June 13, 1902—Litw.); *East. Mong.* (near Hailar town, No. 784, June 20, 1951—Li et al.); *East. Gobi* (Muni-Ula hills, June 1871—Przew.).

IB. **Kashgar:** *West.* (nor. slope of King-Tau mountain range 2 km nor. of Kosh-Kulak settlement, June 10, 1959—Yun.).

IIA. **Junggar:** *Alt. region, Jung. Alt. Tien Shan, Zaisan* (left bank of Ch. Irtysh facing Cherektas hills, June 11; lower course of Belezek river, June 18—1914, Schisch.).

IIIA. **Qinghai:** *Nanshan* (forest belt of Yuzhno-Tetungsk mountain range, 2800 m, July 23, 1880—Przew.).

General distribution: Aral-Casp., Balkh. region, Jung. Tarb.-Nor. and Cent. Tien Shan, Europe, Mediterr. (east.), Balk.-Asia Minor, Near East, Caucasus, Middle Asia, West. Sib., East. Sib., Far East, Nor. Mong. (Fore Hubs., Hent., Hang.), China (Dunbei, Nor., Nor.-West.), Himalayas (Kashmir), Korea, Japan, Nor. Amer.

Note. Highly polymorphic species and, in China and Japan, evidently replaces a whole series of closely related species which have yet to be adequately understood.

24. **P. poiphagorum** Bor in Kew Bull. (1948) 143; ej. Grasses Burma, Ceyl., Ind. and Pakist. (1960) 559.

Described from the Himalayas (Sikkim). Type in London (K).

On rocky and rubble slopes and in riverine pebble beds; in upper mountain belt.

IIIB. **Tibet:** *South.* (Hattan ning daram, 5400 m, Aug. 19, 1892—Rhins).
General distribution: Himalayas.

25. P. relaxa Ovcz. in Izv. Tadzh. bazy AN SSSR, 1, 1 (1933) 20; Roshev. in Fl. SSSR, 2 (1934) 402; Persson in Bot. notiser (1938) 275; Fl. Kirgiz. 2 (1950) 131; Fl. Kazakhst. 1 (1956) 232; Fl. Tadzh. 1 (1957) 146; Pazii in Bot. mater. Gerb. Inst. bot. AN Uzbek.SSR, 17 (1962) 24; Ikonnikov, Opred. rast. Pamira [Key to Plants of Pamir] (1963) 59.—*P. attenuata* var. *stepposa* Kryl. Fl. Alt. and Tomsk. gub. 7 (1914) 1656; ej. Fl. Zap. Sib. 2 (1928) 285.—*P. stepposa* (Kryl.) Roshev. in Fl. SSSR, 2 (1934) 401 and 754; Fl.-Kirgiz. 2 (1950) 131; Fl. Kazakhast. 1 (1956) 232.—*P. urgutina* Drob. in Fl. Uzbek. 1 (1941) 242 and 538.

Described from Middle Asia (Zeravshan mountain range). Type in Leningrad.

In steppes, on rocky slopes and rocks, in riverine pebble beds and sparse forests; up to upper mountain belt.

IA. Mongolia: *Khobd.* (south. peak of Kharkir, July 12, 1879—Pot.; midcourse of Tsagan-Nurin-Gol, Obgor-Ul hill, Aug. 2, 1945—Yun.); *Mong. Alt., Gobi-Alt.* (Dzun-Saikhan mountain range, in Yalo upper creek valley, Aug. 23, 1931—Ik.-Gal.; peak of Dzun-Saikhan mountain range, June 19, 1945—Yun.).

IB. Kashgar: *West., South.* (nor. slope of Kunlun, Tokhtakhon hills, 3300 m, June 18, 1889—Rob.; south of Cherchen, 3400 m, No. 9439, June 5, 1959—Lee et al.).

IIA. Junggar: *Alt. region, Jung. Alt., Tien Shan, jung. Gobi* (east.: Baitak-Bogdo-Nuru mountain range, Takhiltu-Ula, left creek valley of Ulyastu-Gol 4 km from estuary, about 2000 m, Sept. 17; same site, Ulyastu-Gol upper gorge, Sept. 18—1948, Grub.; Beidashan mountain range [Baitak-Bogdo], 1700 m, No. 5223, Sept. 28, 1957—Huang); *Zaisan* (Mai-Kapchagai hill, June 6, 1914—Schisch).

IIIA. Qinghai: *Nanshan* (near Yusun-Khatym river, 3000–3300 m, July 12, 1880—Przew.); *Amdo* (near Dulan-Khit monastery 3650 m, Aug. 8, 1901—Lad.).

IIIB. Tibet: *Weitzan* (nor. slope of Burkhan-Budda mountain range, Khatu gorge, 3650–4000 m, July 11, 1901—Lad.).

IIIC. Pamir

General distribution: Aral-Casp., Balkh. region, Jung.-Tarb., Nor. and Cent. Tien Shan, East. Pam.; Europe (south-east.), Middle Asia (hilly), West. Sib. (south), East. Sib. (rare), Nor. Mong. (Fore Hubs., Hent., Hang.).

Note. Specimens from very low hills and plains, described as an independent species, *P. stepposa* (Kryl.) Roshev., tend to have strongly scabrous stems and just slightly coloured spikelets; still it is extremely difficult to draw any distinct morphological or geographical boundary between them and typical specimens of *P. relaxa*. On the whole, *P. relaxa* occupies a somewhat intermediate position between species of the group *P. ochotensis* Trin. s.l. and the highly polymorphic (variable) species *P. palustris* L., possibly a hybrid.

26. P. reverdattoi Roshev. in Fl. SSSR, 2 (1934) 407 and in Tr. Bot. inst. AN SSSR, ser. 1, 2 (1936) 97; Krylov, Fl. Zap. Sib. 12, 1 (1961) 3107. —Ic.: Fl. SSSR, 2, Plate 30, fig. 13.

Described from East. Siberia (Abakan steppe). Type in Leningrad.

On rocky slopes and rocks; up to midmountain belt.

IA. mongolia: *Khobd.* (valley of Kharkira river, Aug. 9, 1879—Pot.).
General distribution: West. Sib. (Altay), East. Sib. (south-west.).

o P. urssulensis Trin. in Mem. Sav. Etr. Petersb. 2 (1835) 527; Gubanov, Consp. Fl. Outer Mong. (1996) 22.

IA. **Mongolia:** *Cen. Khalkha.*

Section 6. Bolbophorum (Aschers. et Graebn.) Jiras.

27. P. alpina L. Sp. pl. (1753) 76; ? Henderson and Hume, Lahore to Yarkand (1873) 341; ? Deasy, In Tibet and Chin. Turk. (1901) 405; ? Hemsley, Fl. Tibet (1902) 204; Danguy in Bull. Mus.nat. hist. natur. 20 (1914) 147;? Pilger in Hedin, S. Tibet, 6, 3 (1922) 94; Krylov, Fl. Zap. Sib. 2(1928) 287; Pampanini, Fl. Carac. (1930) 75; Roshev. in Fl. SSSR, 2 (1934) 411; Hao in Engler's Bot. Jahrb. 68 (1938) 581; Fl. Kirgiz. 2 (1950) 131; Grubov, Consp. fl. MNR (1955) 70; Fl. Kazakhst. 1(1956) 234; Fl. Tadzh. 1(1957) 162; Keng, Fl. ill. sin., Gram. (1959) 212; Bor, Grasses Burma, Ceyl., Ind. and Pakist. (1960) 555; Pazii in Bot. mater. Gerb. Inst. bot. AN Uzbek.SSR, 17 (1962) 38; Ikonnikov, Opred. rast. Pamira [Key to Plants of Pamir] 1963) 60. —Ic.: Fl. Tadzh. 1, Plate 21, figs. 1–4; Keng, l.c. fig. 165.

Described from Scandinavia (Lapland). Type in London (Linn.).

IA. **Mongolia:** *Mong. Alt.* (Kharagaitu-Daba pass in upper Indertiin-Gol, July 23–24; near summer camp in Bulugun somon on road to Kharagaitu-Khutul, July 27—1947, Yun.).

IIA. **Junggar:** *Alt. region, Jung. Alt.* (Yugantash mountain range, 2000–2400 m, May 25, 1879.—A. Reg.; near nor. bank of Sairam-Nur lake, No. 2137, Aug. 8, 1957—Huang); *Tien Shan.*

IIIA. **Qinghai:** *Amdo* ("Da-ho-ba, bei 4000 m, Sept. 7, 1930" Hao, l.c.).

IIIB. **Tibet:** Chang Tang ("Northern Tibet, camp 17, 5073 m, 1, IX 1896, leg. Hedin"-Pilger, l.c.; "Sarok Tuz valley, 4300 m, 1898"—Deasy, l.c.); *South.* ("Tibet, 5400 m, leg. Thorold"—Hemsley, l.c.).

IIIC. **Pamir** ("Kara-jilga, Basik-kul, 3727 m, 24 VII 1894, leg. Hedin"—Pilger, l.c.).

General distribution: Jung.-Tarb., Nor. and Cent. Tien Shan, East. Pam.; Arct., Europe (hilly), Mediterr. (hilly), Balk.-Asia Minor, Near East, Caucasus, Middle Asia (hilly), West. Sib. (Altay), East. Sib. (west.), Nor. Mong. (Fore Hubs.), Himalayas (west., Kashmir), Korea, Japan, Nor. Amer.

28. P. bulbosa L. Sp. pl. (1753) 70; Krylov, Fl. Zap. Sib. 2 (1928) 284; Roshev. in Fl. SSSR, 2 (1934) 376; Fl. Kirgiz. 2 (1950) 124; Fl. Kazakhst. 1 (1956) 224; Fl. Tadzh. 1 (1957) 187; Bor, Grasses Burma, Ceyl., Ind. and Pakist. (1960) 556; Pazii in Bot. mater. Gerb. Inst. bot. AN Uzbek SSR, 17 (1962) 18. —Ic.: Fl. SSSR, 2, Plate 28, fig. 1.

Described from South. Europe (France). Type in London (Linn.).

In desert steppes, on rocky and rubble slopes, talus, in riverine sand and pebble beds; up to midmountain belt.

IIA. **Junggar:** *Alt. region* (2–3 km south-south-east of Shara-Sume, Kran river valley, July 7, 1959—Yun.); *Tarb.* (Tarbagatai hills near Dachen, 1700 m, No. 1662, Aug. 14, 1957—Huang); *Jung. Gobi* (nor.: along Urungu river, April, 1879—Przew.;

left bank of Ch. Irtysh 38 km east of Shipati crossing on road to Koktogoi, July 8, 1959—Yun.); *Zaisan* (Ch. Irtysh valley, Uzun-Bulak river in foothills of Koksun mountain range, June 5, 1903—Gr.-Grzh.); *Dzhark., Balkh.-Alak.* (30 km west of Durbul'dzhin, No. 1471, Aug. 10, 1957—Huang).

General distribution: Aral-Casp., Balkh. region, Jung.-Tarb., Nor. and Cent. Tien Shan; Europe (cent. and south.), Mediterr., Balk.-Asia Minor, Near East, Caucasus, Middle Asia, West. Sib. (south.), Himalayas (west.).

Note. Report of this species from Qinghai [Hao in Engler's Bot. Jahrb. 68 (1938) 581] is undoubtedly erroneous and probably pertains to *P. albertii* var. *vivipara* (Rendle) Tzvel.

Section 7. O c h l o p o a (Aschers. et Graebn.) Jiras.

29. P. acroleuca Steud. Synops. Pl. Glum. 1 (1854) 256; Forbes and Hemsley, Index Fl. Sin. 3 (1904) 422; Kitag. Lin. Fl. Mansh. (1939) 86; Walker in Contribs U.S. Nat. Herb. 28 (1941) 598; Keng, Fl. ill. sin., Gram (1959) 213. —Ic.: Keng, l.c. fig. 167.

Described from Japan. Type in Leiden.

In riverine sand and pabble beds, in grasslands, along roadsides; up to upper mountain belt.

IIIA. Qinghai: *Nanshan* (alpine belt of Yuzhno-Tetungsk mountain range, July 19–20, 1872—Przew.; "La Ch'iung kou near Sining, No. 638, leg. Ching"—Walker, l.c.).

IIIB. Tibet: *Weitzan* (Yangtze river basin, Nruchu area, 3900 m, July 25, 1900—Lad.).

General distribution: China, Korea, Japan.

30. P. annua L. Sp. pl. (1753) 68; Forbes and Hemsley, Index Fl. Sin. 3 (1904) 422; Krylov, Fl. Zap. Sib. 2 (1928) 282; Roshev. in Fl. SSSR, 2 (1934) 379; Kitag. Lin. Fl. Mansh. (1939) 87; Fl. Kirgiz. 2(1950) 127; Fl. Kazakhst. 1(1956) 226; Fl. Tadzh. 1 (1957) 181; Keng, Fl. ill. sin., Gram. (1959) 224; Bor, Grasses Burma, Ceyl., Ind. and Pakist. (1960) 555; Pazii in Bot. mater. Gerb. Inst. bot. AN Uzbek SSR, 17 (1962) 24. —Ic.: Fl. SSSR, 2, Plate 28, fig. 9; Keng, l.c. fig. 179.

Described from Europe. Type in London (Linn.).

As weed along roadsides, in wastelands, riverine sand and pebble beds; up to midmountain belt.

IA. Mongolia: *East. Mong.* (near Trekhrech'e, 60 m, No. 2003, Aug. 16, 1951—Li et al.).

General distribution: almost cosmopolitan.

31. P. rossbergiana Hao in Engler's Bot. Jahrb. 68 (1938) 581; Keng, Fl. ill. sin., Gram. (1959) 222. —Ic.: Keng, l.c. fig. 178.

Described from Qinghai. Type in Berlin. Isotype in Leningrad.

IIIA. Qinghai: *Amdo* ("In der Nähe des Klosters Ta-schiu-sze, 4200 m, in Jahemari Gebirge, 20 km westlich von Tsi-gi-gan-ba, No. 1197, 9 IX, 1930"—Hao, isotype!).

General distribution: endemic.

32. **P. supina** Schrad. Fl. Germ. 1 (1806) 289; Roshev. in Fl. SSSR, 2 (1934) 379; Fl. Tadzh. 1 (1957) 179; Bor, Grasses Burma, Ceyl., Ind. and Pakist. (1960) 561; Pazii in Bot. mater. Gerb. Inst. bot. AN Uzbek SSR, 17 (1962) 24; Tzvel. in Arkt. fl. SSSR, 2 (1964) 160.

Described from Cent. Europe. Type in Berlin.

In grasslands, riverine pebble beds, often as weed along roadsides and wasteland; in middle and upper mountain belts.

IIA. Junggar: *Alt. region* (near Koktogoi, 2500 m, No. 1912, Aug. 17, 1956—Ching); *Tien Shan.*

General distribution: Nor. and Cent. Tien Shan; Europe (hilly), Caucasus (hilly), Middle Asia (hilly), West. Sib., East. Sib., Himalayas (west.).

33. **P. tibeticola** Bor in Kew Bull. (1948) 139; ej. Grasses Burma, Ceyl., Ind. and Pakist. (1960) 561.—*P. gracillima* Rendle in Forbes and Hemsley, Index Fl. Sin. 3 (1904) 424, non Vasey (1893).

Described from Tibet. Type in London (K).

In grasslands, riverine sand and pebble beds; in upper mountain belt.

IIIB. Tibet: *Weitzan* (Yangtze river basin, Nruchu area, 3900 m, July 25, 1900—Lad.); *South.* ("Khambajong, 7 IX, 1903, leg. Younghusband, type!; Lhasa, IX 1904, leg. Walton"—Bor, l.c. [1948].)

General distribution: China (south-West.), Himalayas.

Note. This species represents an ecogeograpical race of semiweed species *P. acroleuca* Steud. widely found at very high altitudes in East. Asia.

55. Eremopoa Roshev.
in Fl. SSSR, 2 (1934) 429 and 756.

o **E. altaica** (Trin.) Roshev. in Fl. SSSR, 2 (1934) 431; Gubanov, Consp. Fl. Outer Mong. (1996) 19.

IA. Mongolia. *Mong. Alt.*

1. **E. songarica** (Schrenk) Roshev. in Fl. SSSR, 2 (1934) 431; Fl. Kazakhst. 1 (1956) 249; Tzvelev and Grif in Bot. zhurn. SSSR, 50 (1965) 1458, maps 1 and 2.—*E. bellula* (Regle) Roshev. l.c. 431, quoad nom.; Fl. Kirgiz. 2 (1950) 137.—*E. persica* var. *songarica* (Schrenk) Bor, Grasses Burma, Ceyl., Ind. and Pakist. (1960) 532.—*E. glareosa* Gamajun. in Bot. mater. Gerb. inst. bot. AN KazakhSSSR, 2 (1964) 11.—*E. altaica* subsp. *songarica* (Schrenk) Tzvel. in Bot. zhurn. SSSR, 51 (1966) 1104.—*Glyceria songarica* Schrenk, Enum. pl. nov. (1841) 1.—*Festuca bellula* Regel in Acta Horti Petrop. 7 (1880) 594.—*Poa persica* auct. non Trin.: Munro in Henderson and Hume, Lahore to Yarkand (1873) 341; Pampanini, Fl. Carac. (1930) 75.—*Eremopoa persica* auct. non Roshev.: Fl. Tadzh. 1 (1957) 189. —Ic.: Fl. SSSR, 2, Plate 32, fig. 14.

Described from Kazakhstan (from Karatal river). Type in Leningrad.

In riverine pebble beds, on rubble and rocky slopes and talus; in middle and upper mountain belts.

IB. **Kashgar:** *East:* (Karatash river flood plain below Egin. July 2, 1929—Pop.).

IIIB. Tibet : *Chang Tang* ("Lingzi-tang, ca. 5000 m, 1870, leg. Henderson"—Munro, l.c.).

IIIC. Pamir (near Tashkurgan, July 25, 1913—Knorring).

General distribution: Aral-Casp. (south-east.), Balkh. region, Jung.-Tarb., Nor. Tien Shan, East. Pam.; Near East, Caucasus, Middle Asia (hilly), West. Sib. (south-west. Altay), Himalayas (west., Kashmir).

56. Catabrosella (Tzvel.)
in Bot. zhurn. SSSR, 50 (1965) 1320.—*Colpodium* subgen.
Catabrosella Tzvel. in Novosti sist. vyssh. rast. (1964) 12.

1. **C. humilis** (M.B.) Tzvel. in Bot. zhurn. SSSR, 50 (1965) 1320.—*Aira humilis* M.B. Fl. taur.-cauc. 1(1808) 57.—*Colpodium humile* M.B.) Griseb. in Ledeb. Fl. Ross. 4 (1852) 384; Nevski in Fl. SSSR, 2 (1934) 442; Fl. Kirgiz. 2 (1950) 138; Fl. Kazakh. 1 (1956) 253. —Ic.: Fl. SSSR, 2, Plate 33, fig. 2; Fl. Kazakhst. 1, Plate 19, fig. 8.

Described from Fore Caucasus (Beshtau hill). Type in Leningrad.

In desert steppes, on solonchak, rocky slopes, in riverine sand and pebble beds; up to lower mountain belt.

IIA. **Junggar:** *Jung. Gobi.* (south.: near Yantszykhai village, No. 10220, April 21, 1959—Lee et al; nor. foothills of Bogdo-Ula mountain range 10–15 km nor.-east of Urumchi along road to Fukan, April 26, 1959—Yun.); *Dzhark.* (near Kul'dzha town, May, 1877—A. Reg.).

General distribution: Aral-Casp., Balkh. region, Nor. Tien Shan (near Issyk-Kul lake); Europe (far south-east), Near East (nor.), Caucasus, Middle Asia (nor.-east.), West. Sib. (far south).

Note. All specimens of this species from Junggar and Balkhash region belong to a distinct, though very poorly distinguished ecogeographical race—subspecies *C. humilis* subsp. *songorica* Tzvel. [in Spisok rast. Gerb. fl.SSSR [List of Plants in the Herbarium of the Russian Flora], 16 (1967) 49]. They differ from typical specimens of *C. humilis* in very poorly developed and invariably glabrous intermediate ribs of lemma.

57. Paracolpodium (Tzvel.) Tzvel.
in Bot. zhurn. SSSR, 50 (1965) 1320.—*Colpodium* subgen. *Paracolpodium* Tzvel. in Novosti sist. vyssh. rast. (1964) 9.

oP. **altaicum** (Trin.) Tzvel. in Bot. Zhurn. 50, 9 (1965) 1320; Gubanov, Consp. Fl. Outer Mong (1996) 21

IA. **Mongolia.** *Mong. Alt.*

1. **P. leucolepis** (Nevski) Tzvel. comb. nova.—*P. altaicum* subsp. *leucolepis* (Nevski) Tzvel. in Novosti sist. vyssh. rast. (1966) 33.— *Colpodium leucolepis* Nevski in Byull. Mosk. obshch. ispyt. prir. 43 (1934) 224 and in Fl. SSSR, 2 (1934) 437; Fl. Kirgiz. 2 (1950) 138; Fl. Kazkhst. 1 (1956) 252; Fl. Tadzh. 1(1957) 232; Ikonnikov, Opred. rast. Pamira [Key to Plants of Pamir] (1963) 62.

Described from East. Pamir (head of Aksu river). Type in Leningrad.

In grasslands, on rocky slopes, talus, on rocks and in riverine pebble beds; in upper mountain belt.

IIA. Junggar: *Tien Shan.*

IIIC. Pamir (Billuli crossing, June 15, 1909—Divn.; Pil'nen river gorge, about 4700 m, July 1; same site, near glacier, July 16; between Atrakhyr and Tyuzutek rivers, 4500–5000m, July 20; head of Gon-Arek river, 5000–6000 m, July 23—1942, Serp.).

General distribution: Jung.-Tarb., Cent. Tien Shan, East. Pam., Middle Asia (Pam.-Alay).

58. Glyceria R. Br.
Prodr. Fl. Nova Holland. 1 (1810) 179.

1. Panicles with relatively few spikelets, rather secund; spikelets 8–20 mm long, with 4–15 florets; lemma 3–5 mm long. Plant 30–70 cm tall; stems usually ascending, rooting in lower nodes; sheath strongly laterally flattened, with prominent keel ....
.... 1. **G. plicata** (Fries) Fries.

+ Panicles with several spikelets, not secund; spikelets 4–8 m long, with 3–11 florets; lemma 2–4.8 mm long. Plants 30–140 cm tall; stems erect; sheath not flattened, faintly keeled ....2.

2. Glumes broadly lanceolate, lower ones 1–2.5 mm long, upper ones 1.2–3 mm long; lemma 2–3.8 mm long, rather oblong, obtuse; panicles usually widely spreading with somewhat scabrous branches. Leaf blades 4–12 mm broad ....
.... 2. **G. debilior** (Fr. Schmidt) Kudo.

+ Glumes lanceolate, lower ones 3–4 mm long, upper ones 3.5–5 mm long; lemma 3.8–4.8 mm long, lanceolate, gradually narrowed into rather pointed or obtuse tip; panicles weakly spreading, with highly scabrous branches. Leaf blade 2–6 mm broad ....2.

3. Anthers purple about 1 mm long. Ligules up to 1.2 mm long ......
.... 3. **G. longiglumis** Hand.-Mazz.

+ Anthers yellow 1.3–2 mm long. Ligules 1.2–3 mm long ....
....**4. G. spiculosa** (Fr. Schmidt) Roshev.

Section 1. Glyceria

O G. lithuanica (Gorski) Gorski, Icon. Bot. Char. Cyper. Gram. Lith. (1849) tab. 20; Gubanov, Consp. Fl. Outer Mong. (1996) 19.

Cen. Khalkha. IA. Mongolia.

1. G. plicata (Fries) Fries Nova Mant. 3 (1842) 176; Krylov, Fl. Zap. Sib. 2 (1928) 307; Komarov in Fl. SSSR, 2 (1934) 452; Fl. Kirgiz. 2 (1950) 139; Fl. Kazakhst. 1 (1956) 255; Fl. Tadzh. 1 (1957) 222; Keng, Fl. ill. sin., Gram. (1959) 251; Bor, Grasses Burma, Ceyl., Ind. and Pakist. (1960) 570.—*G. fluitans* var. *plicata* Fries, Nova Mant. 2 (1839) 6. —Ic.: Fl. SSSR, 2, Plate 34, fig. 2; Keng, l.c. fig. 204.

Described from Scandinavia. Type in Stockholm.

In marshy meadows, along banks of water reservoirs, often on water; up to midmountain belt.

IA. Mongolia: *Khesi* (Mitszyatsun' nor. of Tszyutsyuan' town, No. 10, July 2, 1956—Ching).

IIA. Junggar: *Tien Shan* (nor. of Kul'dzha, May 3, 1877; Cent. Khorgos, 1000–1700 m, June 15, 1878—A. Reg.).

General distribution: Balkh. region, Jung.-Tarb., Nor. and Cent. Tien Shan; Europe, Balk.-Asia Minor, Near East, Caucasus, Middle Asia, West. Sib. (south.), Afr. (nor.).

Section 2. Hydropoa Dum.

2. G. debilior (Fr. Schmidt) Kudo in J. Coll. Agric. Univ. Sapporo, 11, 2 (1922) 74.—*G. aquatica* var. *debilior* Trin. ex Fr. Schmidt in Mém. Ac. Sci. St.-Pétersb., sér VII, 12, 2 (1868) 201.—*G. aquatica* var. *triflora* Korsh. in Acta Horti Petrop. 12 (1892) 418.—*G. triflora* (Korsh.) Kom. in Fl. SSSR, 2 (1934) 459 and 758; Kitag. Lin. Fl. Mansh. (1939) 77.—*G. kamtschatica* Kom. l.c. 459 and 758.— *G. effusa* Kitag. in Bot. Mag. Tokya, 51 (1937) 152 and l.c. (1939) 77.— *G. angustifolia* Skvortr. in Zap. Kharbin. obshch. estestvoisp. i etnogr. 12 (1954) 29.—*G. maxima* subsp. *triflora* (Korsh.) Hult. in Kungl. Svenska Vetensk.-akad. Handl., ser. V, 8,5 (1962) 184.—*G. aquatica* auct. non Wahl.: Franch. Pl. David. 1(1884) 338; Forbes and Hemsley, Index Fl. Sin. 3 (1904) 428; Grubov, Consp. fl. MNR (1955) 72; Keng, Fl. ill. sin., Gram. (1959) 249. —Ic.: Keng, l.c. fig. 201.

Described from Kamchatka. Type in Leningrad.

Along banks of water reservoirs, in marshes and marshy meadows; up to lower mountain belt.

IA. Mongolia: *Mong. Alt.* (Bulugun river valley, Sept. 17, 1930—Bar.); *Cis-Hing.* (near Trekhrech'e, 700 m, No. 1308, July 11, 1951—Li et al.); *Cent. Khalkha* (in meander scroll of Kerulen, Aug. 6 to 12, 1906—Nov.); *East. Mong.* (near Hailar town, No. 775, June 20; same site, No. 1141, July 4—1951, Li et al.; near Shilin-Khoto, 1959—Ivan.); *Bas. lakes* (near Ulangom, June 20, 1879—Pot.).

0   IIA. **Junggar:** *Alt. region* (near Fuyun' town, 1200 m, No. 1839, Aug. 13; between Barbagai and Burchum, 1400 m, No. 2950, Sept. 4, 1956—Ching).

General distribution: West. Sib. (Altay), East. Sib. (south.), Far East, Nor. Mong., China (Dunbei, Nor., East.), Korea, Japan.

Note. This species is very closely related on the one side to *G. arundinacea* Kunth, found in steppe and forest-steppe belt of Eurasia from Hungary to Altay and, on the other to the North American *G. grandis* S. Wats. ex A. Gray, occupying an intermediate position between the two. The significant rise of polymorphism in *G. debilior* compared to these species and nearly complete similarity of some of its populations sometimes with *G. arundinacea* and sometimes with *G. grandis*, provide every justification to assume the hybrid origin of *G. debilior* from *G. arundinacea* × *G. grandis*. *G. debilior* is also very much isolated from other speices that are quite close to it: predominantly European *G. maxima* (Hartm.) Holmb. [=*G. aquatica* (L.) Wahl.] and East Asian *G. orientalis* Kom. (= *G. leptolepis* Ohwi; = *G. ussuriensis* Kom.).

3. **G. longiglumis** Hand.-Mazz. in Oesterr. bot. Z. 87 (1938) 130.

Described from Inner Mongolia. Type in Vienna.

Along banks of water reservoirs and in marshes.

IA. **Mongolia:** *East. Mong.* ("Gobi, Zwischen dem Bainkure-Nor und dem Shitong-Gol, No. 7456, 18 VI; Zwischen dem Schitong-Gol und Yendo-Sume, No. 7463, 19 VI 1924, leg. Licent, typus!"—Hand.-Mazz., l.c.).

General distribution: endemic.

Note. Judging from the highly imperfect diagnosis of this species, it is mostly similar to *G. spiculosa* (Fr. Schmidt) Roshev. but the possibility of its belonging to another genus cannot be ruled out.

4. **G. spiculosa** (Fr. Schmidt) Roshev. in Fl. Zabaik. 1 (1929) 85; Komarov in Fl. SSSR, 2 (1934) 457.—*Scolochloa spiculosa* Fr. Schmidt in Mém. Ac. Sci. St.-Pétersb., sér. VII, 12, 2 (1868) 201.—*Glyceria paludificans* Kom. in Izv. Bot. sada Petra Vel. 16 (1916) 152. —Ic.: Fl. SSSR, 2, Plate 34, fig. 12.

Described from Sakhalin. Lectotype (Traiziska, Aug. 4, 1860, Fr. Schmidt) in Leningrad.

Along banks of water reservoirs, in marshes and marshy meadows.

IA. **Mongolia:** *Cis-Hing.* (near Yaksha railway station, July 19, 1902—Litw.; near Trekhrech'e, 700 m, No. 1316, July 11, 1951—Wang).

General distribution: East. Sib. (south-east.), Far East (south.), China (Dunbei), Korea, Japan.

59. Puccinellia Parl.
Fl. ital. 1 (1848) 366.

1.   Lemma of lower florets in spikelet (1.5) 1.6–2.2 (2.3) mm long, obtuse at tip, usually broadly rounded or somewhat truncated ..
....2.

+   Lemma of lower florets in spikelet (2.3) 2.4–3.5 (4) mm long, narrowly rounded or somewhat pointed at tip ....6.

2.   Anthers 0.8–1.4 mm long; if less than 1 mm, lemma shorter than 2 mm. Plants of highly saline habitats .... 3.

+   Anthers 0.3–0.8 mm long; if about 0.8 mm, lemma longer than 2 mm. Plants of weakly saline habitats .... 4.

3.   Vegetative shoots numerous, largely covered at base with caps of sheaths of dead leaves; leaf blade longitudinally folded, very narrow (0.2–0.6 mm diameter); ligules 1.2–2 mm long, spinules barely perceptible even on high magnification .... 3. **P. filifolia** (Trin.) Tzvel.

+   Vegetative shoots usually few; sheaths of dead leaves at base only few; leaf blade from flat to longitudinally folded, 0.8–2.3 mm broad; ligules 0.4–1.6 mm long, covered outside with fairly long spinules .... 16. **P. tenuiflora** (Griseb.) Scribn. et Merr.

4.   Lemma glabrous, 1.5–2.1 mm long; palea glabrous along keels, with few very short spinules only in upper third; anthers 0.5–0.8 mm long; branches of relatively dense panicles usually scabrous only in upper part due to very short spinules, occasionally smooth .... 6. **P. himalaica** Tzvel.

+   Lemma rather pilose near base; branches of rather spreading panicles scabrous due to spinules throughout length or almost so .... 5.

5.   Lemma 1.9–2.3 mm long, densely pilose near base; palea with fairly long spinules, in upper half along keels, usually weakly pilose below, sometimes glabrous; anthers 0.5–0.8 mm long .... 2. **P. distans** (L.) Parl.

+   Lemma 1.5–1.9 mm long, weakly pilose near base, occasionally only with some hairs; palea usually with spinules only in upper third along keels, glabrous and smooth below, rarely with few hairs; anthers 0.3–0.5 mm long .... 5. **P. hauptiana** (Krecz.) Kitag.

6(1).   Panicles long and narrow, 8–22 cm long, usually almost as long as rest of stem or not less than two-thirds as long; branches short, longest 1/5–2/5 as long as panicle; lemma with few hairs at base or glabrous. Extravaginal shoots present at base of mats .... 7.

+   Panicles much shorter than rest of stem, invariably shorter by 1/2; branch very long, longest usually 1/2–2/3 as long as panicle .... 8.

7.   Anthers 1.8–2.5 mm long .... 12. **P. przewalskyi** Tzvel.

+   Anthers 0.8–1.2 mm long. .... 15. *P. schischkinii* Tzvel.

8.   Lemma glabrous, sometimes with few hairs at base; palea along keels, with slightly less perceptible few spinules only in upper 1/4 glabrous and smooth below; branches of panicle smooth or rather scabrous only in upper half .... 9.

+ Lemma with numerous hairs near base along ribs and occasionally even between ribs, palea with numerous spinules in upper 1/3–1/2 along keels, glabrous or somewhat pilose below .. .. 12.

9. Panicles spreading at anthesis (dense and compressed only in stunted specimens of *P. nudiflora*). Shortened vegetative shoots almost invariably present; stems usually genuflexed at nodes; leaf blade flat or longitudinally folded loosely, 0.6–2.5 mm broad, slightly scabrous in upper part .... 10.

+ Panicles rather compressed or with slightly laterally declinate branches at anthesis . Shortened vegetative shoots usually absent; stems erect; leaf blade usually longitudinally folded, 0.5–2 mm broad in unfolded state, strongly scabrous in upper part .... 11.

10. Anthers 1.4–2 mm long; panicles greenish or with slight pinkish-violet tinge .... 8. **P. ladyginii** Tzvel.

+ Anthers 0.7–1.2 mm long; panicles pinkish-violet .... 10. **P. nudiflora** (Hack.) Tzvel.

11. Base of mats with arcuately ascending extravaginal shoots covered with scale leaves without blades. Panicles with slight pinkish-violet tinge .... 1. **P. altaica** Tzvel.

+ Base of mats without arcuately ascending extravaginal shoots. Panicles usually pinkish-violet .... 11. **P. pamirica** (Roshev.) Krecz. ex Roshev.

12 (8). Shoots of mat partly extravaginal, covered at base with scale leaves without blades; leaf blade 1.5–3.5 mm broad; ligules 0.6–1.5 mm long, covered with spinules throughout outer surface. Lemma 2.7–3.5 mm long; anthers 1.8–2.2. mm long .... 14. **P. roshevitsiana** (Schisch.) Tzvel.

+ All shoots intravaginal, without scale leaves at base .... 13.

13. Mats lax; shoots covered at base with fairly numerous light brown sheaths of dead leaves; leaf blade 0.5–1.2 mm broad, longitudinally folded, smooth, or almost so, on upper surface. Branches of panicles with 1–4 spikelets; smooth, or almost so, in lower half; lemma 2.8–4.2 mm long; anthers 1–1.3 mm long .... 13. **P. roborovskyi** Tzvel.

+ Mats dense; shoots at base with few sheaths of dead leaves; leaf blade (0.8) 1–2.5 (3) mm broad, flat or longitudinally folded; scabrous on upper surface due to numerous spinules. Branches of panicles usually (barring stunted specimens) with many spikelets .... 14.

14. Ligules covered throughout outer surface with numerous spinules. Anthers 1.4–2 mm long; lemma 2.3–3 mm long; panicles

spreading, with branch scabrous throughout length or almost so
.... 9. **P. macranthera** (Krecz.) Norlindh.

+   Ligules smooth on outer surface or with stray spinules along ribs. Anthers 0.7–1.2 mm long; lemma 2.4–3.4 mm long .... 15.

15.   Plant up to 15 cm tall. Panicles small (usually 2–4 cm long), invariably compressed, with smooth, or nearly smooth, branches ....7. **P. humilis** (Krecz.) Roshev.

+   Plant usually taller than 15 cm. Panicles very large and usually spreading, occasionally compressed, with branches scabrous throughout length or almost so ....
.... 4. **P. hackeliana** (Krecz.) Persson.

1. **P. altaica** Tzvel. sp. nova.—Planta perennis, 20–45 cm alta, sat dense caespitosa, innovationes partim extravaginales foliis squamiformibus stramineis cingentes emittens; culmi et vaginae glabrae et laeves; ligulae 0.6–2 mm lg., glabrae et laeves; laminae 0.6–1.5 mm lt., plus minusve complicatae, subtus laeves, supra scabrae. Paniculae 5–15 cm lt., parum effusae, ramis tenuibus leviter scabris vel laevibus; spiculae vulgo 3–5. florae, leviter roseo-violaceo tinctae; glumae lemmatis mul to breviores, inaequilongae, late lanceolatae, apice acutiusculae vel obtusae; lemmata 2.3–3 mm lg., acutiuscula, glabra vel prope basin vix pilosiuscula; paleae solum in parte superiore secus carinas vix scabris; antherae 1.1–1.6 mm lg.

Typus: Altaj, steppa Czujensis, Dzenishke-Tal, in pratis subsalsis, 10 VIII 1931, B. Schischkin, L. Czilikina et G. Sumnevicz. In Herb. Inst. Bot. Ac. Sci. URSS—LE conservatur.

Affinitas: Haec species a speciebus proximis differt: a *P. pamiricae*—spiculis minus tinctis et innovationibus partim extravaginalibus, a *P. tenuissimae*—lemmatis longioribus glabris vel subglabris et innovationibus partim extravaginalibus.

Described from Altay (Chui steppe). Type in Leningrad. Plate VIII, fig. 1.

In solonetz and saline meadows, riverine pebble beds; in lower and middle mountain belts.

IIA. Junggar: *Jung. Gobi* (nor.: flood plain of Bodonchi river 2–3 km south of Bodonchin-Khure, saline meadow, 1947—Yun.; Tamchi somon, Gun-Tamchi area in Khoni-Usuni-Gobi, saline meadows, Aug. 2, 1947—Yun.).

General distribution: West. Sib. (Altay: Chui steppe).

Note. In general habit and structure of spikelets, this species occupies a somewhat intermediate position beween *P. tenuissima* (Krecz.) Pavl., widely distributed in steppe and forest-steppe zone of Eurasia from Moldavia to Altay foothills, and the Central Asian species *P. pamirica* (Roshev.) Krecz., but is well distinguished from both in the presence of extravaginal shoots covered with coriaceous scale leaves. Besides the type and specimens mentioned above, nine specimens from Chui steppe, also preserved in the herbarium of the Botanical Institute, Academy of Sciences of the USSR, belong to *P. altaica*. It is very likely that variety *P. dolicholepis* var. *paradoxa* Serg. [Krylov, Fl. Zap. Sib. 12, 1 (1961) 3113] has been described from a large specimen of *P. altaica*.

o P. arjinshanensis Cui, in Bull. Bot. Research (Harbin), 17, 2 (1997) 123.

IIA. Junggar. *Tien Shan.*

2. P. distans (L.) Parl. Fl. ital. 1 (1848) 367; Fl. Kazakhst. 1(1956) 247; Keng, Fl. ill. sin., Gram. (1959) 230; Bor, Grasses Burma, Ceyl., Ind. and Pakist. (1960) 562.—*P. filiformis* Keng in J. Wash. Ac. Sci. 28 (1938) 303; Grubov, Consp. fl. MNR (1955) 72.—*P. glauca* (Regel) Krecz. ex Persson in Bot. notiser (1938) 275; Fl. Tadzh. 1 (1957) 228; Ikonnikov, Opred. rast. Pamira [Key to Plants of Pamir] (1963) 64.—*Poa distans* L. Mant. pl.1 (1767) 32.—*Glyceria distans* (L.) Wahl. Fl. Upsal. (1820) 36; ? Hemsley, Fl. Tibet (1902) 204; ? Forbes and Hemsley, Index Fl. Sin. 3 (1904) 428, p.p.—*Atropis distans* (L.) Griseb. in Ledeb. Fl. Ross. 4 (1852) 388; Krylov, Fl. Zap. Sib. 2 (1928) 310, p.p.; Pavlov in Byull. Mosk. obshch. ispyt. prir. 38 (1929) 25, p.p.;? Pampanini, F1. Carac. (1930) 77; Kreczetowicz in Fl. SSSR, 2 (1934) 484.—*A. distans* β. *glauca* Regel in Acta Horti Petrop. 7, 2 (1881) 623.—*A. glauca* (Regel) Krecz. in Fl. SSSR, 2 (1934) 484. —Ic.: Fl. SSSR, 2, Plate 36, figs.18–19; Keng, l.c. (1959) fig. 184.

Described from Europe. Type in London (Linn.).

In grasslands, riverine sand and pebble beds, along banks of water reservoirs, often as weed in oases, along irrigation ditches, along roadsides; up to upper mountain belt.

IA. Mongolia: *Cis-Hing.* (near Yaksha railway station, June 13, 1902—Litw.); *Val. lakes* (Orok-Nor, No. 557, 1925—Chaney); *Gobi-Alt.* (hilly area south of Ikhe-Bain-Ula in region of Tavun-Khobur-Khuduk col., 1709 m, Aug. 2, 1948—Grub.); *East. Gobi* ("Madoni-Ama, Peiling-Miao, Roerich Exped., No. 535, 1935; Batu-Khalkin-Gol, vicinity of Temur-Khoda, Roerich Exped., No. 805, 1935; by side of river Shara-Muren, Roerich Exped., No. 859, 16 VIII 1935"—Keng, l.c. [1938]); *Alash. Gobi* (Edzin-Gol river, Chzhargalante area, June 17, 1909—Czet.).

IB. Kashgar: Nor., West.

IC. Qaidam: *hilly* (Kurlyk-Nor, 3000 m, May 31, 1895—Rob.).

IIA. Junggar: *Tien Shan* (B. Yuldus basin, 2460 m, No. 6470, Aug. 10, 1958—Lee and Chu); *Jung. Gobi* (Baitak-Bogdo-Nuru mountain range, Ulyastu-Gola gorge 3–4 km from estuary, 1948—Grub.).

IIIC. Pamir.

General distribution: Aral-Casp., Balkh. region, Jung.-Tarb., Nor. and Cent. Tien Shan, East. Pam.; Europe, Mediterr., Balk.-Asia Minor, Near East, Caucasus, Middle Asia (hilly), West. Sib. (south.), East. Sib. and Far East (only as introduced plant), China (Dunbei, Nor.), Himalayas (West., Kashmir); introduced in many other countries of both hemispheres.

Note. Southern populations of this fairly polymorphic species with very stiff greyish-green leaf blades and slightly smaller (0.5–0.6 and not 0.6–0.8 mm long) anthers are sometimes treated as an independent species, *P. glauca* (Regel) Krecz. ex Persson (l.c.). These populations differ somewhat from typical specimens of *P. distans* (L.) Parl. and specimens described from Inner Mongolia, *P. filiformis* Keng (l.c.), which also have small (0.5–0.6 mm long) anthers and less spreading panicles.

3. **P. filifolia** (Trin.) Tzvel. in Novosti sist. vyssh. rast.(1964) 18.—
Colpodium filifolium Trin. in Bull. Ac. Sci. St.-Pétersb. 1(1836) 69.

Described from Mongolia (East. Gobi). Type in Leningrad.
In solonchak with sandy and loamy soils.

IA. **Mongolia:** *Cent. Khalkha* (25–30 km from Sorgol-Khairkhan hill along old road to Dalan-Dzadagad, July 16; 15 km south of Erdeni-Dalai somon, July 17—1943, Yun.); *East. Gobi* (in salsis Mongholiae mediae, 1831—Bunge, type !; sand hillocks near Sain-Usu, Aug. 20, 1926—Lis.; south-east. side of Baishintin-Sume, Orle area, Aug. 17, 1927—Zam.; Sain-Usu col. On Kalgan road between sand knolls, Aug. 10, 1931—Pob).

General distribution: endemic.

Note. This species is very close to *P. tenuiflora* (Griseb.) Scribn. et Merr. described recently but widely known.

o **P. florida** Cui, in Bull. Bot. Research (Harbin), 17, 2 (1997) 122.

IIA. **Junggar.** *Tien Shan.*

4. **P. hackeliana** (Krecz.) Persson in Bot. notiser (1938) 275; Fl. Kirgiz. 2 (1950) 143; Fl. Kazakhst. 1 (1956) 248; Fl. Tadzh. 1 (1957) 230; Ikonnikov, Opred. rast. Pamira [Key to Plants of Pamir] (1963) 64.—*Atropis hackeliana* Krecz. in Fl. SSSR, 2 (1934) 484 and 762. —Ic.: Fl. SSSR, 2, Plate 36, fig. 20.

Described from Pamir (Sasyk-Kul' lake). Type in Leningrad.

In riverine sand and pebble beds, on rocky and rubble slopes, sometimes along roadsides and wastelands; in middle and upper mountain belts.

IA. **Mongolia:** *Mong.-Alt.* (Adzi-Bogdo mountain range, Ara-Tszusylan-Ikhe-Gol water divide, rubble-stone talus, Aug. 7, 1947—Yun.).

IB. **Kashgar:** *Nor.* (around Bai town, 1580 m, No. 8160, Sept. 2, 1958—Lee and Chu; west. rim of Bai basin 4–5 km south of Kzyl-Bulak settlement, Sept. 13, 1958—Yun.); *West.* (Jerzil, 3200 m, Aug. 14, 1930—Persson).

IIIB. **Tibet:** *Chang Tang* (nor. slope of Russky mountain range, upper course of Aksu river, 3500–4000 m, June 3, 1890—Rob.).

IIIC. **Pamir** (Toili-Bulung area on Pas-Rabat river, Aug. 2, 1909—Divn.).

General distribution: Cent. Tien Shan, East. Pam.; Middle Asia (Pam.-Alay), Himalayas (West.).

Note. Like *P. distans* (L.) Parl., this species is very easily propagated by man and animal, which probably explains the rather frequent reports of these species together, alongwith the formation of hybrids.

5. **P. hauptiana** (Krecz.) Kitag. Lin. Fl. Mansh. (1939) 90; Grubov, Consp. fl. MNR (1955) 72; Fl. Kazakhst. 1 (1956) 247.—*P. kobayaschii* Ohwi in Acta Phytotax. et Geobot. 4 (1935) 31; Keng, Fl. ill. sin., Gram. (1959) 230.— *P. distans* var. *micrandra* Keng in Sunyatsenia, 6, 1 (1941) 58.—? *P. tenuiflora* var. *multiflora* Norlindh, Fl. mong. steppe, 1 (1949) 104.—*P. micrandra* (Keng) Keng, l.c. (1959) 232.—*P. iliensis* (Krecz.) Serg. in Krylov, Fl. Zap. Sib. 12, 1 (1961) 3116.—*Atropis hauptiana* Krecz. in Fl. SSSR, 2

(1934) 485.—*A. iliensis* Krecz. l.c. 485. —Ic.: Fl. SSSR, 2, Plate 36, figs. 21–22; Keng, l.c. (1959) fig. 185 (sub. *P. kobayaschii* Ohwi) and 186 (sub. *P. micrandra* Keng).

Described from West. Siberia (Kurgan district). Type in Leningrad.

In solonetz meadows, riverine sand and pebble beds, along banks of water reservoirs, often as week along roadsides, around irrigation ditches, and in wastelands; up to midmountain belt.

IA. Mongolia: *Mong. Alt.* (Bulugun river valley, Sept. 18, 1930—Bar.); *Cent. Khalkha* (Bain-Ula hill along old road to Dalan-Dzadagad, under rocks, July 12, 1948—Grub.); *East. Mong., Bas, lakes* (Kharkira river valley, June 9, 1879—Pot.); *Val. lakes* (Tatsin-Gol river, July 26, 1923—Pisarev; "Kolobolchi-Nor, No 288; Tsagan-Nor, No. 503, 1925, leg. Chaney"—Keng, l.c. [1941]); *Gobi-Alt., East. Gobi* (Batu-Khalka, Roerich Exped., No. 513, July 25; Bailingmiao, Roerich Exped., No. 690, Aug. 5—1935, Keng; near Beilinmyao, 1959—Ivan.); *Alash. Gobi* (Alashan hills, Tszosto gorge, May 15, 1908—Czet.); *Khesi* (Ninyanlu settlement 40 km nor. of Yunchan town, July 1, 1935—Petr.).

IB. Kashgar: *Nor.* (on Shakh'yar-Kucha town road, No. 8747, Sept. 13; near Shakh'yar settlement, No. 8776. Sept. 19—1958, Lee and Chu); *South* (8–10 km southeast of Karasai settlement along road to Karakash, May 25, 1959—Yun.); *East.* (near Bagrashkul' lake, No. 6175, July 25, 1958—Lee and Chu).

IIA. Junggar: *Alt. region* (Daban crossing near Qinhe, 2200 m, No. 1595, Aug. 8, 1956—Ching); *Tien Shan* (Tsagan-Tyunge river, 1600–2000 m, June 2, 1879—A. Reg.); *Zaisan* (Ch. Irtysh bank below Burchum river estuary, June 15, 1914—Schisch.); *Dzhark.* (Khoyur-Sumun village south of Kul'dzha, May 27, 1877—A. Reg.).

General distribution: Aral-Casp. (rare), Balkh. region, Jung.-Tarb.; Europe (east.), West. Sib., East. Sib., Far East, Nor. Mong., China (Dunbei, Nor., Nor.-West.), Korea, Nor. Amer.; introduced in many other countries.

Note. Compared to the very close species *P. distans* (L.) Parl., this species is a more eastern and older ecogeographical race. In the past, these species were no doubt more isolated and the present-day extensive overlapping of their distribution ranges is largely a consequence of their rapid dispersal by human agencies. Like *P. distans*, more southern populations of *P. hauptiana* have a more xeromorphic habit and slightly smaller anthers and lemma. *P. iliensis* (Krecz.) Serg. (l.c.) and *P. micrandra* (Keng) Keng (l.c.) have been described from such specimens.

6. P. himalaica Tzvel. in Bot. mater. Gerb. Bot. inst. AN SSSR, 17 (1955) 67; Bor, Grasses Burma, Ceyl., Ind. and Pakist. (1960) 562.— ? *Atropis convoluta* var. *glaberrima* Hack. in Ann. Naturhist. Hofmus. Wien, 22 (1907) 32.— ? *A. distans* var. *convoluta* f. *glaberrima* (Hack.) Pampanini, Fl. Carac. (1930) 77.

Described from Kashmir. Type in Leningrad.

In grasslands, riverine pebble beds, along banks of water reservoirs; in upper mountain belt.

IIIB. Tibet: *Chang Tang* (Karakash valley, 4500 m, Juy 31, 1870—Henderson; hills south of Khotan and Kerii, 5570 m, July 20, 1892—Rhins).

IIC. Pamir (Gëz-Darya river valley 60–65 km south of Upal village on Kashgar-Tashkurgan road, June 15, 1959—Yun.).

General distribution: Himalayas (west., Kashmir).

Note. This species is probably a high-altitude derivative of *P. hauptiana* (Krecz.) Kitag. with which it is closely related.

.7. **P. humilis** (Krecz.) Roshev. in Fl. Kirgiz. 2 (1950) 140; Bor in Nytt Mag. Bot. Oslo, 1 (1952) 19; Fl. Kazakhst. 1 (1956) 248; Fl. Tadzh. 1(1957) 224; Ikonnikov, Opred. rast. Pamira [Key to Plants of Pamir] (1963) 64.— *Atropis humilis* Krecz. in Fl. SSSR, 2 (1934) 473 and 759. —Ic. : Fl. SSSR, 2, Plate 35, fig. 5.

Described from Middle Asia (Alay mountain range). Type in Leningrad.

In solonetz grasslands, riverine pebble beds, on rocky and melkozem slopes; in upper mountain belt.

IB. Kashgar: *West.* (Torugart settlement nor.-west of Kashgar, 3500 m, June 20, 1959—Yun.

IIIC. Pamir (Chicheklik river valley, July 28, 1909—Divn.).

General distribution: Nor. and Cent. Tien Shan, East. Pam.; Middle Asia (Pamir-Alay), Himalayas (west.).

Note. There is no doubt about the very close relationship of this species with *P. hackeliana* (Krecz.) Persson inhabiting relatively low hills.

O **P. kreczetoviczii** Bubnova, in Bot. Zhurn. 73, 9 (1998) 1334; Gubanov, Consp. Fl. Outer Mong. (1996) 23.

IA. Mongolia. *Mong. Alt.*

8. **P. ladyginii** Tzvel. in Bot. mater. Gerb. Bot. inst. AN SSSR, 17 (1955) 65. —Ic.: Tzvel. l.c. fig. 3.

Described from Tibet. Type in Leningrad.

In grasslands, on rocky slopes and rocks; in upper mountain belt.

IIIB. Tibet: *Weitzan* (nor. slope of Burkhan-Budda mountain range, Khatu gorge, 4300–4600 m, No. 268, July 18, 1901—Lad., type!).

General distribution: endemic.

9. **P. macranthera** (Krecz.) Norlindh, Fl. mong. steppe, 1 (1949) 102; Grubov, Consp. fl. MNR (1955) 72.—*P. jeholensis* Kitag. in Rep. First. Sci. Exped. Manch., sect. 4, 4 (1936) 102; ej. Lin. Fl. Mansh. (1939) 90; Keng, Fl. ill. sin., Gram. (1959) 226.—*P. poaeoides* Keng in J. Wash. Ac. Sci. 28 (1938) 301.—*P. palustris* subsp. *jeholensis* (Kitag.) Norlindh, l.c. 105.—*Atropis macranthera* Krecz. in Fl. SSSR, 2 (1934) 471 and 758.—*Glyceria convoluta* auct. non Fries: Franch. Pl. David 1 (1884) 338. —Ic.: Keng, l.c. (1938) fig. 2 and l.c. (1959) fig. 181.

Described from East. Siberia (Shiro lake). Type in Leningrad.

In solonchak, saline meadows, along banks of brackish water reservoirs; up to lower mountain belt.

IA. Mongolia: *Mong. Alt.* (2–3 km south-east of Yusun-Bulak, July 13, 1947—Tuvanzhab; east. bank of Tonkhil'-Nur lake, July 16, 1947—Yun.; south. bank of Tonkhil'-Nur lake, Sept. 7, 1948—Grub.); *Cis-Hing.* (near Siguitutsi settlement, No.

996, June 29, 1951—Li et al.); *Cent. Khalkha* (Dzaragiin-Gol valley along road to Khadasan 40 km from Khashiat somon camp, June 24, 1948—Grub.); *East. Mong., East. Gobi* ("Djatono-Khuduk, 18 VII 1920, leg. Andersson; Khujurtu-Gol, 22 VI and 3 VII, 1927, leg. Hummel"—Norlindh, l.c.; 'A shown Coop, about 5 miles north-east of Naran-Obo, Peiling-Miao, No. 3395,VIII, 1935"—Keng, l.c. [1938]).

General distribution: East. Sib. (south.), Nor. Mong., China (Dunbei, Nor., Nor.-West.).

Note. Notwithstanding the absence of a very close affinity between *P. macranthera* and *P. tenuiflora* (Griseb.) Scribn. et Merr., these species hybridise readily and, as a result, are related through a large number of transitional specimens and populations.

o P. microanthera Cui, in Bull. Bot. Research (Harbin), 17, 2 (1997) 125.

IIA. Junggar: *Cis-Alt. Tara., bag., Tian Schan.* IIIB. Tibet. *Chang Tang.*

o P. mongolica (Norlindh) Bubnova, in Bot. Zhurn. 73, 9 (1998) 1336; Gubanov, Consp. fl. Outer Mong (1996) 23.

IA. Mongolia. *Val. Lakes.*

10. P. nudiflora (Hack.) Tzvel. in Bot. mater. Gerb. Inst. bot. AN Uzbek SSR, 17 (1962) 76.—*P. pauciramea* (Hack.) Krecz. ex Roshev. in Fl. Kirgiz. 2 (1950) 140; Fl. Kazakhst. 1(1956) 246; Fl. Tadzh. 1 (1957) 227; Ikonnikov, Opred. rast. Pamira [Key to Plants of Pamir] (1963) 65.—*Poa nudiflora* Hack. in Oesterr. bot. Z. 52 (1901) 453.—*Atropis distans* f. *pauciramea* Hack. in Acta Horti Petrop. 21 (1903) 442.—*A. pauciramea* (Hack.) Krecz. in Fl. SSSR, 2 (1934) 477 and 760.—Ic.: Fl. SSSR, 2, Plate 36, fig. 8.

Described from Cent. Tien Shan (Aksu river basin). Type in Vienna. Map 9.

In rather solonetzic grasslands, riverine pebble beds, on rocky and rubble slopes; in upper mountain belt.

IA. Mongolia: *Mong. Alt.*—(Adzhi-Bogdo mountain range, Ara-Tszusylan—Ikhe-Gol water divide, rubble-stone talus, Aug. 7, 1947—Yun.); *Gobi-Alt.* (stone talus in upper belt of Ikhe-Bogdo mountain range, June 29, 1945—Yun.).

IB. Kashgar: *West.* (between Kashgar and Torugart, 3500 m, No. 9741, June 20, 1959—Lee et al.).

IC. Qaidam: *hilly* (rock bed of Ichegyn-Gol river, 3300 m, June 20, 1895—Rob.).

IIIB. Tibet: *Chang Tang* (Raskem- and Tiznaf-Darya water divide 3–4 km south of Saryk crossing, 5000 m, July 4, 1959—Yun.).

IIIC. Pamir (Billuli river at its discharge into Gumbus river, June 11, 1909—Divn.; upper Kaplyk river, 4500–5000 m, July 14, 1942—Serp.).

General distribution: Cent. Tien Shan, East. Pam.; Middle Asia (Alay).

Note. Typical specimens of *P. nudiflora* are small-sized with dense and compressed panicle while those of *P. pauciramea* (Hack.) Krecz. ex Roshev. are much larger with panicles spreading at anthesis. Evidently, no significant taxonomic importance should be attached to this feature.

11. P. pamirica (Roshev.) Krecz. ex Roshev. in Fl. Kirgiz. 2(1950) 140; Fl. Kazakhst. 1 (1956) 245; Fl. Tadzh. 1 (1957) 224; Ikonnikov, Opred. rast.

Pamira [Key to Plants of Pamir] (1963) 65.—*Atropis distans* f. *pamirica* Roshev. in Acta Horti Petrop. 38 (1924) 121.—*A. pamirica* (Roshev.) Krecz. in Fl. SSSR, 2(1934) 474 and 759.

Described from Pamir (Kara Kul' lake). Type in Leningrad.

In grasslands and riverine pebble beds; in upper mountain belt.

IIIA. Qinghai: *Nanshan* (Sharagol'dzhin river, Buglu-Tologoi area, June 14, 1894—Rob.).

IIIB. Tibet: *Chang Tang* (Karakash valley, July 31, 1870—Henderson).

IIIC. Pamir.

General distribution: Cent. Tien Shan, East. Pam.; Middle Asia (Pam.-Alay), Himalayas (Kashmir).

12. **P. przewalskyi** Tzvel. in Bot. mater. Gerb. Bot. inst. AN SSSR, 17 (1955) 63.

Described from Qaidam. Type in Leningrad.

On moist solonchak, along banks of salt lakes; up to upper mountain belt.

IA. Mongolia: *Bas. lakes* (Dzeron-Hur lake, July 25, 1879—Pot.).

IB. Qaidam: *plains* (south-east. Qaidam, 3000 m, No. 511, Aug. 16, 1884—Przew., type!).

General distribution: endemic.

13. **P. roborovskyi** Tzvel. sp. nova.—Planta perennis, 20–40 cm alta, laxe caespitosa; vaginae glabrae et laeves; ligulae 0.8–2 mm lg.; laminae 0.5–1.2 mm lt., convolutae, glauco-virides, subtus laeves, supra laeves vel sublaeves. Paniculae 5–10 cm lg., effusae, sat paucispiculae, ramis longis (ad 6 cm lg.), in parte suprema scabris et spiculas 1–4 gerentibus, in parte inferiore laevibus vel sublaevibus; spiculae 3–5-florae, plus minusve roseoviolaceo tinctae; glumae lemmatis mul to breviores, inaequilongae, lanceolatae, apice acutae vel obtusiusculae; lemmata 2.8–4.2 mm lg., lanceolata, apice longe acutata, prope basin (praesertim secus nervos) pilosa; paleae secus carinas scabris, in parte inferiore saepe plus minusve pilosae; antherae 1–1.3 mm lg.

Typus: Tibet bor.-occid., in declivitate boreali jugi Rossici, fl. Aksu superior, Tasch-Bulak, in solo arenoso humido, 17 VI 1890, W. Roborovsky. In Herb. Inst. Bot. Acad., sci. URSS—LE conservatur.

Affinitas: Haec species a speciebus proximis differt: a *P. hackeliana* (Krecz.) Drob.—*foliorum* laminis angustioribus (0.5–1mm lt.) supra laevibus vel sublaevibus et caespitibus mul to laxioribus, a *P. nudiflora* (Hack.) Tzvel.—lemmatis prope basin pilosis et paleis in parte superiore secus carinos scabris.

Described from Tibet. Type in Leningrad. Plate VIII, fig. 2.

IIIB. Tibet: *Chang Tang* (nor. slope of Russky mountain range, Tash-Bulak on Upper Aksu river, in wet sandy soil, June 17, 1890—Rob., type!).

General distribution: endemic.

Note. In the structure of spikelets, this species is more similar to large specimens of *P. hackeliana* (Krecz.) Persson. Its mat is very lax and very narrow, however; leaf blade almost smooth on upper surface and branches of panicles somewhat scabrous. Concomitantly, fairly profuse pubescence on lemma as well as general habit clearly distinguish it from the other relatively close speices *P. nudiflora* (Hack.) Tzvel. It is quite likely that *P. roborovskyi* is a hybrid of these two species (one of them, *P. hackeliana*, is also known from the nor. slope of Russky mountain range) but this assumption requires substantiation.

14. **P. roshevitsiana** (Schischk.) Krecz. ex Tzvel. in Bot. mater. Gerb. Bot. inst. AN SSSR, 17 (1955) 60; Fl. Kazakhst. 1 (1956) 244; Krylov, Fl. Zap. Sib. 12 (1961) 3114.—*Atropis roshevitsiana* Schischk. in Sist. zam. Gerb. Tomsk. univ. 3 (1929) 1. —Ic.: Tzvelev, l.c. fig. 2.

Described from Junggar. Type in Tomsk, isotype in Leningrad.

In solonetz and saline meadows and chee grass thickets.

IIA. **Junggar:** *Zaisan* (right bank of Ch. Irtysh river below Burchum river estuary between Sary-Dzhasykarea and Kiikpai collective, June 15, 1914—Schisch., Type!). General distribution: Balkh. region (east.).

Note. This species occupies a somewhat intermediate position between two species lacking close affinity—*P. schischkinii* Tzvel. and *P. dolicholepis* (Krecz.) Pavl., suggesting its possible hybrid origin. However, *P. dolicholepis* widely distributed in solonchak in Kazakhstan, has not been reported thus far from Chinese Junggar.

15. **P. schischkinii** Tzvel. in Bot. mater. Gerb. Bot. inst. AN SSSR, 17 (1955) 57 and 20 (1960) 415; Fl. Kazakhst. 1 (1956) 245; Ikonnikov, Opred. rast. Pamira [Key to Plants of Pamir] (1963) 64.—*P. palustris* subsp. *filiformis* (Keng) Norlindh, Fl. mong. steppe, 1 (1949) 111, quoad pl.—*P. kuenlunica* Tzvel. l.c. (1955) 62.—*Atropis roshevitsiana* auct. non Schischk. : Kreczetowicz in Fl. SSSR, 2 (1934) 493.—*Puccinellia roshevitsiana* auct. non Krecz. ex Tzvel.: Grubov, Consp. fl. MNR (1955) 72. —Ic.: Tzvelev, l.c. (1955) fig. 1.

Described from East. Kazakhstan (Zaisan lake region). Type in Leningrad. Map 9.

In wet solonchak, along banks of salt lakes; up to upper mountain belt.

IA. **Mongolia:** *Cent. Khalkha* (nor. of Delger-Khangai, Aug. 1, 1931—Ik.-Gal.); *Gobi-Alt.* (Bain Tukhum area, Aug. 12, 1931—Ik.-Gal.; same site, July, 1993—Simukova; south. trail of Artsa-Bogdo mountain range, Dzhirgalant-Khuduk collective, July 20, 1948—Grub.); *West. Gobi* (Nusha, June–July, 1879—Przew.; Noyan somon, Bilgekhu-Bulak area 30–35 km east of Tsagan-Bogdo,—July 31; Bain-Gobi somon, Tsagan-Burgasun area, Aug. 8; Tseel'somon, Tszakhoi-Tszaram area in foothill plain south of Mong. Altay, Aug. 18—1943, Yun.); *Alash. Gobi* ("Bayan-Bogdo Camp, Wan-Tsun-Hai-Tze, 18 VI 1929, leg. Söderbom; Khoburin-Nor, Aug. 29; Tukhumin-Gol, 1, IX, 1927, leg. Hummel"—Norlindh, l.c.).

IB. **Kashgar:** *East.* (east of Irakhu village in Toksun settlement zone, No. 1270, June 15; nor.-east of Toksun settlement, No. 1328, June 19; 5 km east of Botakhu settlement on road to Kurl', 1000 m, No. 6832, July 16; near Yuili west of Bagrashkul' lake, No. 8588, Aug. 11—1958, Lee and Chu).

IC. **Qaidam:** *plains* (south-east. Qaidam, 3000 m, Aug. 16, 1884—Przew.; Syrtyn valley, 3000 m, July 20, 1895—Rob.).

IIA. Junggar : *Tien Shan* (near Ulausu, No. 321, July 3; south-west. bank of Barkul 'lake, No. 2146, Sept. 25—1957, Huang); *Jung. Gobi* (nor.: between Barbagai and Burchum 540 m, No. 2997, Sept. 6, 1959—Lee et al.; nor.-west.: Darbaty river valley at its intersection with Karamai-Altay road, June 20; west.: Kzyl-Tuz meteorological station near Junggar outlet, Aug. 16—1957, Yun.).

IIIA. Qinghai: *Amdo* (Dulan-Khit monastery, 3100 m, Aug. 5, 1901—Lad.).

General distribution: Balkh. region (east.), Nor. and Cent. Tien shan, East. Pam.; East. Sib. (only near Tussykh lake in Tuva Autonomous Region).

16. P. tenuiflora (Griseb.) Scribn. et Merr. in Contribs. U.S. Nat. Herb. 13 (1910) 78, quoad nom.; Kitag. Lin. Fl. Mansh. (1939) 90; Norlindh, Fl. mong. steppe, 1 (1949) 103, p.p.; Fl. Kirgiz. 2 (1950) 144; Grubov, Consp. fl. MNR (1955) 73; Fl. Kazakhst. 1 (1956) 244, p.p.; Keng, Fl. ill. sin., Gram. (1959) 230; Tzvelev in Bot. mater. Gerb. Bot. inst. AN SSSR, 20 (1960) 416; Krylov, Fl. Zap. Sib. 12, 1 (1961) 3114; Ikonnikov, Opred. rast. Pamira [Key to Plants of Pamir] (1963) 64.—*P. chinampoensis* Ohwi in Acta Phytotax. et Geobot. 4 (1935) 31; Kitag. l.c. 89.—*P. manchuriensis* Ohwi, l.c. 31.—*Atropis tenuiflora* Turcz. ex Griseb. in Ledeb. Fl. Ross. 4 (1852) 389; Danguy in Bull. Mus. nat. hist. natur. 20 (1914) 147; Pavlov in Byull. Mosk. Obshch. ispyt. prir. 38 (1929) 25; Kreczetowicz in Fl. SSSR, 2 (1934) 493.— *A. distans* var. *tenuiflora* (Griseb.) Kryl. Fl. Zap. Sib. 2 (1928) 310.—**Ic.**: Fl. SSSR, 2, Plate 38, fig. 33; Norlindh, l.c. Table 15 (var. *mongolica* Norlindh); Keng, l.c. fig. 183.

Described from Transbaikal (Selenga river basin). Type in Leningrad. Map. 9.

In solonchak meadows, along banks of brackish water reservoirs, in riverine sand and pebble beds; up to upper mountain belt.

IA. Mongolia: *Mong. Alt., Cis-Hing.* (Khuntu somon 18 km south-east of Bain-Tsagan, Aug. 6, 1949—Yun.); *Cent. Khalkha, East. Mong., Bas. lakes, Val. lakes, Gobi-Alt., East. Gobi, West. Gobi* (Tsagan-Bogdo mountain range, Suchzhi-Bulak col., July 4, 1943—Yun.); *Alash. Gobi, Khesi* (between Gaotai town and Fuitin village, June 5, 1886—Pot.).

IIA. *Junggar: Jung. Alt.* (south. slope of Barlyk mountain range, Arba-Kezen' river valley, July 9, 1905—Obruchev; Maili mountain range 4 km south of Dzhirmas, 1300 m, July 18, 1953—Mois.); *Tien Shan* (from Barkul' lake to Beisyan, No. 4910, Sept. 26, 1957—Huang; Muzart river upper course, Sazlik area, Sept. 8, 1958—Yun.); *Zaisan* (between Karoi area and village on Kabe river, July 16, 1914—Schisch.).

General distribution: Jung.-Tarb., Cent. Tien Shan, East. Pam.; West. Sib. (Altay), East. Sib. (south., Cent. Yakutia), Nor. Mong., China (Dunbei, Nor.), Korea.

Note. Typical specimens of this species have fairly small (1.5–1.9 mm long) glabrous lemma while the latter in most specimens from Siberia as well as Cent. Asia are rather pilose at the base and often larger (up to 2.2 mm long). Such specimens can be regarded as hybrids of *P. tenuiflora* (Griseb.) Scribn. et Merr. × *P. macranthera* (Krecz.) Norlindh. There is no doubt of the existence of these hybrids. Nevertheless, it is quite possible that they belong to an independent, although very poorly isolated ecogeographical race on plains already described under the name *P. chinampoenis* Ohwi (l.c.) from Korea and *P. manchuriensis* Ohwi (l.c.) from Harbin district. Variety *P. tenuiflora* var. *mongolica* Norlindh (l.c. 105, Table 15) described from a quite typical specimen of *P. tenuiflora*, but with lemma about 2 mm long, hardly deserves this taxonomic rank.

60. Festuca L.

Sp. pl. (1753) 73.

1. Ligules 4–7 mm long; leaf blade flat or somewhat longitudinally convoluted, 2.5–6 mm broad, strongly scabrous in lower part, without auricles at base. Panicles 15–30 cm long, spreading; lemma 8–10 mm long, without awn ovary puberulent at tip .... 6. **F. ladyginii** Tzvel.

+ Ligules up to 2.5 mm long .... 2.

2. Leaf blade 3–15 broad, flat, sometimes partly convoluted longitudinally; lanceolate auricles at base. Panicles 15–35 cm long, rather spreading at anthesis; spikelets 8–15 mm long; lemma 6–8 mm long .... 3.

+ Leaf blade up to 3 mm broad, usually longitudinally folded or convoluted, occasionally partly flat, without auricles at base .... 4.

3. Tip of lemma with 10–18 mm long awn. Leaf blade 7–15 mm broad; ligule 1–2.5 mm long .... 20. **F. gigantea** (L.) Vill.

+ Tip of lemma with up to 2.5 mm long awn or cusp leaf blade 3–8 mm broad; ligule 0.5–1.3 mm long .... 21. **F. orientalis** (Hack.) Krecz. et Bobr.

4. Anthers 0.5–1.4 mm long. Plants monoecious with bisexual flowers. Leaf blade longitudinally folded, 0.4–1 mm in diameter, smooth in lower part .... 5.

+ Anthers 1.6–4 (5) mm long. Plants sometimes dioecious .... 8.

5. Anthers 1–1.4 mm long; lemma 3.5–4.5 mm long, dull or slightly lustrous, with 0.6–1.8 mm long awn at tip; ovary glabrous. Plant 8–20 cm tall, densely caespitose .... 9. **F. coelestis** (St.-Yves) Krecz. et Bobr.

+ Anthers 0.5–0.8 mm long; lemma 4–5.5 mm long, lustrous .... 6.

6. Ovary glabrous at tip; lemma in lower half smooth or almost so, with 0.8–2 mm long awn at tip; panicles dense and rather compressed. Plant 8–20 cm tall, densely caespitose .... 8. **F. brachyphylla** Schult. et Schult.f.

+ Tip of ovary covered with short stiff hairs; lemma with scattered spinules even in lower half .... 7.

7. Plant 10–30 cm tall, densely caespitose; ligules 0.3–1 mm long. Panicles 4–7 cm long, fairly dense; lemma with 1.5–3 mm long awn at tip .... 15. **F. nitidula** Stapf.

+ Plant 20–40 cm tall, loosely caespitose; ligules 0.1–0.3 mm long. Panicles 5–8 cm long, spreading at anthesis; lemma with 0.5–1.5 mm long awn or cusp at tip .... **F. venusta** St.-Yves.

8(4). Plant 15–50 cm tall, with short-creeping subsurface shoots; blade of cauline leaves often flat, up to 3 mm broad; sheath of leaves of vegetative shoots closed almost throughout length. Panicles 4–10 cm long, somewhat spreading .... 18. **F. rubra** L.

+ Plant forming dense mats, without creeping subsurface shoots .. .. 9.

9. Leaf blade longitudinally folded or convoluted, 1–2 mm in diameter, occasionally partly flat and then up to 3 mm broad, Lemma (5.5) 6–9 (9.5) mm long; ovary with short stiff hairs at tip, but occasionally with only few hairs .... 10.

+ Leaf blade of all leaves longitudinally folded, less than 1 mm in diameter. Lemma (2.5) 3–6 (6.5) mm long; ovary usually glabrous at tip, less often puberulent .... 13.

10. Plant dioecious. Panicles usually dense and short, 3–8, less often upto 12 cm long, very lax; branches short ascending; lemma pale green, less often with faint pinkish-violet fringe, a unawned ovary densely pilose at tip .... 11.

+ Plant monoecious. Panicles widely spreading, 7–20 cm long, with long branches; lemma almost invariably pinkish-violet, without awn or with short awn; ovary with only few hairs at tip .... 12.

11. All vegetative shoots intravaginal, covered at base with numerous lustrous straw-coloured sheaths of dead leaves not disintegrating into fibres; ligule of upper cauline leaf 0.5–1.7 mm long. Panicles 3–12 cm long, usually with rather interrupted spikelets; lemma 5.5–9 mm long .... 1. **F. olgae** (Regel) Krivot.

+ Vegetative shoots partly extravaginal, similarly covered at base with numerous but duller light grey sheaths of dead leaves which disintegrate into fibres on breakage; ligules of upper cauline leaf 0.1–0.3 mm long. Panicles 3–7 cm long, with close-set spikelets; lemma 5.5–7 mm long .... 2. **F. sibirica** Hack. ex Boiss.

12. Lemma 6–7 mm long, smooth or almost so, lustrous in lower half. Sheath smooth; leaf blade smooth in lower part, less often somewhat scabrous .... 3. **F. alatavica** (St.-Yves) Roshev.

+ Lemma 6.5–9.5 mm long, surface entirely covered with short spinules, duller. Sheath and leaf blade rather scabrous in lower part due to sparse spinules .... 4. **F. altaica** Trin.

13 (9). Panicles 5–12 cm long, with long, nearly smooth branches, lemma 4.5–6.5 mm long, with 0.6–2 mm long cusp or awn at tip; ovary with few short hairs at tip. Leaf blade 0.3–0.6 mm in diameter, green, with poorly developed strands of mechanical tissue ..... 14.

+ Panicles 2–8 cm long, with short, scabrous branches throughout length; lemma 3–5.8 mm long, with or without awn at tip; ovary usually glabrous, less often (in *F. forrestii* St.-Yves) with few hairs at tip .... 15.

14. Panicles widely spreading; spikelets dark-coloured, pinkish-violet. Leaf blade with 5 ribs in lower part and scabrous along ribs due to short spinules .... 7. **F. tristis** Kryl. et Ivanitzk.

+ Panicles with fairly long, ascending branches; spikelets greenish or with slight pinkish-violet tinge. Leaf blade in lower part with indistinct ribs, smooth .... 5. **F. erectiflora** Pavl.

15. Lemma without awn at tip but sometimes with up to 0.5 mm long cusp ....16.

+ Lemma at tip with 0.8–3.5 mm long awn ....17.

16. Plant densely caespitose. Leaf blade very stiff, smooth in lower part (outer surface); 3(5) well-developed strands of mechanical tissue in cross-section. Panicles dense, erect; lemma 4.5–5.8 mm long .... 10. **F. dahurica** (St.-Yves) Krecz. et Bobr.

+ Plant loosely caespitose. Leaf blade less stiff, somewhat scabrous in lower part, with 5–7 relatively poorly developed strands of mechanical tissue in cross-section; intermediate tissues fall opposite ribs. Panicles very lax, often pendent at tip; lemma 3–4.5 mm long ....12. **F. jacutica** Drob.

17. Cross-section of leaf blade with continuous subepidermal sheath of mechanical tissue of uniform thickness; leaf blade rather scabrous or smooth on outer surface. Lemma 3–4 mm long ....16. **F. ovina** L.

+ Cross-section of leaf blade usually with strands of mechanical tissue: central and marginal; less often, 2 or 4 more intermediate strands present ....18.

18. Lemma 4.5–5.5 mm long, brownish-green, often with golden hue; palea along keels, almost right up to base, with slender elongated cappilliform spinules. Leaf blade 0.4–0.8 mm in diameter, green, with weakly developed strands of mechanical tissue; sheath of leaves of vegetative shoots closed for 1/3–1/2 its length ....13. **F. kryloviana** Reverd.

+ Lemma green, occasionally with pinkish-violet tinge or glaucescent bloom; palea with very short and stubby spinules along keels.... ....19.

19. Palea at tip with 2 fairly long sharp awn-like teeth; awn of lemma 2–3.5 mm long. Panicles 3–8 cm long, fairly lax; ovary at tip with few short hairs. Plant 25–40 cm tall. Leaf blade green, with 5–7 highly prominent ribs in upper part and 5–7 poorly formed strands of mechanical tissue in cross-section. Intermediate strands fall opposite ribs .... 11. **F. forrestii** St.-Yves.

+ Palea at tip with 2 very short sharp teeth; awn of lemma usually 1–2.5 mm long. Panicles very dense; ovary glabrous. Cross-section of leaf blade usually with 3 well-developed strands of mechanical tissue, less often with 2 more poorly formed intermediate strands .... 20.

20. Sheath of leaves of vegetative shoots closed for 1/3–1/2 its length. Mats dense but shoots arising individually or in tufts surrounded by common cap of well-separated sheaths of dead leaves; shoot formation mixed. Plant 8–30 cm tall. Leaf blade with 3(5) barely distinguishable ribs in upper part. Panicles 2–3 cm long, very dense; lemma broadly lanceolate .... 14. **F. lenensis** Drob.

+ Sheath of leaves of vegetative shoots laciniate up to, or almost up to base. Mats dense, with shoots or shoot tufts barely separated from each other; shoot formation intravaginal. Plants 8–50 cm tall. Panicles 2–8 cm long, dense or lax; lemma lanceolate .... 21.

21. Sheath of leaves of vegetative shoots connate near base. Leaf blade 0.5–0.8 mm in diameter; (3) 5 (7) less prominent ribs in upper part. Lemma 4.5–5.5 mm long .... 17. **P. pseudosulcata** Drob.

+ Sheath of leaves of vegetative shoots laciniate up to base. Leaf blade 0.3–0.6 mm in diameter; (3) 5 highly prominent ribs in upper part. Lemma 3–5 mm long .... 19. **F. valesiaca** Schleich. ex Gaud.

Section 1. L e u c o p o a (Griseb.) Krivot.

1. **F. olgae** (Regel) Krivot. in Bot. mater. Gerb. Bot. inst. AN SSSR, 20 (1960) 56.—*F. deasyi* Rendle in J. Bot. (London) 38 (1900) 429; Deasy, In Tibet and Chin. Turk. (1901) 405; Hemsley, Fl. Tibet (1902) 204.—*F. sibirica* var. *deasyi* (Rendle) St.-Yves in Bull. Soc. Bot. France, 71 (1924) 132.—*Molinia olgae* Regel in Acta Horti Petrop. 7, 2 (1881) 615.—*Leucopoa* olgae (Regel) Krecz. et Bobr. In Fl. SSSR, 2 (1934) 495; Fl. Kirgiz. 2 (1950) 145; Fl. Kazakhst. 1 (1956) 256; Fl. Tadzh. 1 (1957) 203; Ikonnikov, Opred. rast. Pamira [Key to Plants of Pamir] (1963) 66.—*Festuca sibirica* auct. non Boiss.: Hemsley, l.c. 205; Pampanini, Fl. Carac. (1930) 78.—*Leucopoa albida* auct. non Krecz. et Bobr.: Bor, Grasses Burma, Ceyl., Ind. and Pakist. (1960) 544.

Described from Middle Asia (Alay). Type in Leningrad.

On talus, rocky and rubble slopes and rocks, in pebble beds of river valleys; in upper mountain belt.

IB. Kashgar: *Nor.* (nor.-west. slope of Sogdyn-Tau mountain range 10–12 km south-east of Akchii hydropower station on Kokshaal-Darya river, 2600 m, Sept. 19, 1958—Yun.); *West., South.*

IIA. Junggar: *Tien Shan.*

IIIB. Tibet: *Chang Tang* (Kerii mountain range, near Khan-Yut pass, 4000 m, July 13, 1885—Przew.; 35–38 km beyond Kyude settlement near bridge on road to Saryk pass, 4180 m, June 1, 1959—Yun.; south of Cherchen settlement, 4190 m, No. 9420, June 4, 1959—Lee et al.); *South* ("Tisum, 5000 m, no. 1848, leg. Strachey and Witterbottom"—Hemsley, l.c.).

IIIC. Pamir.

General distribution: Nor. and Cent. Tien Shan, East. Pam.; Middle Asia (West. Tien Shan, Pam.-Alay, Giss-Darv.), Himalayas (west.).

Note: All specimens from Tibet and a significant proportion of them from other regions of Cent. Asia within China have panicles with spikelets more interrupted than in typical specimens of this species while lemma are longer (7–9 mm long) and acuminate for longer length. However, as it has not been possible to draw a distinct morphogeo graphical boundary between these specimens and the typical specimen of *F. olgae*, we have deranked this species to a variety, *F. olgae* var. *deasyi* (Rendle) Tzvel. comb. nova (=*F. deasyi* Rendle, l.c.).

2. F. sibirica Hack. ex Boiss. Fl. or 5 (1884) 626; Pavlov in Byull. Mosk. obshch. ispyt. prir. 38 (1929) 26; Grubov, Consp. fl. MNR (1955) 74.—*Poa albida* Turcz. ex Trin. in Mém. Ac. Sci. St.-Pétersb., sér. VI, 1 (1831) 387, non *Festuca albida* Lowe (1831).—*Leucopoa sibirica* Griseb. in Ledeb. Fl. Ross. 4 (1852) 282, nom. illeg.—*L. albida* (Trin.) Krecz. et Bobr. in Fl. SSSR, 2 (1934) 495; Kitag. Lin. Fl. Mansh. (1939) 82; Norlindh, Fl. mong. steppe, 1 (1949) 113; Keng, Fl. ill. sin., gram. (1959) 134.—*L. kreczetoviczii* K. Sobol. in Bot. mater. Gerb. Bot. inst. AN SSSR, 14 (1951) 75.—Ic.: Fl. SSSR, 2, Plate 39, fig. 1; Keng, l.c. fig. 99.

Described from Baikal region. Lectotype ("In sabulosis ad Baicalem prope thermas Turrenses, 1829, leg. Turczaninov") in Leningrad.

On rocky and rubble slopes, rocks, talus, and in sand and pebble beds of river and lake valleys; up to midmountain belt.

IA. Mongolia: *Cis-Hing.* (Khalkhin-Gol somon, near Salkit hill, June 9, 1956—Dashnyam); *Cent. Khalkha* (8–10 km south of Sergulen somon on road to Dalan-Dzadagad, June 4, 1949—Yun.; hill ridge intersecting road to Dalan-Dzadagad, 72 km south of Ulan Bator, Aug. 8, 1951—Kal.); *East. Mong.* (near Manchuria railway station, June 6, 1902—Litw.; same site, May 6, 1908—Komarov; same site, 1915—Nechaeva; same site, 800 m, No. 907, July 24, 1951—Li et al.; Ara-Dzhargalant somon, slopes of hillocks, Aug. 25, 1954; Choibalsan somon 10 km west of Khatorgi, June 2, 1955—Dashnyam); *Gobi-Alt.* (Ikhe-Bogdo mountain range, Bityut en-Ama area, Sept. 12; same site, Pakhoituin-Khundei creek valley, Sept. 14; same site, near Puntsuk-Obo, Sept. 14; middle belt of Baga-Bogdo mountain range, Sept. 18—1943; Ikhe-Bogdo mountain range, Narin-Khurimt creek valley, June 28; midsection of Shishkhid creek valley, south. Steep talus slope, 2620 m, June 30—1945, Yun.).

General distribution: East. Sib. (south.), Nor. Mong., China (Dunbei).

Notes. 1. We regard the establishment of this section with this species as the type of the independent genus *Leucopoa* Griseb. Inadequately substantiated since species belonging to it are very intimately associated with species of the section Breviaristatae Krivot. (especially with *F. altaica* Trin.) and, through it, with species of section Festuca. of these three sections, very closely related, section *Leucopoa* (Griseb.) Krivot. should be regarded as more primitive, not withstanding the presence of dioecism among many of its species, and section Festuca as more advanced in evolution. Sections Bromoides Rouy and Bovinae Fries are far more isolated, perhaps justifying their placement in an independent genus, *Schedonorus* Beauv.

2. Two more species of section Leucopoa close to *F. sibirica* Hack. ex Boiss., namely *F. komarovii* Krivot. [in Bot. mater. Gerb. Bot. inst. AN SSSR, 17 (1955) 80] and *F. hubsugulica* Krivot. (l.c. 77), are reported from Nor. Mongolia (Fore Hubs.).

Section 2. Breviaristatae Krivot.

3. **F. alatavica** (St.-Yves) Roshev. in Fl. SSSR, 2 (1934) 528;? Fl. Kirgiz. 2 (1950) 156;? Fl. Kazakhst. 1 (1956) 264; Bor, Grasses Burma, Ceyl., Ind. and Pakist. (1960) 538.—*F. rubra* subsp. *alatavica* Hack. ex St.-Yves in Candollea, 3 (1928) 393.—*F. tianshanica* Roshev. l.c. 526 and 772; Fl. Kirgiz. 2 (1950) 156; Fl. Kazakhst. 1 (1956) 263; Ikonnikov, Opred. rast. Pamira [Key to Plants of Pamir] (1963) 66.—*F. altaica* auct. non Trin.:? Pampanini, Fl. Carac. (1930) 78; ? Bor, l.c. 538.—Ic.: Fl. SSSR, 2, Plate 39, fig. 6.

Described from Nor. Tien Shan. Lectotype ("Kungei Alatau, Kokoirak ad fontes fl. Kebin majoris, 13 VII 1896, leg. Brotherus") in Geneva; isotype in Leningrad.

In meadows and grasslands, on rocky and rubble slopes and rocks; in upper mountain belt.

IIA. **Junggar:** *Jung. Alt.* (20 km south of Ven'tsyuan', 2900 m, No. 1548, Aug. 14; 30 km west of Ven'tsyuan', No. 2037, Aug. 25—1957, Huang); *Tien Shan.*

IIIC. **Pamir** (Billuli river near its discharge into Chumbus river, june 12, 1909—Divn.; Yashilkan-Gunyt' settlement, 4000 m, July 21, 1941—Serp.).

General distribution: Jung. Tarb., Nor. and Cent. Tien Shan, East. Pam.; Middle Asia (West. Tien Shan, Pam.-Alay), Himalayas (west.).

Note. Of the two simultaneously published names for this species, undoubtedly synonyms, we accept the name *F. alatavica* (St.-Yves) Roshev. Since it is based on much earlier and more complete diagnosis by St.-Yves.

4. **F. altaica** Trin. in Ledeb. Fl. alt. 1 (1829) 109; Krylov, Fl. Zap. Sib. 2 (1928) 330; Pavlov in Byull. Mosk. obshch. ispyt. prir. 38 (1929) 26; Krecz. and Bobr. in Fl. SSSR, 2 (1934) 528; Grubov, consp. fl. MNR (1955) 73; Fl. Kazakhst. 1 (1956) 264. —Ic.: Ledeb. Ic. pl. fl. ross. 3 (1831) Table 288.

Described from altay. Type in Leningrad.

In meadows, on rocky and rubble slopes, talus, in larch forests; in middle and upper mountain belts.

IA. **Mongolia:** *Khobd.* (south. peak of Kharkira, June 11, 1879—Pot.); *Mong. Alt.* (Taishiri-01 mountain range, in forest, July 3, 1877—Pot.; 8 km south-east of Dzhasaktu-Khan station, larch forest, Aug. 9, 1930—Pob.; Bulugun river basin, Kharagaitu Khutul' pass, rubble scree, July 24, 1947—Yun.).

IIA. **Junggar:** *Alt. region* (near Qinhe, 2630 m, No. 1037, Aug. 4, 1956—Ching; in Altay hills, 2450 m, No. 10758, July 18, 1959—Lee et al.).

General distribution: Jung.-Tarb. (indicated for Jung. Alatau); Arct. (Asiat.), West. Sib. (Altay), East. sib., Far East (excluding south.), Nor. Mong. (Fore Hubs., Hang.), Nor. Amer. (only in Mackenzie basin).

5. **F. erectiflora** Pavl. in Byull. Mosk. obshch. ispyt. prir. 47 (1938) 79; Fl. Kazakhst. 1 (1956) 264.—Ic.: Fl. Kazakhst. 1, Plate 20, fig. 3.

Described from Nor. Tien Shan (Transili Alatau). Type in Alma Ata.

In meadows and grasslands among rock screes and rocks; in upper mountain belt.

IIA. Junggar: *Tien Shan* (upper part of Manas river basin, left bank of Danu-Gol river, sedge-carex short grass meadow among rocks, July 21, 1957—Yun.).

General distribution: Nor. Tien Shan.

6. **F. ladyginii** Tzvel. sp. nova.—Planta perennis, 50–100 cm alta, laxe caespitosa, innovationes breviter repentes emittens; culmi sub panicula scaberuli; vaginae praesertim in parte superiore scabre; ligulae 4–7 mm lg., hyalinae, margine minute ciliatae; laminae 2.5–6 mm lg., planae vel laxe convolutae, glabrae, intus valde scabrae, supra sublaeves, prope basin sine auriculis. Paniculae 15–30 cm lg., effusae, ramis longis scabris vulgo binis; spiculae 10–15 mm lg., 3–7 florae, pallide-virides; glumae anguste-lanceolatae, trinerviae, secus nervis scabrae, inferior 5.5–8, superior 7–9 mm lg.; lemmata 8–10 mm lg., lanceolata, 5-nervia, dorso scabra, apice acuta, inermia; paleae lemmatis leviter breviores, secus carinis scabris; germen apice breviter pilosum; caryopsis et antherae ignotae.

Typus: Tibet orientali, systema fl. Jangtze, secus fl. Iczu in silva juniperina, ca. 4200 m, No. 420, 28 VII 1900, V. Ladygin. In Herb. Inst. Bot. Ac. Sci. URSS—LE conservatur.

Affinitas. Haec species ut videtur speciei himalaicae *F. lucidae* Stapf proxima, sed lemmatis et glumis longioribus et angustioribus bene differt.

Described from Tibet. Type in Leningrad.

IIIB. Tibet: *Weitzan* (Yangtze river basin, along Ichu river in juniper forest, about 4200 m, No. 420, July 28, 1900—Lad., type!).

General distribution: endemic.

Note. The affinity of this wholly isolated species to section *Breviaristatae* Krivot. is not quite clear since it exhibits many features common to species of section *Phaeochloa* Griseb. emend. Krivot. [in Bot. mater. Gerb. Bot. inst. AN SSSR, 20 (1960) 59] and *Leucopoa* (Griseb.) Krivot. as well. In any case, together with the relatively close species Himalayan *F. lucida* Stapf and Himalayan-south. Chinese *F. modesta* Steud., it stands among more primitive species of the genus.

7. **F. tristis** Kryl. et Ivanitzk. in Sist. zam. Gerb. Tomsk. univ. 1 (1928) 1; Krylov, Fl. Zap. Sib. 2 (1928) 331; ? Pavlov in Byull. Mosk. Obshch. ispyt. prir. 38 (1929) 26; Krecz. and Bobr. in Fl. SSSR, 2 (1934) 527; Grubov, consp. fl. MNR (1955) 74; Fl. Kazakhst. 1 (1956) 263.—*F. sajanensis* Roshev. in Izv. Glavn. bot. sada SSSR, 38 (1929) 383.—Ic.: Fl. SSSR, 2, Plate 39, fig. 4.

Described from Altay. Lectotype (Narym mountain range, near Katon-Karagai, Sukhoi river upper course, No. 235, July 16, 1927, V. Vereshchagin) in Tomsk; isotype in Leningrad.

In meadows, on rocky slopes and rocks; in upper mountain belt.

IA. Mongolia: *Mong. Alt.* (Kharagaitu-Daba pass in upper Indertiin-Gol, alpine meadow, July 23 and 24; Bulugun river basin, in upper course of Ketsu-Sairin-Gol river, alpine meadow, July 26—1947, Yun.).

General distribution: West. Sib. (Altay), East. Sib. (Sayans), Nor. Mong. (? Hang.).

Section 3. Festuca

8. **F. brachyphylla** Schult. et Schult. f. Add. ad Mant. 3 (1827) 646; Tzvelev in Bot. mater. Gerb. Bot. inst. AN SSSR, 20 (1960) 424; Skvortsov in Arkt. fl. SSSR, 2 (1964) 221, map 72.—*F. brevifolia* R. Br. Suppl. to App. Parry's First Voy., Bot. (1824) 289, non Muhl. (1817); Krylov, Fl. Zap. Sib. 2 (1928) 313; Krecz. and Bobr. in Fl. SSSR, 2 (1934) 514; Grubov, Consp. fl. MNR (1955) 73; Fl. Kazakhst. 2 (1956) 260.—Ic.: Fl. SSSR, 2, Plate 40, figs. 10 and 11.:

Described from Arctic America. Type in London (K).

In meadows and grasslands, among carex thickets, on rocky and rubble slopes, rocks and rock screes; in upper (baldpeak) mountain belt.

IA. Mongolia: *Khobd.* (near source of Kharkira river, July 12, 1879—Pot.); *Mong. Alt.* (Tolbo-Kungei-Ala Tau mountain range, 3200 m, carex meadow, Aug. 5, 1945; nor. trail of Bus-Khairkhan mountain range, July 17, 1947; Adzhi-Bogdo mountain range, Burgasin-Daba pass between Indertiin-Gol and Dzuslangin-Gol, rubble scree, Aug. 6, 1947—Yun.), *Gobi-Alt.* (Ikhe-Bogdo mountain range, flat peak among granite outliers, 3500 m, Sept. 6; same site, carex meadow, Sept. 11—1943; same site, in upper creek valley of Narin-Khurimt, June 28; same site, above Ketsu creek valley, carex thicket, June 29—1945, Yun.; same site, ridge of mountain range, alpine meadow among granite outliers, about 3700 m, July 29, 1948—Grub.).

IIA. Junggar: *Jung. Alt.* (30 km west of Ven'tsyuan', 2700 m, No. 2035, Aug. 25; 10 km east of Taldy, 2300 m, No. 2089, Aug. 26—1957; Huang); *Tien Shan* (3 km south of Danu settlement, 3150 m, No. 431, July 18; Ulanusu river 30 km south of Nyutsyuan'tsz village, 3150 m, No. 291, July 19—1957, Huang; Manas river basin, valley of Ulanusu river near discharge of Koisu into it on road to Danu-Gol, July 19, 1957—Yun.).

IIIB. Tibet: *Weitzan* (nor. slope of Burkhan-Budda mountain range, Khatu gorge, 5000 m, July 12, 1901—Lad.).

General distribution: East. Pam., Arct., West. Sib. (Altay), East. Sib. (bald peaks), Far East (Kamchatka), Nor. Mong. (Fore Hubs., Hang.).

Note. The fairly extensive distribution of this species in Cent. Asia including its report in Pamir (Tzvelev, l.c.) suggests beyond doubt its habitation in hills of Asia even before the evolution of Circumpolar arctic flora, which it entered subsequently together with many other Asian high-altitude species. This is also supported by the presence of two more genetically related but relatively more primitive species, *F. nitidula* Stapf and *F. venusta* St. Yves, in mountains of Asia.

9. **F. coelestis** (St.-Yves) Krecz. et Bobr. in Fl. SSSR, 2 (1934) 514 and 770; Fl. Kirgiz. 2 (1950) 152; Fl. Kazakhst. 1 (1956) 262.—*F. ovina* subsp. *coelestis* St.-Yves in Candollea, 3 (1928) 376; Bor, Grasses Burma, Ceyl., Ind. and Pakist. (1960) 539.— *F. alaica* auct. non Drob. : Fl. Tadzh. 1 (1957) 198; Ikonnikov, Opred. rast. Pamira [Key to Plants of Pamir] (1963) 67.— Ic.: Fl. SSSR, 2, Plate 40, fig. 11.

Described from Cent. Tien Shan. Lectotype ("Thian-schan, mts. Célestes, valée de Djoukoutchiak, 2100–2300 m, No. 392, leg. Brocherel") in Geneva.

In meadows, on rocky, rubble and melkozem slopes, riverine pebble beds; in upper mountain belt.

IIA. Junggar: *Tien Shan.*

III B. Tibet: *Chang Tang* (upper course of Tiznaf river 3–4 km east of Saryk pass, 4800 m, June 4; Raskem-Darya and Tiznaf water divide 1–4 km south of Saryk pass, 5000 m, June—1959, Yun.).

IIIC. Pamir (Mia river gorge, 4000 m, July 21, 1941—Serp.; Kingtau mountain range 3–4 km south-east of Kosh-Kulak settlement, June 10, 1959—Yun.).

General distribution: Jung.-Tarb., Nor. and Cent. Tien Shan, East. Pam., Middle Asia (Pam.-Alay), Himalayas (west., Kashmir).

Note. This wholly isolated species was evidently of hybrid origin from *F. brachyphylla* Schult. et Schult. f. x *F. valesiaca* Gaud. The "type" of *F. coelestis* (St.-Yves) Krecz. et Bobr. from Shugnan cited by authors of Flora SSSR [Flora of the USSR] (l.c.) cannot be accepted as the type of this species due to considerations of nomenclature and also because it belongs to an altogether different species, *F. valesiaca* Gaud.

10. **F. dahurica** (St.-Yves) Krecz. et Bobr. in Fl. SSSR, 2 (1934) 517 and 771; Grubov, Consp. fl. MNR (1955) 73.—*F. ovina* subsp. *laevis* var. *dahurica* St.-Yves in Bull. Soc. Bot. France, 71 (1924) 40 and in Candollea, 3 (1929) 358.—F. ovina auct. non L.: ? Danguy in Bull. Mus. nat. hist. natur. 20 (1914) 147.—*F. ovina* var. *duriuscula* auct. non Koch: Kitag. Lin. Fl. Mansh. (1939) 76.—Ic.: Fl. SSSR, 2, Plate 40, fig. 13.

Described from Transbaikal (Selenga river basin). Type in Paris; isotype in Leningrad.

In riverine sand, steppes with sandy and loamy soil; up to lower mountain belt.

IA. Mongolia: *Cis-Hing.* (17–20 km east-south-east of Bain-Tsagan, Aug. 6; 5 km west of Toge-Gol, Aug. 7—1949, Yun.; Khalkha-Gol somon, near Khamar-Daban, June 18, 1954—Dashnyam); *Cent. Khalkha* (water divide between Ubur- and Ara-Dzhirgalante rivers, July 2, 1949—Yun.); *East. Mong.* (nor.).

General distribution: East. Sib. (south.), China (Dunbei).

11. **F. forrestii** St.-Yves in Candollea, 3 (1929) 383. —Ic.: St.-Yves, l.c. fig. 29.

Described from Kam upland. Lectotype ("N.W. Yunnan, Thali Range, ca. 3400 m, No. 2574, leg. Forrest") in Paris.

In meadows and forest glades; in middle and upper mountain belts.

IIIB. Tibet: *Weitzan* (Yangtze river basin, around Rkhombo-Mtso lake and in Ichu river valley, 4100–4400 m, No. 435, Aug. 1, 1900—Lad.).

General distribution: China (South-West.).

Note. Specimens of this species collected by V. Ladygin differ from its typical specimens (judging from the diagnosis of St.-Yves) in narrower (0.4–0.6, not 0.7–0.8 mm in diameter) leaf blade and shorter (4–5.2, not 5.5–7 mm long) lemma with 2–3.5, not 4–7 mm long awn and belong at least to a distinct variety *F. forrestii* var. *kozlovii* Tzvel. var. nova.—A varietate typica foliorum laminis angustioribus, lemmatis et aristis brevioribus differt. Typus: "Tibet occid., in systemate fl. Jangtze, in viciniis lac. Rchombo-Mtzo et in valle fl. Iczu, 4100–4400 m, No. 435, 1, VIII 1900, V. Ladygin" in Herb. Inst. Bot. Ac. Sci. URSS—LE conservatur.

12. **F. jacutica** Drob. in Tr. Bot. muzeya AN, 14 (1915) 163; Krecz. and Bobr. in Fl. SSSR, 2 (1934) 416.—Ic.: Drob. l.c. Plate 6, figs. 14–17.

Described from East. Siberia. Lectotype (Amgin road, upper course of Krestyakh river, Amga river tributary, No. 238, June 27, 1912, V. Drobov) in Leningrad.

In larch forests, among shrubs, on rocky slopes and rocks; up to lower mountain belt.

IA. Mongolia: *Cis-Hing.* (hill slope near Yaksha railway station, June 13, 1902—Litw.).

General distribution: East., Sib., Far East (south.), China (Dunbei).

13. **F. Kryloviana** Reverd. in Sist. zam. Gerb., Tomsk. univ. 2 (1927) 3; Krylov, Fl. Zap. Sib. 2 (1928) 321; Krecz. and Bobr. in Fl. SSSR, 2 (1934) 512; Fl. Kirgiz. 2 (1950) 151; Grubov, Consp. fl. MNR (1955) 73; Fl. Kazakhst. 1 (1956) 259.—Ic.: Fl. SSSR, 2, Plate 40, fig. 22.

Described from Altay. Lectotype (mountain range between Berezovka and Khapsyn rivers, Aug. 1, 1920—V. Sapozhnikov) in Tomsk; isotype in Leningrad.

In meadows, on rocky and rubble slopes, in riverine pebble beds; in upper and, less often, middle mountain belts.

IA. Mongolia: *Khobd.* (peak of Kharkira, 2600 m, July 10; near Kharkira river source, July 12—1879, Pot.); *Mong. Alt.*

IIA. Junggar: *Alt. region, Tarb,* (south. Slope of Saur mountain range, valley of Karagaitu river, Bain-Tsagan creek valley, subalpine meadow, June 23, 1957—Yun.); *Jung. Alt.* (south. Slope of Junggar Alatau, below Koketau pass, July 21, 1909—Lipsky); *Tien Shan, Jung. Gobi* (pasture on Beidashan [Baitak-Bogdo], 1700 m, No. 2420, Sept. 28, 1957—Huang.

General distribution: Jung. Tarb., Nor. and Cent. Tien Shan; West. Sib. (Altay), East. Sib. (Sayans), Nor. Mong. (Fore Hubs., Hent., Hang.).

O**F. kurtschumica** E. Alexeev, in Novit. Syst. Pl. Vasc. 13 (1976) 30; Gubanov, Consp. Fl. Outer Mong. (1996) 19.

IA. Mongolia. *Mong. Alt.*

14. **F. lenensis** Drob. in Tr. Bot. muzeya AN, 14 (1915) 158; Krecz. and Bobr. in Fl. SSSR, 2 (1934) 513; Grubov, Consp. fl. MNR (1955) 73. —? *F.*

auriculata Drob. l.c. 159; Krecz. and Bobr. l.c. 511; Skvortsov in Arkt. fl. SSSR, 2 (1964) 219, map 70.—*F. tschuiensis* Reverd. in Sist. zam. Gerb. Tomsk. Univ. 3 (1936) l.—*F. albifolia* Reverd. l.c. 2; Krylov, Fl. Zap. Sib. 12, 1 (1961) 3119.—*F. albifolia* var. *tschuiensis* (Reverd.) Serg. in Krylov, l.c. 3119.—*Ic.*: Drob. l.c. Plate 5, figs. 5 and 9.

Described from East. Siberia. Lectotype (slope of Lena river bank near Kyatchin village, No. 103, June 6, 1914, G. Dolenko) in Leningrad.

On rubble and rocky slopes, rocks and in upland steppes; in middle and upper (bald peak) mountain belts.

IA. **Mongolia:** *Khobd.* (Kharkira hill group, Burtu area, July 16, 1903—Gr.-Grzh.; Khutiyn-Khutul' pass south of Tsagan-Nur on road to Khobdo, Aug. 4, 1945—Yun.); *Mong. Alt., Cent.* Khalkha (hill ridge intersecting road to Dalan-Dzadagad 72 km south of Ulan Bator in Arbai-Khere, upper part of rocky slope of knoll, June 19, 1952—Davazhamts); *Gobi-alt.* (nor. Slope of Dzun-Saikhan mountain range near peak, Oct.8, 1940; pass between Dundu- and Dzun-Saikhan, steep rocky slope toward gorge, July, 22; Ikhe-Bogdo mountain range, Bityuten-Ama creek valley, hilly subalpine steppe, Sept. 12—1943, Yun.; south-east. slope of Ikhe-Bogdo mountain range, Narin-Khurimt gorge, on rocks, about 2900 m, June 28, 1948—Grub.).

General distribution: Arct. (Asian), West. Sib. (Altay), East. Sib., Far East (nor.), Nor. Mong. (Fore Hubs., Hang.), China (Dunbei: Greater Hinggan).

Note. Under *F. lenensis* Drob. we include some morphologically poorly differentiated ecogeographical races representing East. Siberian, highly typical bald and sub-bald peak plants. The type of *F. auriculata* Drob. has unfortunately, been lost but, most probably, this name should pertain to more mesophilic specimens of *F. lenensis* with poorly formed sclerenchymatous strands of leaf blades, as treated by A.K. Skvortsov (l.c.). Such specimens have not been reported within Cent. Asia. Specimens with abundant glaucous bloom on the leaf blades, described as an independent species *F. albifolia* Regverd. (l.c.) and, in fact, meriting recognition at least as a variety—*F. lenensis* var. *albifolia* (Reverd.) Tzvel. comb. nova—are quite widely distributed in Mongolia but occasionally found together with specimens of the very similar *F. lenensis* bearing green leaves but otherwise very similar to them in all other respects. Our specimens of *F. lenensis*, in external structure of mat, differ quite well from all other forms of polymorphic species *F. valesiaca* Gaud.

15. **F. nitidula** Stapf in Hook. f. Fl. Brit. Ind. 7 (1896) 350; Hemsley, Fl. Tibet (1902) 205; Pampanini, Fl. Carac. (1930) 78; Bor, Grasses Burma, Ceyl., Ind. and Pakist. (1960) 539.

Described from West. Himalayas. Type in London (K).

In meadows, on rocky and rubble slopes, in riverine pebble beds; in upper mountain belt.

IIA. **Junggar:** *Tien Shan* (Taskhan canal east of Khomote, 3880 m, No. 7113, Aug. 3; same site, 3300 m, No. 7130, Aug. 6—1958, Lee and Chu.).

IIIB. **Tibet:** *South.* ("Tisum, 5000 m, 1848, leg. Strachey and Witterbottom"-Hemsley, l.c.).

General distribution: Himalayas (west., Kashmir).

16. **F. ovina** L. Sp. pl. (1753) 73, s.s.;? Forbes and Hemsley, Index Fl. Sin. 3 (1904) 429; Krylov, Fl. Zap. Sib. 2 (1928) 314, p.p.; Pavlov in Byull. Mosk.

obshch. ispyt. prir. 38 (1929) 25; Krecz. and Bobr. in Fl. SSSR, 2 (1934) 503; Kitag. Lin. Fl. Mansh. (1939) 76, p.p.; Keng, Fl. ill. sin., Gram. (1959) 131, p.p.; Skvortsov in Arkt. fl. SSSR, 2 (1964) 216, map 67.—? *F. supina* Schur, Enum. Pl. Transs. (1866) 784; Krylov, l.c. 317; Krecz. and Bobr. l.c. 504; Grubov, Consp. Fl. MNR (1955) 74; Fl. Kazakhst. 1(1956) 258.—? *F. ovina* var. *supina* (Schur) Hack. Monogr. Fest. Eur. (1882) 88; Kitag. l.c. 77.—Ic.: Keng, l.c. fig. 97.

Described from Nor. Europe. Type in London (Linn.).

In meadows, moss-lichen tundras, on rocky slopes and rocks, in larch forests, among shrubs; in middle and upper mountain belts.

IA. Mongolia: *Mong. Alt.* (lichen tundra in Gul'cha river upper course, Aug. 3, 1930—Bar.; Khargatiin-Daba near summer camp in Bulugun somon, alpine meadow, July 23, 1947—Yun.); *Gobi Alt.* (cent. massif of Nemegetu-Nuru mountain range near its main peak, about 2700 m, on nor. slope under rocks, Aug. 8, 1948—Grub.).

IIA. Junggar: *Tarb.* (in hills near Koktubai, 2250 m, No. 1594, Aug. 13, 1957—Huang); *Tien Shan* (Manas river basin, valley of Ulanusu river near confluence with Dzhartas, nor. Slope at upper forest boundary, July 18; same site, upper course of Danu-Gol river near ascent to Se-Daban pass, sedge-carex high-altitude meadow, July 21 and 22, 1957—Yun.; same site, in Danu-Gol river gorge south of Danu settlement, 3100 m, No. 370, July 21, 1957—Huang).

General distribution: Jung.- Tarb., Nor. Tien Shan; Arct. (Europ.), Europe (nor. and cent.), Balk., West. Sib., East. Sib., Far East. Nor. Mong. (Fore Hubs., Hent., Hang.), China (Dunbei, Nor., Nor.-West.,? South-West.).

Note. After A.K. Skvortsov (l.c.), we have appended to this species its more high-altitude form known under the name *F. supina* Schur, partly due to problems of their differentiation (only difference between high-altitude form and type *F. ovina* L. is the leaf sheath of vegetative shoots closed near base) and partly due to lack of reliability in the similarity of east. Asian high-altitude forms of *F. ovina* with typical Carpathian *F. supina*.

17. **F. pseudosulcata** Drob. in Tr. Bot. muzeya AN, 14 (1915) 156; Krecz. and Bobr. in Fl. SSSR, 2 (1934) 504; Grubov, Consp. fl. MNR (1955) 73.—*F. kolymensis* Drob. l.c. 155; Krecz. and Bobr. l.c. 511; Skvorzov in Arkt. fl. SSSR, 2 (1964) 217.—Ic.: Drob. l.c., Plate 5, figs. 7–8.

Described from East. Siberia. Lectotype (bay in Vilyuisk region, Chona river, 35 km beyond estuary of its right tributary, Igody river, No. 556, July 30, 1914—V. Drobov) in Leningrad.

On rocky slopes and rocks, in larch forests, sometimes in sandy steppes; up to midmountain belt.

IA. Mongolia: *Khobd.* (east. descent of Ulan-Daba pass on Ulangom—Tsagan-Nur road, Aug. 29, 1945—Yun.); *Mong. Alt.* (valley of Khentulyasutai river, larch forest, July 30; valley of Urkhu-Gol river, larch forest, Aug. 19—1930, Pob.; nor. Slope of Khan Taishiri mountain range, larch forest, July 12, 1945—Yun.; same site 15 km south-east of Yusun-Bulak near fringe of larch forest, Sept. 1, 1948—Grub.); *Cis-Hing.* (near Yaksha railway station, rocks, June 11, 1902—Litw.); *East. Mong.* (Garbuni—Bulak area near Kerulen [Choibalsan] settlement, 1870—Lomonosov; near Manchuria railway station, June 6, 1902—Litw.; near Trekhrech'e, 750–850 m, No. 1236, July 10,

1951—Li et al.); *East. Gobi* (sandy exposed steppe near Batu-halka, Roerich Exped., No. 668, Aug. 5, 1935—Keng).

IIA. Junggar: *Jung. Gobi* (Beidashan' mountain range [Baitak Bogdo], 2400 m, No. 2408, Sept. 28, 1957—Huang).

General distribution: East. Sib., Far East (south.), Nor. Mong., China (Dunbei, Nor.).

Note. Barring a characteristic such as connate leaf sheaths of vegetative shoots near base which is difficult to detect, this species differs from the closely related *F. valesiaca* Gaud. Only in very broad leaf blade and very large spikelets, i.e., the same features that distinguish the European steppe species *F. rupicola* Heuff. [=*F. sulcata* (Hack.) Nym., nom. illeg.] from *F. valesiaca*. The similarity of *F. pseudosulcata* Drob. with *F. rupicola* is in fact so great that we retain *F. pseudosulcata* as an idependent species, only on the basis of the vast gap between their distribution ranges. The independently formed, though very poorly isolated, ecogeographical race *F. kolymensis* Drob. (unfortunately, authentic specimens of this species have been lost) was evidently described from specimens with very stiff leaves and well-developed sclerenchymatous strands of leaf blades and probably does not differ from *F. jenisseiensis* Reverd. and *F. malzewii* (Litv.) Reverd. described later from South. Siberia. The specimen noted above from Cis-Hinggan can be placed in this race. Our specimens cited from East. Mongolia and East. Gobi, also having very stiff leaf blades but altogether smooth in the lower part, probably belong to a special ecogeographical race very close to *F. kolymensis* and may be treated as a variety: *F. pseudosulcata* var. *litvinovii* Tzvel. var. nova.—A varietate typica foliorum laminis rigidis subtus laevibus differt.

Typus: "Prope st. viae fer. Manczshuria, 6 VI 1902, D. Litvinov" in Herb. Inst. Bot. Ac. Sci. URSS—LE conservatur.

18. F. rubra L. Sp. pl. (1753) 74; Krylov, Fl. Zap. Sib. 2 (1928) 323; Pavlov in Byull. Mosk. Obshch. ispyt. prir. 38 (1929) 26; Pampanini, Fl. Carac. (1930) 78; Krecz. and Bobr. in Fl. SSSR, 2 (1934) 517; ? Hao in Engler's Bot. Jahrb. 68 (1938) 582; Fl. Kirgiz. 2 (1950) 152; Grubov, Consp. fl. MNR (1955) 73; Fl. Kazakhst. 1 (1956) 262; Fl. Tadzh. 1 (1957) 199; Keng, Fl. ill. sin., Gram. (1959) 129; Bor, Grasses Burma, Ceyl., Ind. and Pakist. (1960) 540; Skvorzov in Arkt. fl. SSSR, 2 (1964) 213, map 66.—*F. kirilovii* Steud. Synops. pl. glum. 1 (1854) 306; Krecz. and Bobr. l.c. 524; Persson in Bot. notiser (1938) 275; Fl. Kirgiz. 2 (1950) 155; Fl. Kazakhst. 1 (1956) 263; Ikonnikov, Opred. rast. Pamira [Key to Plants of Pamir] (1963) 66.—Ic.: Fl. SSSR, 2, Plate 40, fig. 21; Keng, l.c. fig. 96.

Described from Europe. Type in London (Linn.).

In meadows, riverine sand and pebble beds, on rocky slopes, in sparse forests; up to upper mountain belt.

IA. Mongolia: *Khobd.* (valley of Kharkira river, July 9; Kharkira peak, July 10; near Kharkira river source, July 12—1879, Pot.); *Mong. Alt., Gobi-Alt.* (Dzun-Saikhan mountain range, top of Yalo creek valley, Aug. 26, 1931—Ik.-Gal.).

IIA. Junggar: *Tarb.* (along Khobuk river, July 20, 1914—Sap.); *Jung. Alt., Tien Shan.*

IIIA. Qinghai: *Nanshan* (108 km west of Xining and 6 km west of Daudankhe settlement, 3400 m, Aug. 5, 1959—Petr.); *Amdo* ("auf dem Ming-ke Gebirge, sudlich vom Kloster Taschin-sze, 3900 m, 25 VIII 1930"—Hao, l.c.).

IIIB. Tibet: *Chang Tang* (upper course of Tiznaf river near Kyude settlement, June 4, 1959—Yun.); *Weitzan* (nor. slope of Burkhan-Budda mountain range, 3800–4300 m, Aug. 2, 1884—Przew.; same site, Khatu gorge, 3500 m, June 17, 1901—Lad.).

IIIC. Pamir (Sarykol'sk mountain range, Pistan gorge, 4500 m, July 15; Tagdumbash-Pamir, near confluence of Kara-Chukur and Ilyksu rivers, July 16—1901, Alekseenko; Ulug-Tuz gorge in Charlym river basin, June 20, 1909 — Divn.; Minteke Pass, about 4150 m, July 3, 1935.—Persson; in midcourse of Kaplyk river, 3200 m, July 12, 1942—Serp.).

General distribution: Jung.—Tarb., Nor. and Cent. Tien Shan, East. Pamir, Europe, Mediterr., Near East, Balk.-Asia Minor, Caucasus, Middle Asia (hilly), West. Sib., East. Sib., Far East, Nor. Mong. (Fore Hubs., Hent., Hang.), China (Dunbei, Nor.-West.,? South-West.), Himalayas, Korea, Japan, Nor. Amer.

Note. Like A.K. Skvorzov — (l.c.), we cover in this highly polymorphic (variable) species a broad range mainly because of its poorly investigated status to date and difficulties of nomenclature. In Tibet as well as in Pamir, only one morphologically wholly justified variety — *F. rubra* var. *alaica* Drob. [in Tr. Bot. muzeya AN, 16 (1916) 135] has been found. This variety, also known under the species name *F. kirilovii* Steud. (l.c.), differs from typical specimens of *F. rubra* in profuse and fairly long pubescence of lemma. It is interesting that it is extremely similar to many other 'minor' species genetically related to *F. rubra*, viz., *F. arenaria* Osbeck, *F. richardsonii* Hook., *F. cryophila* Krecz. et Bobr. and others, whose differences are not yet clearly understood. In the hills of Tien Shan, Junggar Alatau, Tarbagatai and Altay, this variety is even rare and occasionally found at the same site as typical specimens of *F. rubra*. The above mentioned specimen from Qinghai probably belongs to a distinct variety *F. rubra* s.l.

19. F. valesiaca Schleich. ex Gaud. Agrost. Helv. 1 (1811) 242; Deasy. In Tibet. and Chin. Turk. (1901) 398; Hemsley, Fl. Tibet (1902) 205; Krylov, Fl. Zap. Sib. 2 (1928) 318; Pampanini, Fl. Carac. (1930) 77; ? Norlindh, Fl. mong. steppe, 1 (1949) 112; Bor, Grasses Burma, Ceyl., Ind. and Pakist. (1960) 542.—F. ovina var. valesiaca (Gaud.) Koch, Synops. Fl. Germ. et Helv. (1837) 812; Pilger in Hedin, S. Tibet, 6, 3 (1922) 94—F. pseudoovina Hack. ex Wiesb. in Oesterr. bot. Z. 30 (1880) 126; Krylov, l.c. 319; Krecz. and Bobr. l.c. 508; Fl. Kazakhst. 1 (1956) 259 —F. ganeschinii Drob. in Tr. Bot. muzeya AN, 14 (1915) 105; Krecz. and Borb. l.c. 509; Krylov, l.c. 12, 1 (1961) 3118.— F. recognita Reverd. in Sist. zam. Gerb. Tomsk. univ. 3–4 (1928) 7; Krylov, l.c. (1928) 321. — F. ovina auct. non L. : Hemsley in J. (1894) 140; Simpson in J. Linn. Soc. London (Bot.) Linn. Soc. London (Bot.) 30 (1894) 140; Simpson in J. Linn. Soc. London (Bot.) 41 (1913) 453; ? Rehder in J. Arnold Arb. 14 (1933) 4; ? Hao in Engler's Bot. Jahrb. 68 (1938) 582; Keng, Fl. ill. sin., Gram. (1959) 131, p.p.— F. sulcata auct. non Nym.: Pavlov in Byull. Mosk. obşhch. ispyt. prir. 38 (1929) 26; Fl. Kirgiz. 2 (1950) 149; Fl. Tadzh. 1 (1957) 195; Ikonnikov, Opred. rast. Pamira [Key to Plants of Pamir] (1963) 67.—Ic.: Fl. SSSR, 2, Plate 40, fig. 5.

Described from Europe (Alps). Type in Paris.

In plains and hilly steppes, on rocky and rubble slopes, rocks, in riverine pebble beds; characteristic species of various varieties found in fescue steppes; up to upper mountain belt.

IA. **Mongolia:** *Mong. Alt., Cent. Khalkha* (Tsinkhir-Mandal somon, near Tsinkhiriin-Dugang, July 23, 1949—Yun.; hill ridge intersecting the road to Dalan-Dzadagad 72 km south of Ulan Bator, Aug. 8, 1957—Kal.); *East. Mong.* (near Hailar town, in stone crevices on hill slope, No. 918, June 24, 1951 — Li et al.; 20 km north of Khukh-Khoto, pass through Datsin'shan' mountain range, 1900 m, June 4, 1958 — Petr.); *Bas. lakes* (foothill plain north of Khan-Khukhei mountain range, July 24, 1945 — Yun.); *Gobi-Alt.* (Ikhe-Bogdo mountain range, Bityuten-Ama creek valley, Aug. 12, 1927 — Simukova; nor. slope of Ikhe-Bogdo mountain range, Tsagan-Burgas creek valley, subalpine steppe, Sept. 15, 1943; same site, Shishkhid upper creek valley, June 30, 1945—Yun.); *East. Gobi* ("Khujirtu-Gol, 2 VI 1927, leg. Hummel"—Norlindh, l.c.).

IB. **Kashgar:** *Nor.* (nor.-west. slope of Sogdyn-Tau mountain range 10 to 12 km south-east of Akchii hydropower station on Kokshaal-Darya, 2500 m, Sept. 8; valley of Muzart river before its discharge into Baisk basin, Lyangar peak, Sept. 12 — 1958, Yun.); *West.*

IIA. **Junggar:** *Altay region, Tarb., Jung. Alt., Tien Shan, Jung. Gobi* (south.: between Tien Shan-Laoba and Myaoergou, No. 2370, Aug. 9, 1957—Huang; Kuitun river basin, mouth of Bain-Gol creek valley south of Tushandzy settlement, June 29, 1957—Yun.; near Gan'khedza settlement north of Urumchi, 1200 m, No. 2260, Sept. 23, 1957—Huang; east.: Baitak-Bogdo-Nur mountain range, Takhiltu-Ula, left of Ulyastu-Gol creek valley 7 km from mouth, 2600 m, Sept. 17, 1948—Grub.; Beidashan' [Baitak-Bogdo] mountain range, 2300 m, No. 2395, Sept. 8, 1957—Huang); Zaisan (left bank of Ch. Irtysh 19–20 km south-west of Burchum on road to Zimunai, July 10, 1959—Yun.).

IIIA. **Qinghai:** *Nanshan* (66 km west of Xining, 2800 m, July 5, 1959 — Petr.).

IIIB. **Tibet:** *Chang Tang* ("Tibet, 34° 46', 81° 15' 20", 5300 m, 28 VI 1896"—Deasy, l.c.; "Northern Tibet, 5073 m, 1 IX 1896, leg. Hedin"—Pilger, l.c.); *Weitzan* ("two miles north of Murus river, headwaters of Yangtsekiang, 91° 31', 33° 53', 4900 m, leg, Rockhill"—Hemsley, l.c.); *South.* ("Tibet, 5500 m, leg. Thorold; Tisum, 5000 m, 1848, leg. Strachey and Witterbottom" —Hemsley, l.c.).

IIIC. **Pamir** ("Taghdumbash, 4700 m" — Deasy, l.c.; Ulug-Tuz gorge in Charlym river basin, June 20, 1909 — Divn.; pass from Arpalyk river gorge to Kizyl-Bazar, 3500 m, June 13, 1941 — Serp.).

General distribution: Aral-Casp., Balkh. region, Nor. and Cent. Tien Shan, East. Pam.; Europe (cent. and south.), Mediterr., Balk.-Asia Minor, Near East, Caucasus, Middle Asia (hills and foothills), West. Sib. (south.), East. Sib. (south-west.), Nor. Mong. (Fore Hubs, Hang., Mong.-Daur.), China (? Nor.-West.,? South-West.), Himalayas.

Note. Typical specimens of *F. valesiaca* are covered with glaucescent bloom and have spikelets 5.5–7 mm long, lemma 3.5–5 mm long and anthers 2–2.9 mm long. Specimens of *F. pseudoovina* Hack. ex Wiesb. grouped by us with *F. valesiaca* differ only slightly from those of the latter. For the former, spikelets 4–5.5 mm long, lemma 2.3–4 mm long and anthers 1.6–2 mm long have been recorded, while their leaf blades may be green or glaucescent due to the presence of bloom, and the sclerenchymatous strands are usually more poorly formed than in *F. valesiaca* s.s. Although we could not satisfactorily differentiate these two taxa for Cent. Asia, it is wholly possible that *F. pseudoovina* nevertheless merits recognition as a subspecies — *F. valesiaca* subsp.

pseudoovina (Wiesb.) Hegi or variety — *F. valesiaca* var. *pseudoovina* (Wiesb.) Stohr. Another ecogeographical race related to *F. valesiaca* s.l., namely *F. rupicola* Heuff. [-*F. sulcata* (Hack.) Nym. nom. illeg.], evidently not found in Cent. Asia, is more distinct. Compared to *F. valesiaca* s.l. (including *F. pseudoovina*), it possesses very broad (usually 0.6–0.8 mm in diameter), invariably green leaf blades with 5–7 ribs and slightly larger spikelets.

O **F. valesiaca** Gaud. subsp. **hypsophila** (St.-Yves) Tzvel. in Bot. Zhurn. 56, 9 (1971) 1255; Gubanov, Consp. Fl. Outer Mong. (1996) 19.

IA. Mongolia. *Mong. Alt., East. Mong., Depr. lakes, Val. lakes.*

F. venusta St.-Yves in Izv. Glavn. bot. sada SSSR, 28 (1929) 383; Krecz. and Bobr. in Fl. SSSR, 2 (1934) 527; Grubov, Consp. fl. MNR (1955) 74; Malyshev in Bot. mater. Gerb. Bot. inst. AN SSSR, 21 (1961) 464.

Described from Nor. Mongolia (Hang.). Type in Moscow; isotype in Leningrad.

In meadows, on rocky slopes, along banks of rivers and brooks; in middle and upper mountain belts.

Found in border region of Nor. Mongolia (Hang.).

General distribution: East. Sib. (south.), Nor. Mong. (Hang.).

Section 4. Bromoides Rouy

20. **F. gigantea** (L.) Vill. Hist. Pl. Dauph. 2 (1787) 110; Krylov, Fl. Zap. Sib. 2 (1928) 329; Krecz. and Bobr. in Fl. SSSR, 2 (1934) 534; Fl. Kirgiz. 2 (1950) 160; Fl. Kazakhst. 1 (1956) 266; Fl. Tadzh. 1 (1957) 202; Bor, Grasses Burma, Ceyl., Ind. and Pakist. (1960) 538.—*Bromus giganteus* L. Sp. pl. (1753) 77.—Ic.: Fl. SSSR, 2, Plate 40, fig. 23.

Described from Europe. Type in London (Linn.).

In forests, among shrubs along banks of rivers and brooks, in gorges; in midmountain belt.

IIA. Junggar: *Tien Shan* (near Sairam-Nur lake, June; Talki gorge, July — 1877; Borgaty brook in Kash river basin, July 8, 1879—A. Reg.).

General distribution: Jung.-Tarb., Nor. and Cent. Tien Shan; Europe, Mediterr., Balk.-Asia Minor, Near East (nor.), Caucasus, Middle Asia (hilly), West. Sib., East. Sib. (south-west.), Himalayas (west., Kashmir).

Section 5. Bovinae Fries

21. **F. orientalis** (Hack.) Krecz. et Bobr. in Fl. SSSR, 2 (1934) 531; Fl. Kirgiz. 2 (1950) 159; Fl. Kazakhst. 1 (1956) 265; Fl. Tadzh. 1 (1957) 201.— *F. elatior* subsp. *arundinacea* var. *genuina* subvar. *orientalis* Hack. Monogr. Fest. Eur. (1882) 154.— *F. arundinacea* auct. non Schreb.: Krylov, Fl. Zap. Sib. 2 (1928) 329; Keng, Fl. ill. sin., Gram. (1959) 127; Bor, Grasses Burma, Ceyl.,

Ind. and Pakist. (1960) 538. —Ic.: Fl. SSSR, 2, Plate 39, fig. 11; Keng, l.c. fig. 92.

Described from Cent. Europe. Type in Vienna.

In rather saline meadows, riverine meadows and pebble beds; up to lower mountain belt.

IIA. Junggar: *Jung.-Alt.* (between Gunlyu and Shakhe, No. 3659, Aug. 18, 1957 — Huang); *Dzhark.*

General distribution: Aral-Casp., Balkh. region, Jung.-Tarb., Nor. and Cent. Tien Shan; Europe (cent. and south-east.), Balk.-Asia Minor, Near East (nor.) Caucasus (Fore Caucasus), Middle Asia, West. Sib. (south.).

Note. This species is closely related to the widely distributed *F. arundinacea* Schreb. and, compared to it, represents a more eastern (desert-steppe) ecogeographical race.

61. Littledalea Hemsl.

in Hook. Icon. Pl., ser. IV, 5 (1896) Tab. 2472 and in Kew Bull. (1896) 215.

1. Stalks of spikelets glabrous and smooth, less often with rare spinules; lemma 1.8–3 cm long, glabrous, smooth or scabrous due to presence of very short spinules; palea usually half as long as lemma. Leaf blade 1.5–5 mm broad, flat or loosely longitudinally convoluted, scabrous on upper surface due to spinules, smooth or somewhat scabrous on under .... 2. **L. racemosa** Keng.

+ Stalks of spikelets densely covered with spinules or very short hairs; lemma 1.1–1.8 cm long; palea usually 2/3 length of lemma .... 2.

2. Leaf blade 1.5–4 mm broad, flat or loosely, longitudinally convoluted, with very short but profuse hairs on both sides; ligules up to 5 mm long. Lemma covered with very short hairs or elongated spinules along outer surface .... 3. **L. tibetica** Hemsl.

+ Leaf blade 0.8–2.5 mm broad, usually longitudinally convoluted bristle-like, covered with short or rather elongated spinules on upper surface, glabrous and smooth on under; less often with scattered very short spinules; ligules up to 2.5 mm long. Lemma in upper part and along lateral ribs near their base scabrous due to short spinules, glabrous and smooth over rest of surface .... 1. **L. przevalskyi** Tzvel.

1. **L. przevalskyi** Tzvel. sp. nova—? *L. tibetica* var. *paucispica* Keng, Fl. ill. sin., Gram. (1959) 254, diagn. sin.—Planta perennis, 25–50 cm alta; culmi erecti, glabri et laeves; vaginae glabrae; ligulae ad 2.5 mm lg.; laminae 0.8–2.5 mm lt., convolutae, griseo-virides, supra scabrae, subtus laeves vel sub-laeves. Panicula 5–15 cm lg., sat effusae, ramis scabris, saepe flexuosis, spiculas 1–2 gerentibus; spiculae 1.7–3 cm lg., 5–11-

florae; glumae lanceolatae, inferior 5–9 mm, superior 9–14 mm lg.; lemmata 1.1–1.8 cm lg., apice obtusata vel emarginata, saepe brevissime aristata (arista ad 2.5 mm lg.), glabra, in parte superiore at secus nervos laterales scabriuscula; paleae quam lemmata subsequi breviores, carinis scabris; antherae 5.2–6.7 mm lg.

Typus: Prov. Kansu, in fluxu superiore fl. Hoangho (jugum Sjanj-Si-Bej), in abruptis argillosis, 16 (28) V 1880, No. 244, N. Przevalsky. In Herb. Inst. Bot. Acad. Sci. URSS—LE conservatur.

Affinitas. Haec species a *L. alaicae* (Korsh.) V. Petrov ex Nevski—habitu altiore et paniculae ramis longioribus saepe flexuosis spiculas 1–2 gerentibus, a *L. tibeticae* Hemsl.—laminis angustioribus, supra scabris, subtus laevibus vel sublaevibus (non utrinque puberulis) et lemmatis sublaevibus, a *L. racemosae* Keng—paniculae ramis scabris et lemmatis brevioribus differt.

Described from Qinghai. Type in Leningrad. Plate VII, fig. 3; map 1.

On rocky and clayey slopes, talus, pebble beds; in middle and upper mountain belts.

IIIA. Qinghai: *Nanshan* (high Nanshan foothills 60 km south-east of Chzhan"e town, 2200 m, July 12, 1958—Petr.); *Amdo* (Syan'sibei mountain range in upper course of Huang He river, on clay cliffs, No. 244, May 16, 1880—Przew., type!).

IIIB. Tibet: *Weitzan* (Burkhan-Budda mountain range, 3300 to 5000 m, Sept., 1879—Przew.).

General distribution: endemic.

Note. Although we are not aware of *L. tibetica* Hemsl., it can be judged from its sketch (Hook, Icon, Pl., No. 2472) and original diagnosis that the above-cited specimens from Qinghai and Nor.-East. Tibet differ fairly well from this species and deserve recognition as a new independent species. Our species, of course, is closer to *L. alaica* (Korsh.) V. Petrov ex Nevski found in the USSR (Alay, Nor.-West. Pamir, Chu-Ili hills) but differs in very large plant size as a whole and very much longer, often flexuose branches of panicles.

2. **L. racemosa** Keng in Contribs, Biol. Lab. Sci. Soc. China, Bot. 9 (1934) 136; ej. Fl. ill. sin., Gram. (1959) 257. — Ic.: Keng, l.c. (1934) fig. 15 and (1959) fig. 207.

Described from China (Sichuan province). Type in Nanking. Map 1.

On rocky and stony slopes, talus, in riverine pebble beds; in upper mountain belt.

IIIA. Qinghai: *Amdo* (South. Kukunor mountain range, near Ulan-Khutan crossing, 3300–3700 m, July 27, 1894—Rob.).

IIIB. Tibet: *Weitzan* (upper course of Huang He river, Dzhagyn-Gol river, 4500–4800 m, July 1884 — Przew.; nor. slope of Burkhan Budda mountain range, Khatu gorge, 4300–4700 m, July 17, 1901—Lad.).

General distribution: China (South-West.).

3. **L. tibetica** Hemsl. in Hook. Icon. Pl., ser. IV, 5 (1896) Table 2472 and in Kew Bull. (1896) 215; ej. Fl. Tibet (1902) 204; Keng, Fl. ill. sin., Gram. (1959) 254, excl. var.—Ic.: Hemsl. l.c.; Keng, l.c. fig. 207.

Described from Tibet. Type in London (K).

On rocky slopes, talus and in riverine pebble beds; in upper mountain belt.

IIIB. Tibet: *South.* ("Gooring valley, 90° 25', 30° 12', about 5500 m, leg. G.R. Littledale" — Hemsl., l.c., type!).

General distribution: endemic.

62. Z e r n a Panz.

in Denkschr. Akad. München (1814) 296, s.s.

1. Tip of lemma with 10–15 mm long awn; panicles 15–30 cm long, widely spreading, with long scabrous branches. Forest plant, 40–120 cm tall, forming loose mats without long creeping subsurface shoots; leaf blade 2–5 mm broad, flat, scabrous with scattered hairs above .... 3. **Z. plurinodis** (Keng) Tzvel.

+ Tip of lemma unawned or with up to 4 (5) mm long awn. Plants of open habitats, not forming mats, with long creeping subsurface shoots .... 2.

2. Spikelets usually greenish, rarely with pinkish-violet tinge; rachial segments scabrous due to short spinules; lemma glabrous or puberulent only in lower third; panicles spreading or rather compressed, with scabrous branches ....
.... 1. **Z. inermis** (Leyss.) Lindm.

+ Spikelets usually rather pinkish-violet, less often greenish; rachial segments pilose or with highly elongated (piliform) spinules; lemma hairy along sides and usually even along midrib, for more than half length from base .... 3.

3. Branches of panicles hairy, usually short compared to overall length of panicle; lemma with dense, long hairs along sides. Plant inhabiting coastal sand and pebble beds with long creeping subsurface shoots .... 2. **Z. korotkyi** (Drob.) Nevski.

+ Branches of panicles scabrous, long or rather shortened; lemma with less abundant and less long pubescence along sides. Meadow plant with very short creeping subsurface shoots ....
.... 4. **Z. pumpelliana** (Scribn.) Tzvel.

1. **Z. inermis** (Leyss.) Lindm. Svensk Fanerogamfl. (1918) 101; Fl. Tadzh. 1 (1957) 238. —*Bromus inermis* Leyss. Fl. Hal. (1761) 16; Danguy in Bull. Mus. nat. hist. natur. 20 (1914) 148; Krylov, Fl. Zap. Sib. 2 (1928) 335; Pavlov in Byull. Mosk. obshch. ispyt. prir. 38 (1929) 27; Nevski and Sochava in Fl. SSSR, 2 (1934) 558; Kitag. Lin. Fl. Mansh. (1939) 64; Norlindh, Fl. mong. steppe, 1 (1949) 115; Fl. Kirgiz. 2 (1950) 171; Grubov, Consp. fl. MNR (1955) 74; Fl. Kazakhst. 1 (1956) 275; Keng, Fl. ill. sin.,

Gram. (1959) 266; Bor, Grasses Burma, Ceyl., Ind. and Pakist. (1960) 454. —Ic.: Fl. Tadzh. 1, Plate 31, figs. 6–9; Keng, l.c. Fig. 216.

Described from Cent. Europe. Type in Galle or in Berlin.

In meadows, riverine sand and pebble beds, along roadsides; up to midmountain belt.

IA. **Mongolia:** *Cis-Hing.* (near Yaksha railway station, June 13, 1902—Litw.; same site, No. 2226, 1954—Wang); *Cent. Khalkha, East. Mong., Bas. lakes, Gobi-Alt.* (Dundu-Saikhan, Aug. 19, 1931—Ik.-Gal.); *East. Gobi* (Muni-Ul hills, June 30, 1871 —Przew.; 1 km south-east of Dalan-Dzadagad, test plantation, July 28, 1951—Kal.).

IIA. **Junggar:** *Alt. region* (20 km nor.-west of Shara-Sume, July 7, 1959—Yun.); *Tarb.* (floodplain of Khobuk river between Saur and Semistei mountain ranges 5–6 km west-nor.-west of Kosh-Tologoi settlement, June 22, 1957—Yun.); *Jung. Alt.* (valley of Kepeli river in west. spurs of Barlyk, June 25, 1905—Obruchev); *Tien Shan, Zaisan* (right bank of Ch. Irtysh river below Burchum river estuary, Sary-Dzhasyk collective, June 15, 1914—Schisch.).

General distribution: Aral-Casp., Balkh. region, Jung.-Tarb., Nor. and Cent. Tien Shan; Europe, Mediterr., Balk.-Asia Minor, Near East, Caucasus, Middle Asia, West. and East. Sib., Far East (south.), Nor. Mong., China (Dunbei, Nor., Nor.-West., Cent., East.), Himalayas, Korea, Japan; introduced in many other countries.

2. **Z. korotkyi** (Drob.) Nevski in Tr. Sredneaz. gos. univ., ser. 8b, 17 (1934) 17. — *Z. ircutensis* (Kom.) Nevski, l.c. 17.— *Bromus korotkyi* Drob. in Tr. Bot. muzeya AN, 12 (1914) 238; Nevski and Sochava in Fl. SSSR, 2 (1934) 560.— *B. ircutensis* Kom. in Bot. mater. Gerb. Glavn. bot. Sada RSFSR, 2 (1921) 130; Nevski and Sochava, l.c. 560; Norlindh, Fl. mong. steppe, 1 (1949) 115; Grubov, Consp. fl. MNR (1955) 74; Keng, Fl. ill. sin., Gram. (1959) 268.— *B. pavlovii* Roshev. in: Nor. Mongolia, 1 (1926) 162; Pavlov in Byull. Mosk, obshch. ispyt. prir. 38 (1929) 27.—Ic.: Drobov, l.c. Plate 18; Fl. SSSR, 2, Plate 43, fig. 5; Keng, l.c. fig. 217.

Described from Transbaikal (Barguzin river valley). Type in Leningrad.

In sand and pebble beds of river and lake valleys; up to lower mountain belt.

IA. **Mongolia:** *Cent. Khalkha, East. Mong.* ("Gul-Chaghan, 26 VII; 12.5 km ad orient. versus a Dayen, 31 VII; Dayen, 17 VIII, leg. Eriksson" — Norlindh, l.c.); *Bas. lakes* (valley of Shuryk river, July 11, 1877 — Pot.; Kharanur Khunguisk area 30 km west of Santu-Margats somon, sand, Aug. 18, 1944; 40 km south of Ulyasutai on road to Tsagan-Ol, Shurygin-Gol area, July 17, 1947—Yun.).

General distribution: East. Sib. (south.), Nor. Mong. (Hang., Mong.-Daur.).

Note. Typical specimens of *Bromus korotkyi* and *B. pavlovii* Roshev. (described from Hangay) are entirely similar while the typical specimen of *B. ircutensis* Kom. differs from them in profuse pubescence of all leaves—evidently, a characteristic with no significant taxonomic importance in the entire group of rhizomatous species of the genus *Zerna*.

3. **Z. plurinodis** (Keng) Tzvel. comb. nova.—*Bromus plurinodis* Keng, Fl. ill. sin., Gram. (1959) 260, diagn. sin.—Ic.: Keng, l.c. fig. 211 (type!).— Planta perennis, 40–120 cm alta, laxa caespitosa; culmi glabri et laeves,

plurinodes; vaginae scaberulae, inferiores patule pilosae; ligulae 2–5 mm lg., glabrae; laminae 2–5 mm lt., planae utrinque scaberulae, supra sparsiuscule pilosae, prope basin auriculis lanceolatis praeditae. Paniculae 15–30 cm lg., effusae, ramis longis, tenuibus et scabris; spiculae 9–15 mm lg., 3–6-florae, pallide virides; rachillae breviter pilosae; glumae anguste lanceolatae, apice subuliformiter acutatae, nervis scaberulis, inferior 3.5–6.5 mm lg., uninervia, superior 7–10 mm lg., trinervia; lemmata 7–10 mm lg., lanceolata, trinervia, nervis scabris, prope basin pilosiuscula, apice in arista 10–15 mm lg., recta vel vix curvata transeuntia; paleae lemmatis subaequilongae, carinis scabris; antherae 2.2–3.2 mm lg.

Described from China. Type ("Gansu province, No. 682" — Keng, l.c.) in Nanking.

In forests and among shrubs; in midmountain belt.

IIIA. **Qinghai:** *Nanshan* (in forest belt of Yuzhno-Tetungsk mountain range, maple forest, 2500 m, July 28, 1880—Przew.).

General distribution: China (Nor.-West., South-West.).

Note. This wholly isolated species is evidently very close to the Himalayan species *Z. himalaica* (Stapf.) Henr.

4. **Z. pumpelliana** (Scribn.) Tzvel. in Arkt. fl. SSSR, 2 (1964) 225.—*Z. occidentalis* Nevski in Tr. Sredneaz. gos. univ., ser. 8b, 17 (1934) 18.— *Bromus pumpellianus* Scribn. in Bull. Torr. Bot. Club, 15 (1889) 9.—*B. sibiricus* Drob. in Tr. Bot. muzeya AN, 12 (1914) 229; Nevski and Sochava in Fl. SSSR, 2 (1934) 561; Grubov, Consp. fl. MNR (1955) 75.—*B. inermis* var. *sibiricus* (Drob.) Kryl. Fl. Zap. Sib. 2 (1928) 335.—*B. inermis* subsp. *pumpellianus* (Scribn.) Wagnon in Rhodora, 52 (1950) 211.—*B. richardsonii* auct. non Link: Pavlov in Byull. Mosk. obshch. ispyt. prir. 38 (1929) 27; Nevski and Sochava, l.c. 562; Grubov, l.c. 74; Keng, Fl. ill. sin., Gram. (1952) 261.—*B. ciliatus* auct. non L.: Kitag. Lin. Fl. Mansh. (1939) 63. — Ic.: Keng, l.c. fig. 213.

Described from the USA (Montana state). Type in Washington.

In meadows, riverine sand and pebble beds; in lower and middle mountain belts.

IA. **Mongolia:** *Cis-Hing., Cent. Khalkha* (water divide between Ara- and Ubur-Dzhargalante rivers, 1925—Krasch. and Zam.; valley of Tsinkhiriin-Gol river near Tsinkhiriin-Duganga, July 22, 1949—Yun.); *East. Mong.* (near Lukh-Sume monastery, June 30, 1899—Pot. and Sold.; near Hailar town, No. 759, June 20, 1951—Li et al.); *Bas. lakes* (Chara-Khargai-Kharga crossing, Aug. 10, 1909—Sap.; south. bank of Khara-Nur lake 20 km west of Santu-Margats somon, Aug. 18, 1944—Yun.); Gobi-Alt., East. Gobi (Muni-ul hills, June 30, 1871—Przew.).

General distribution: Arct., Europe (Urals), West. Sib. (nor., Altay), East. Sib., Far East, Nor. Mong., China (Dunbei, Nor.), Korea, Nor. Amer. (Alaska, Nor. Cordillera).

Note. American species *Z. ciliata* (L.) Henr. and *Z. richardsonii* (Link) Nevski, similar in structure of the spikelets to *Z. pumpelliana* (Scribn.) Tzvel., are easily differentiated from this species in the absence of creeping subsurface shoots.

○**Bromopsis altaica** Peschk. in Novit. Syst. Pl. Vasc. 23 (1986) 26: Gubanov, Consp. Fl. Outer Mong. (1996) 16.

IA. Mongolia: *Mong. Alt.*

○**B. ubsunurica** Tzvel. in Bot. Zhurn. 76, 4 (1991) 609.

IA. Mongolia: *East. Mong.*

63. Anisantha C. Koch
in Linnaea, 21 (1848) 394.

1. **A. tectorum** (L.) Nevski in Tr. Sredneaz. gos. univ., ser. 8b, 17 (1934) 22; Fl. Tadzh. 1 (1957) 254. — *Bromus tectorum* L. Sp. pl. (1753) 77; Forbes and Hemsley, Index Fl. Sin. 3 (1904) 431; Krylov, Fl. Zap. Sib. 2 (1928) 338; Pampanini, Fl. Carac. (1930) 79; Kreczetowicz and Vvedensky in Fl. SSSR, 2 (1934) 573; Hao in Engler's Bot. Jahrb. 68 (1938) 582; Walker in Contribs. U.S. Nat. Herb. 28 (1941) 596; Fl. Kirgiz. 2 (1950) 175; Fl. Kazakhst. 1 (1956) 278; Keng, Fl. ill. sin., Gram. (1959) 275; Bor, Grasses Burma, Ceyl., Ind. and Pakist. (1960) 456. —Ic.: Fl. Tadzh. 1, Plate 33, figs. 1–3; Keng, l.c. fig. 224.

Described from South. Europe. Type in London (Linn.).

In desert steppes, on rubble and rocky slopes, talus, in riverine pebble beds and sand, along roadsides; up to upper mountain belt.

IIA. Junggar: *Dzhark.* (valley of Khorgos river 2–3 km east of Santokhodze on road to Manas, July 4, 1957—Yun.; Pilyuchi near Kul'dzha, May 17, 1877—A. Reg.).

IIIA. Qinghai: *Nanshan* (valley of Datungkhe river, 2800 m, Aug. 20, 1958—Dolgushin; "Pa-Yen-Jung-Ke, No. 741, leg. Ching"—Walker, l.c.); *Amdo* (along Yusun-Khatym river, 3000–3300 m, July 11, 1880—Przw.; "Scha-Chu-Yi, am Ufer des Flusses Kia-Po-Kia-Ho, 2900 m"—Hao, l.c.).

IIIB. Tibet: *Weitzan* (Yangtze basin, Kabchzha-Kamba village on Khichu river, 4000 m, July 21, 1900—Lad.; nor slope of Burkan Budda mountain range, Khatu gorge, 3500 m, July 28, 1901—Lad.).; *South.* (Khamba, July 8–10, 1903—Younghusband).

General distribution: Aral-Casp., Balkh. region, Jung.-Tarb., Nor. and Cent. Tien Shan, East. Pam.; Europe (cent. and south.), Mediterr., Balk.-Asia Minor, Near East, Caucasus, Middle Asia, West. Sib. (south.), China (Dunbei, Nor., Cent., South.-West.), Himalayas; introduced in other countries.

64. Bromus L.
Sp. pl. (1753) 76.

1. Lemma 12–15 mm long, lanceolate, with 2 acute 0.6–1.6 mm long teeth at tip; awns of all lemmas, except lemma of lowest flower,

14–20 mm long, laterally recurved and diverging 3.5-6 mm from tip; palea 4–6 mm shorter than lemma; panicles usually rather spreading .... 3. **B. oxyodon** Schrenk.

+ Lemma 7–12 mm long, with or without (may have been broken!) teeth up to 0.6 mm long, at tip; awns of all lemmas, except lemma of lowest flower, 5–14 mm long and diverging 0.5–2.5 mm from tip; palea 0.5–2.8 mm shorter than lemma .... 2.

2. Lemma lanceolate-oblong, usually with uniformly rounded lateral margin, with 2 small acute teeth at tip and a notch 0.4–0.6 mm deep between teeth; awns not recurved, or very slightly so, and diverging 0.5–1.5 mm from tip of lemma in central and upper florets; panicles compressed or poorly spreading ....3.

+ Lemma obovate-lanceolate, without acute teeth at tip, obtuse or slightly notched (with notch up to 0.2 mm deep); awns in fruit and in dry state laterally recurved, diverging 1.2–2.5 mm from tip of lemma in central and upper florets; panicles spreading; ligules puberulent on back ....4.

3. Panicles lax, usually rather spreading; lemma scabrous with very narrow membranous margin; palea 0.5–1.5 mm shorter than lemma. Stems glabrous; sheath of lower leaves puberulent or glabrous, of upper ones glabrous; ligules glabrous ....
.... 1. **B. gedrosianus** Pénzes.

+ Panicles dense, rather compressed; lemma scabrous or hairy, relatively broadly membranous along margin; palea 1.5–2.3 mm shorter than lemma. Stems with very short hairs at, or often under, nodes; sheath of lower and often even upper leaves puberulent; ligules with very short hairs on back ...... 4. **B. sewerzowii** Regel.

4. Lemma rather rounded laterally, almost double breadth of palea, with indistinct teeth at tip; branches of panicles often with 2–3 spikelets ....2. **B. japonicus** Thunb.

+ Lemma with prominent corners laterally, almost thrice broader than palea; teeth distinct about 0.3 mm long at tip; branches of panicles with 1, occasionally 2 spikelets .... 5. **B. squarrosus** L.

1. **B. gedrosianus** Pénzes in Bot. Kozlemén. 33 (1936) 111; Bor, Grasses Burma, Ceyl., Ind. and Pakist. (1960) 454.— Ic.: Pénzes, l.c. figs. 48 and 39. Described from India. Type in Vienna; isotype in Leningrad.

As weed in farms and plantations of various crops, along roadsides and around irrigation ditches, in wastelands; up to upper mountain belt.

IB. **Kashgar:** *Nor.* (Uchturfan, June 16, 1908—Divn.); *West.* (around Yangigissar, May 24, 1909—Divn.); *South.* (Karasai village on nor. slope of Russky mountain range, 3300 m, June 13, 1890—Rob.).

IC. Qaidam: *plains* (Qaidam, July, 1879—Przew.; lower course of Nomokhun-Gol river, about 3300 m, Aug. 3; near Nomokhun-Gol river, Aug. 30—1884, Przew.).

IIIB. Tibet: *Weitzan* (nor. slope of Burkhan-Budda mountain range, Khatu gorge, 3500 m, July 3 and 25, 1901—Lad.).

IIIC. Pamir (near Pas-Rabat village, July 3, 1909—Divn.; Tagarma valley, July 23, 1913—Knorring; Issyksu river, 3100 m, July 3; near discharge of Tyna river into Issyksu, 3200 m, July 19—1942, Serp.).

General distribution: Near East, China (Cent., South-West.), Himalayas, Afr. (nor.-east. part); possibly more widely distributed.

Note. Like the European *B. secalinus* L., this species is a specialised weed found mainly in cereal crops.

2. **B. japonicus** Thunb. Fl. Jap. (1784) 52; Forbes and Hemsley, Index Fl. Sin. 3 (1904) 430; Krylov, Fl. Zap. Sib. 2 (1929) 339; Pampanini, Fl. Carac. (1930) 79; Kreczetowicz and Vvedensky in Fl. SSSR, 2 (1934) 578; ? Persson in Bot. notiser (1938) 275; Kitag. Lin. Fl. Mansh. (1939) 64; Fl. Kirgiz. 2 (1950) 179; Grubov, Consp. fl. MNR (1955) 74; Fl. Kazakhst. 1 (1956) 280; Fl. Tadzh. 1 (1957) 246; Keng, Fl. ill. sin., Gram. (1959) 273; Bor, Grasses Burma, Ceyl., Ind. and Pakist. (1960) 455.— Ic.: Fl. SSSR, 2, Plate 42, fig. 10; Keng, l.c. fig. 221.

Described from Japan. Type in Uppsala.

On rocky slopes, in riverine pebble beds and sand, along banks of water reservoirs, often as weed in farms and various crops, along roadsides; up to upper mountain belt.

IA. Mongolia: *Mong. Alt.* (Buyantu river valley, Aug. 27; Buyantu station, Aug. 28 — 1930, Bar.; Bombotu-Khairkhan hills, El'river, Oct. 10, 1930—Pob.); *Bas. lakes* (nor. bank of Khirgis-Nur lake, Shitsirgan-Bulak area, Aug. 21, 1944—Yun.); *East. Gobi* (Dalan-Dzadagad, July 21, 1943—Yun.; near Dalan-Dzadagad, Sept. 7, 1950— Lawr.; same site, July 28 and Sept. 12, 1951.—Kal.); *Alash. Gobi* (Chagan-Beli area at Edzin-Gol, July 5, 1886—Pot.).

IB. Kashgar: *West.* ("Jerzil, 2800 m, 8 VII 1930; Kentalek, 2700 m, 16 VII 1931"— Persson, l.c.).

IIA. Junggar: *Alt. region, Jung. Alt.* (near Toli, No. 976, Aug. 4, 1957—Huang; nor. rim of Dzhair mountain range 24 km nor.-east of Toli along road to Temirtam, Modun-Obo brook, Aug. 5, 1957—Yun.); *Tien Shan, Jung. Gobi, Zaisan* (Kaba river near Kaba village, June 16, 1914—Schisch.); *Dzhark.*

General distribution: Aral-Casp., Balkh. region, Jung.-Tarb., Nor. and Cent. Tien Shan,? East. Pam.; Europe, Mediterr., Balk.-Asia Minor, Near East, Caucasus, Middle Asia, West. Sib. (south.), China (south. Dunbei, Nor., Nor.-West., Cent., East.), Himalayas, Korea, Japan; introduced in many other countries of both hemispheres.

3. **B. oxyodon** Schrenk in Bull. Ac. Sci. St. Pétersb. 10 (1842) 355; Pampanini, Fl. Carac. (1930) 79; Kreczetowicz and Vvedensky in Fl. SSSR, 2 (1934) 581; Norlindh, Fl. mong. steppe, 1 (1949) 114; Fl. Kirgiz. 2 (1950) 179; Grubov, Consp. fl. MNR (1955) 74; Fl. Kazakhst. 1 (1956) 281; Fl. Tadzh. 1 (1957) 249; Bor, Grasses Burma, Ceyl., Ind. and Pakist. (1960) 455. —*B. macrostachys* var. *oxyodon* (Schrenk) Griseb. in Ledeb. Fl. Ross. 4

(1852) 363.—*B. krausei* Regel in Acta Horti Petrop. 7 (1881) 600.—Ic.: Fl. SSSR, 2, Plate 42, fig. 11; Fl. Kazakhst. 1, Plate 21. fig. 6.

Described from Junggar (Tarbagatai). Type in Leningrad.

On rubbly and rocky slopes, rocks, talus, in riverine pebble beds; sometimes as weed along roadsides, in farms; up to mid-mountain belt.

IA. **Mongolia:** *Mong. Alt., Alash. Gobi* ("prope fl. Edsen-Gol, circa 4 km ad merid. versus a Camp 59, 24 V 1930, leg. Bohlin"— Norlindh, l.c.).

IIA. **Junggar:** *Jung. Alt.* (near Toli, No. 976, Aug. 4, 1957—Huang); *Tien Shan* (35 km east of bridge on Kash river, 980 m, Aug. 21; Borgaty in Sairam-Nur lake region, 1150 m, No. 1265, Aug. 30—1957, Huang); *Jung. Gobi* (left bank of Ch. Irtysh 38 km east of Shipati crossing on road to Koktogoi, July 8, 1959—Yun.); *Dzhark.* (Chimpansi village, May 8; Aktyube village, May 13—1877, A. Reg.).

General distribution: *Balkh. region* (south. and east.), Jung.-Tarb., Nor. and Cent. Tien Shan; Middle Asia (east.), Himalayas (west.).

Note. In Cent. Asia, this species replaces the very closely related Mediterranean species *B. macrostachys* Desf.

o **B. scoparius** L. 1755, Cent. Pl. 1 (1775) 6; Gubanov, Consp. Fl. Outer Mong. (1996) 17.

IIA. **Junggar:** *Jung. Gobi.*

4. **B. sewerzowii** Regel in Acta Horti Petrop. 7 (1881) 601; Kreczetowicz and Vvedensky in Fl. SSSR, 2 (1934) 580; Fl. Kirgiz. 2 (1950) 179; Fl. Kazakhst. 1 (1956) 280; Fl. Tadzh. 1 (1957) 247.—Ic.: Fl. Kazakhst. 1, Plate 21, fig. 9.

Described from Middle Asia. Type in Leningrad.

On rocky slopes, talus, in riverine pebble beds, rocks; up to midmountain belt.

IB. **Kashgar:** *West.* (Egin village, near fort ruins, Aug. 11, 1913—Knorring).

General distribution: Aral-Casp. (south-east.), Balkh. region (south.); Near East (east.), Middle Asia.

5. **B. squarrosus** L. Sp. pl. (1753) 76; ? Forbes and Hemsley, Index Fl. Sin. 3 (1904) 431; Danguy in Bull. Mus. nat. hist. natur. 20 (1914) 148; Krylov, Fl. Zap. Sib. 2 (1928) 339; Kreczetowicz and Vvedensky in Fl. SSSR, 2 (1934) 577; ? Norlindh, Fl. mong. steppe, 1 (1949) 114; Fl. Kirgiz. 2 (1950) 180; Grubov, Consp. fl. MNR (1955) 75; Fl. Kazakhst. 1 (1956) 279; Fl. Tadzh. 1 (1957) 252.—Ic.: Fl. Kazakhst. 1, Plate 21, fig. 10.

Described from Europe. Type in London (Linn.).

On rubbly and rocky slopes, in riverine sand and pebble beds, sometimes as weed along roadsides, in field borders; up to midmountain belt.

IA. **Mongolia:** *Bas. lakes* ("Environs de Kobdo, 25 IX 1895, leg. Chaffanjon" — Danguy, l.c.); *Alash. Gobi* ("ad fl. Edsen-Gol, circa 4 km ad merid. versus a Camp 59, 24 V 1930, leg. Bohlin"—Norlindh, l.c.).

IIA. Junggar: *Tarb.* (nor. of Dachen-Chuguchak road, Aug. 13, 1957—Huang); *Jung. Alt.* (Barlyk hills, valley of Tasty river, June 23, 1905—Obruchev); *Zaisan* (Mai-Kapchagai hill, June 6, 1914—Schisch.; near Burchum settlement, 1200 m, No. 3244, Sept. 20, 1956—Ching).

General distribution: Aral-Casp., Balkh. region, Jung.-Tarb., Nor. Tien Shan; Europe (cent. and south.), Mediterr., Balk.-Asia Minor, Near East, Caucasus, Middle Asia, West. Sib. (south.), Afr. (nor.).

65. Brachypodium Beauv.
Ess. Agrost. (1812) 100.

1. Plant with long creeping subsurface shoots, not forming mats. Spikes not spreading; awn of lower lemma in upper part of spikelets 1.5–5 mm long; anthers 3–4.5 mm long. .... 1. **B. pinnatum** (L.) Beauv.

+ Plant with short creeping subsurface shoots, forming mats. Spikes usually spreading, with very slender rachis; awn of lemma in upper part of spikelets 7–12 mm long; anthers 1.8–3 mm long .... 2. **B. silvaticum** (Huds.) Beauv.

1. **B. pinnatum** (L.) Beauv. Ess. Agrost. (1812) 155; Krylov, Fl. Zap. Sib. 2 (1928) 340; Nevski in Fl. SSSR, 2 (1934) 594; Kitag. Lin. Fl. Mansh. (1939) 63; Fl. Kirgiz. 2 (1950) 184; Grubov, Consp. fl. MNR (1955) 75; Fl. Kazakhst. 1 (1956) 285; Bor. Grasses Burma, Ceyl., Ind. and Pakist. (1960) 450. — *Bromus pinnatus* L. Sp. pl. (1753) 78.— Ic.: Fl. SSSR, 2, Plate 44, fig. 16; Fl. Kazakhst. 1, Plate 23, fig. 1.

Described from Europe. Type in London (Linn.).

In forests, among shrubs, in forest glades, subalpine meadows; in middle and upper mountain belts.

IIA. **Junggar:** *Tien Shan* (south. bank of Sairam-Nur lake, July 22; Talki river gorge, July, 1877; Kokkamyr hills, 2300 m, July 27, 1878; Bargaty river, 2000 m, July 4, 1879 — A. Reg.; near Turfan town, 1886—Krasnov; east. part of Ketmen' mountain range, Sarbushin pass on Ili-Kzyl-Kure road, Aug. 23; slope on Tekes river valley, 4–5 km south-east of Aksu settlement, Aug. 24, 1957—Yun.).

General distribution: Aral-Casp. (Mugodzhary), Balkh. region (Karkaralin hills), Jung.-Tarb., Nor. and Cent. Tien Shan; Europe, Mediterr., Balk.-Asia Minor, Near East, Caucasus, Middle Asia (Chatkal' mountain range), West. Sib., East. Sib., Nor. Mong. (Mong.-Daur.), China (Greater Hinggan).

2. **B. silvaticum** (Huds.) Beauv. Ess. Agrost. (1812) 101; Forbes and Hemsley, Index Fl. Sin. 3 (1904) 431; Krylov, Fl. Zap. Sib. 2 (1928) 342; Nevski in Fl. SSSR, 2 (1934) 594; Walker in Contribs. U.S. Nat. Herb. 28, 4 (1941) 596; Keng in Sunyatsenia, 6 (1941) 54; Fl. Kirgiz. 2 (1950) 184; Fl. Kazakhst. 1 (1956) 285; Fl. Tadzh. 1 (1957) 262; Keng, Fl. ill. sin., Gram. (1959) (279); Bor, Grasses Burma, Ceyl., Ind. and Pakist. (1960) 450.—*B. manshuricum* Kitag. in J. Jap. Bot. 9, 2 (1933) 117; Kitag. Lin. Fl. Mansh.

(1939) 63. — *Festuca silvatica* Huds. Fl. Angl. (1762) 38.—*Brevipodium silvaticum* (Huds.) A. et D. Löve in Bot. notiser, 114, 1 (1961) 36. —Ic.: Fl. SSSR, 2, Plate 44, fig. 15; Keng, l.c. (1959) fig. 226.

Described from British Isles. Type in London (K).

In forests, among shrubs, in forest glades; up to midmountain belt.

IIIA. Qinghai: *Nanshan* (between Choibsen monastery and Yuzhno-Tetungsk mountain range, July 17, 1880—Przew.; "Near Sining, No. 626, leg. R.C. Ching"— Keng, l.c. 1941).

General distribution: Jung.-Tarb., Nor. Tien Shan; Europe, Mediterr., Balk.-Asia Minor, Near East, Caucasus, Middle Asia (hilly), West. Sib. (south-east.), East. Sib. (south.), Far East (Sakhalin and Kuril islands), China (Dunbei, Nor., Nor.-West., Cent., East., South-West., South.), Himalayas, Japan, Indo-Malay.

66. L o l i u m L.
Sp. pl. (1753) 83.

1. **L. persicum** Boiss. and Hohen. in Boiss. Diagn. pl. or., sér. 1, 13 (1853) 66; Nevski in Fl. SSSR, 2 (1934) 548; Persson in Bot. notiser (1938) 275; Fl. Kirgiz. 2 (1950) 164; Fl. Kazakhst. 1 (1956) 272; Fl. Tadzh. 1 (1957) 215; Keng, Fl. ill. sin., Gram. (1959) 449; Bor, Grasses Burma, Ceyl., Ind. and Pakist. (1960) 545.—Ic.: Fl. SSSR, 2, Plate 44, fig. 5; Keng, l.c. Fig. 382.

Described from Near East (Iran). Type in Geneva; isotype in Leningrad.

As weed in farms and various crops, along roadsides and around irrigation ditches, in wastelands; up to upper mountain belt.

IA. Mongolia: *Alash. Gobi* (Tengeri sand near Min'tsin town, July 12, 1958—Petr.).
IB. Kashgar: *Nor.* (in Aksu town region, 2080 m, No. 8908, Sept. 16, 1958—Lee and Chu); *West.* ("Kentalek, 2700 m, 13 VII 1931"—Persson, l.c.).

General distribution: Nor. and Cent. Tien Shan, East. Pam.; Asia Minor, Near East, Caucasus, Middle Asia, China (Nor.-West., South-West.), Himalayas.

67. A g r o p y r o n Gaertn.
in Novi Comm. Ac. Sci. Petrop. 14, 1 (1970) 539, s.s.

1. Glumes lanceolate, laterally compressed boat-like, with prominent midrib like keel and undistinct lateral ribs, asymmetrical; spikes linear to broadly ovate with close-packed spikelets, glabrous or rather pilose; anthers (2.5) 3–4.5 (5) mm long. Plants forming mats, without creeping subsurface shoots, less often (in *A. michnoi* Roshev.) with creeping subsurface shoots .... 2.

+ Glumes lanceolate or elliptical, with ribs rather equally developed, without keel but occasionally with midrib, slightly better developed than lateral ribs .... 6.

2. Plants inhabiting sand, with creeping subsurface shoots, not forming mats or forming small mats joined with rhizome Spikes

2–5 cm long, dense and broad, but usually only slightly pectinate, with obliquely ascending spikelets; lemma densely pilose to subglabrous .... 21. **A. michnoi** Roshev.

+ Plants without creeping subsurface shoots forming rather dense mats .... 3.

3. Spikes oblong to broadly ovate, 1–6 cm long and 0.8–2.5 cm broad, rather dense, pectinate, with spikelets, strongly divergent from axis of spikes .... 4.

+ Spikes linear, (2) 3–10 (15) cm long and 0.5–1 cm broad, with wider-spaced spikelets, not pectinate (spikelets adhere to axis of spikes or only slightly divergent from it) .... 5.

4. Spikes with very close-packed spikelets—no gaps between spikelet bases; glumes and lemma usually rather pilose, less often subglabrous or even totally glabrous ....
.... 18. **A. cristatum** (L.) Beauv.

+ Spikes with more interrupted spikelets—gaps distinctly visible between bases of spikelets; glumes and lemma usually glabrous or subglabrous; less often somewhat pilose ....
.... 2. **A. pectinatum** (M.B.) Beauv.

5. Spikes 2–7 cm long; lemma glabrous, with 1–3 mm long awn at tip; palea along keels with interrupted and relatively few spinules (up to 25 on each keel). Plant 20–60 cm tall, usually inhabiting clayey and stony soils, with roots without distinct caps of sand particles .... 19. **A. desertorum** (Link) Schult. et Schult. f.

+ Spikes 3–15 cm long; lemma glabrous or with scattered hairs along sides, acute at tip or with up to 1 mm long cusp; palea with densely disposed numerous (more than 30 on each keel) spinules along keels. Plant 30–80 cm tall, inhabiting sandy soil, with caps of sand particles on roots .... 20. **A. fragile** (Roth.) Nevski.

6(1). Glumes and lemma glabrous but occasionally with few short spinules in upper part of ribs; spikes invariably linear with rather interrupted spikelets; anthers (3.2) 3.6–5 (5.5) mm long .... 7.

+ Lemma pilose; glumes glabrous or pilose; spikes linear to broadly ovate and then with close-packed spikelets; anthers (1.5) 1.8–3.5 (4.2) mm long .... 14.

7. Plants 20–100 cm tall, not forming mats, with long creeping subsurface shoots; leaf blade 3–10 mm broad, usually flat. Glumes broadly lanceolate, not more than 2/3 as long as lemma .... 8.

+ Plants 15–70 cm tall, forming fairly dense mats, without creeping subsurface shoots or with few short-creeping subsurface shoots; leaf blade 1.5–3.5 mm broad, usually longitudinally convoluted, less often flat. Glumes ellipitcal or broadly lanceolate .... 9.

8. Glumes rather obtuse at tip; lemma also usually obtuse at tip .... 1. **A. elongatiforme** Drob.

+ Glumes gradually acuminate at tip, often with cusp or short awn; lemma acute at tip, often with up to 6 mm long awn .... 2. **A. repens** (L.) Beauv.

9. Lemma with rather laterally recurved awn 13–20 mm long at tip; glumes 2/5–2/3 as long as lemma, with 3–5 ribs .... 3. **A. aegilopoides** Drob.

+ Lemma acute at tip, occasionally with cusp or awn up to 3 (5) mm long .... 10.

10. Sheath and leaf blades, usually also stems, densely covered with very short hairs. Glumes 1/2–2/3 as long as lemma, acute or with up to 1.2 mm long cusp ....8. **A. nevskii** Ivanova ex Grub.

+ Stems and sheath glabrous, less often subglabrous; leaf blade glabrous in lower part, covered in upper part with short, sometimes also very long hairs .... 11.

11. Rachis subglabrous, only in upper part with 2 lateral ribs and spinules indistinct even on high magnification; glumes rather obtuse at tip, usually 2/3 as long as lemma .... 4. **A. dshungaricum** (Nevski) Nevski.

+ Rachis with 2 lateral ribs and very distinct spinules; glumes usually acute or with cusp .... 12.

12. Glumes abruptly acuminate at tip or with up to 1 mm long cusp; more than 2/3 as long as contiguous lemma .... 6. **A. geniculatum** (Trin.) C. Koch.

+ Glumes gradually acuminate at tip 2/3 or less as long as contiguous lemma .... 13.

13. Palea with short and stubby spinules along keels ....5. **A. ferganense** Drob.

+ Palea with long and slender spinules along keels .... 7. **A. kanashiroi** Ohwi.

14(6). Lemma villous-haired due to rather flexuose long hairs, acute or with up to 3 (4) mm long cusp at tip; spikes strongly shortened, slightly pectinate; anthers 1.5–2.3 mm long. Plant growing in sand, usually with leaf blade 1.5–4 mm broad, rather loosely convoluted longitudinally ....15.

+ Lemma puberulent .... 16.

15. Plants 40–100 cm tall, not forming mats, with long creeping subsurface shoots; sheath of upper cauline leaves not swollen. Spikes 4–9 cm long, relatively lax, often with slightly flexuose rachis; Dumes 5–8 mm long, usually glabrous.............................. .... 11. **A. grandiglume** (Keng) Tzvel.

+   Plant 15–40 cm tall, usually forming lax mats with short creeping subsurface shoots; sheath of upper cauline leaves swollen. Spikes 2–6 cm long, very dense and somewhat secund; glumes 4–5 mm long, usually pilose .... 17. **A. thoroldianum** Oliver.

16.   Spikes linear, 8–15 cm long, with greatly interrupted spikelets; glumes 7–11 mm long, with 5 ribs, glabrous, with up to 2 mm long cusp at tip; lemma 8–11 mm long, with up to 3.5 mm long cusp or awn at tip .... 9. **A. alatavicum** Drob.

+   Spikes broadly linear or oblong, 2.5–8 cm long, very dense, occasionally with spikelets arranged subpectinately; glumes 3.5–7 (8) mm long, with (1) 3–5 ribs; lemma 4.5–8.5 mm long .... 17.

17.   Lemma acute or with up to 1 mm long cusp at tip; glumes glabrous, usually 2/3 as long as contiguous lemma; anthers 2.3–2.8 mm long .... 15. **A. muticum** (Keng) Tzvel.

+   Lemma with 1.5–14 mm long awn at tip .... 18.

18.   Glumes 3.5–5 mm long, puberulent, gradually narrowing at tip into awn almost as long as glume lemma with 1–3 and palea with 3–5 ribs; lemma 4.5–6 mm long, with 3–6 mm long awn at tip ....
.... 13. **A. kokonoricum** (Keng) Tzvel.

+   Glumes acute or with up to 1.5 mm long cusp at tip, with 3–5 ribs
.... 19.

19.   Plant of rocky slopes and rocks, forming fairly dense mats, without creeping subsurface shoots. Spikes broadly linear, relatively lax; glumes usually pilose, 5–7.5 mm long, less than 2/3 as long as lemma; latter with (3) 5–10 (14) mm long awn at tip; anthers 2.3–3.2 mm long, with reddish tinge ....
.... 10. **A. batalinii** (Krasn.) Roshev.

+   Plant growing in sand, forming lax mats, with creeping subsurface shoots. Spikes oblong, very dense; glumes usually glabrous, 3.5–6.5 mm long; 2/3 or more as long as lemma; tip of latter with 1.5–5 mm long awn anthers 1.8–2.3 mm long, with blackish tinge .... 20.

20.   Lemma covered with short stiff hairs; glumes with highly prominent ribs .... 12. **A. kengii** Tzvel.

+   Lemma covered with very long soft hairs; glumes with barely discernible ribs .... 14. **A. melantherum** Keng.

Section 1. Stolonifera Rouy

1. **A. elongatiforme** Drob. in Vvedensky et al., Opred. rast. okr. Tashkenta [Key to Plants of Tashkent Region], 1 (1923) 42 and in Feddes repert. 21 (1925) 44; Nevski in Fl. SSSR, 2 (1934) 651; Fl. Kirgiz. 2 (1950) 206.—

Elytrigia elongatiformis (Drob.) Nevski in Tr. Sredneaz. gos. univ., ser. 8b, 17 (1934) 61.

Described from Middle Asia (around Tashkent). Type in Tashkent (TAK).

In meadows and as weed along roadsides and around irrigation ditches, in farms and various crops; up to midmountain belt.

IIA. Junggar: *Jung. Gobi* (nor.-west.: in Kran river valley east of Barbagai, No. 10595, July 8, 1959—Lee et al.).

General distribution: Balkh. region (south), Nor. and Cent. Tien Shan; Caucasus (South. and East. Transcaucasus), Middle Asia.

Note. This species is very close to *A. lolioides* (Kar. et Kir.) Roshev. widely distributed in south-eastern European USSR and Kazakhstan, and represents a relatively more southern and predominantly weedy ecogeographical race. It would perhaps be more correct to regard the two species as subspecies of *A. lolioides* (Kar. et Kir.) Roshev. s. l.

2. **A. repens** (L.) Beauv. Ess. Agrost. (1812) 102; Forbes and Hemsley, Index Fl. Sin. 3 (1904) 432, p.p.; Krylov, Fl. Zap. Sib. 2 (1928) 351, p.p.; Pavlov in Byull. Mosk, obshch. ispyt. prir. 38 (1929) 28; Pampanini, Fl. Carac. (1930) 79; Nevski in Fl. SSSR, 2 (1934) 652; Fl. Kirgiz. 2 (1950) 200; Grubov, Consp. fl. MNR (1955) 76; Fl. Kazakhst. 1 (1956) 297; Melderis in Bor, Grasses Burma, Ceyl., Ind. and Pakist. (1960) 664.—*Triticum repens* L. Sp. pl. (1753) 86.—*Elytrigia repens* (L.) Desv. ex Nevski in Tr. Bot. inst. AN SSSR, ser. 1, 1 (1933) 14, in adnot.; Kitag. Lin. Fl. Mansh. (1939) 75; Norlindh, Fl. mong. steppe, 1 (1949) 122; Fl. Tadzh. 1 (1957) 317; Keng, Fl. ill. sin., Gram. (1959) 410.—Ic.: Fl. SSSR, 2, Plate 47, fig. 6; Kang, l.c. fig. 339.

Described from Europe. Type in London (Linn.).

In meadows, grassy slopes, sand and pebble beds, sparse forests and shrubs, often as weed in farms and various crops, along roadsides, in wastelands; up to midmountain belt.

IA. Mongolia: *Khobd., Mong. Alt. Cis-Hing.* (Khalkha-Gol somon, valley of Khalkhin-Gol river 13 km south-east of Khamar-Daban, Aug. 11, 1949—Yun.); *Cent. Khalkha* (Kerulen river 150 km below Tsetsen-Khan, Aug. 12, 1928—Tug.; Choiren-Ula, 1940—Sanzha); *East. Mong.* (near Hailar town, No. 362, July 4, 1951—Li et al.); *Bas. lakes, Val. lakes* (Saikhan-Obo somon, valley of Ongiin-Gol river 10 km beyond Khoshu-Khida, July 8, 1941—Tsatsenkin); *Gobi-Alt.* (along Leg river, Aug. 16, 1886—Pot.; Dundu-Saikhan hills, in Ulan-Khunde creek valley, Aug. 17 and 18; hills of Dzun-Saikhan, Aug. 27—1931, Ik.-Gal.).

IB. Kashgar: *Nor., West., South.* (nor. foothills of Kunlun, 1700 m, June 13, 1889—Rob.; Khotanskii circle, Sandzhu oasis, May 27, 1959—Yun.; Guma district, near Sandzhubazar village, No. 162, May 28, 1959—Lee et al.).

IIA. Junggar: all regions.

General distribution: Aral-Casp., Balkh. region, Jung.-Tarb., Nor. and Cent. Tien Shan; Europe, Mediterr., Balk.-Asia Minor, Near East, Caucasus, Middle Asia, West. Sib., East. Sib., Far East (evidently, only as introduced plant), Nor. Mong., China

(Altay, Dunbei, Nor.), Himalayas, Korea, Japan, Indo-Malay. (nor.-west. part); introduced in many other countries of both hemispheres.

Note. In most of Mongolia and Kashgar, as also in countries of East. Asia, this species is evidently found only as an introduced but widely distributed plant.

Section 2. Caespitosa Rouy

3. A. aegilopoides Drob. in Tr. Bot. muzeya AN, 12 (1914) 46, emend. Tzvel. h.l.; Pavlov in Byull. Mosk. obshch. ispyt. prir. 38 (1929) 27.—*A. propinquum* Nevski in Izv. Bot. sada AN SSSR, 30 (1932) 498 and in Fl. SSSR, 2 (1934) 634; ? Fl. Kirgiz. 2 (1950) 202; Fl. Kazakhst. 1 (1956) 296.— *A. stenophyllum* Nevski, l.c. (1932) 500.—*A. roshevitzii* Nevski, l.c. (1932) 503 and l.c. (1934) 635; Grubov, Consp. fl. MNR (1955) 76.—*A. gmelinii* (Trin.) Nevski, l.c. (1934) 635, non Scribn. et J.G. Smith (1897).—*Triticum gmelinii* Trin. in Linnaea, 12 (1838) 467.—*Elytrigia propinqua* (Nevski) Nevski in Tr. Sredneaz. gos. univ., ser. 8b, 17 (1934) 61.—*E. stenophylla* (Nevski) Nevski, l.c. (1934) 61.—*E. gmelinii* (Trin.) Nevski in Tr. Bot. inst. AN SSSR, ser, 1, 2 (1936) 78; Melderis in Norlindh, Fl. mong. steppe, 1 (1949) 122.—*Agropyron strigosum* auct. non Boiss.: Krylov, Fl. Zap. Sib. 2 (1928) 346.—Ic.: Nevski, l.c. (1932) fig. 3.

Described from Baikal region. Type in Leningrad.

On rubble and rocky slopes, rocks, talus, pebble beds; up to midmountain belt.

IA. Mongolia: *Khobd.* (upper course of Kharkira river, July 10; south. peak of Kharkira, July 11—1879, Pot.); *Mong. Alt., Gobi-Alt.* (south of Tostu mountain range, Aug. 4, 1886—Pot.).

IIA. Junggar: *Altay region* (Altay district, 1400 m, No. 2548, Aug. 27; near Shara-Sume, No. 2626, Aug. 29—1956, Ching); *Tien Shan* (Bogdashan' hills near Turfan, 2000 m, No. 5739, June 18, 1958—Lee and Chu; Bogdashan' hills south-east of Urumchi, Tyan'tazy lake, 1900 m, Aug. 29, 1959—Petr.).

General distribution: Jung.-Tarb.; West. Sib. (Altay), East. Sib. (south), Nor. Mong. (Fore Hubs, Hent., Mong.-Daur.).

Note. Like L.P. Sergievskaja [Sist. zam. Gerb. Tomsk, univ. 81 (1957) 7–10], we have not been able to detect persistent differences among species described by S.A. Nevski cited by us in synonyms. Further, *A. stenophyllum* Nevski and *A. roshevitzii* Nevski were treated in the much later works of the author himself as synonyms of *A. gmelinii* (Trin.) Nevski. It would perhaps be more correct to treat *A. aegilopoides* and the very close species *A. reflexiaristatum* Nevski from South. Urals, *A. jacutorum* Nevski from Yakutia and *A. amgunense* Nevski from Amur basin as subspecies of the same species namely *A. aegilopoides* Drob. s.l.—which in turn is very closely related to the Crimean endemic species *A. strigosum* (M.B.) Boiss. and some species of Nor. American prairies grouped around *A. spicatum* (Pursh) Scribn. et J.G. Smith. All the species listed above are very close in ecology and their contemporary distribution ranges undoubtedly represent remnants of the far wider range covered by a single Tertiary Eurasian-American species.

4. **A. dshungaricum** (Nevski) Nevski in Fl. SSSR, 2 (1934) 641; Fl. Kirgiz. 2 (1950) 205; Fl. Kazakhst. 1 (1956) 299.—*Elytrigia dshungarica* Nevski in Tr. Sredneaz. gos. univ., ser. 8b, 17 (1934) 61 and Tr. Bot. inst. AN SSSR, ser. 1, 2 (1936) 81.

Described from Junggar Alatau. Type in Leningrad.

On rocky slopes, rocks and talus; in lower and middle mountain belts.

IIA. Junggar: *Jung. Alt.* (south-west. slopes of Maili mountain range 40–42 km nor.-east of Junggar outlets on road to Karagaity pass, Aug. 14, 1957—Yun.).

General distribution: Jung.-Tarb., Nor. and Cent. Tien Shan.

5. **A. ferganense** Drob. in Tr. Bot. muzeya AN, 16 (1916) 138; Nevski in Fl. SSSR, 2 (1934) 641; Fl. Kirgiz. 2 (1950) 202.—? *A. cognatum* Hack. in Allgem. Bot. Z. (1905) 22, in nota; Melderis in Bor, Grasses Burma, Ceyl., Ind. and Pakist. (1960) 660. —*Elytrigia ferganensis* (Drob.) Nevski in Tr. Sredneaz. gos. univ., ser. 8b, 17 (1934) 61; Fl. Tadzh. 1 (1958) 314.

Described from Middle Asia (nor. slope of Altay mountain range). Type in Tashkent (TAK); isotype in Leningrad.

On rocky slopes, rocks and talus; in lower and middle mountain belts.

IIA. Junggar: *Tien Shan* (south. slope of Boro-Khoro mountain range 4 km beyond Nizhn. Ortai village on Sairam-Nur—Kul'dzha road, Aug. 19, 1957—Yun.).

General distribution: Middle Asia (Fergana, Alay, Transalay, Turkestan and Zeravshan mountain ranges), ? Himalayas (Kashmir).

Note. This and the preceding species are very closely related and may be treated as subspecies of *A. ferganense* Drob. s.l. Melderis (l.c.) combines the two species with *A. congnatum* Hack. described from Kashmir but reliable specimens of the latter have eluded us and its original diagnosis is very brief.

6. **A. geniculatum** (Trin.) C. Koch in Linnaea, 21 (1848) 425; Nevski in Fl. SSSR, 2 (1934) 645; Fl. Kazakhst. 1 (1956) 298; Krylov, Fl. Zap. Sib. 12, 1 (1961) 3132.—*A. repens* var. *geniculatum* (Trin.) Kryl. Fl. Zap. Sib. 2 (1928) 353.—*Triticum geniculatum* Trin. in Ledeb. Fl. alt. 1 (1829) 117.— *T. bungeanum* Trin. in Mém. Ac. Sci. St.-Pétersb. Sav. Etrang. 2 (1835) 529.— *Elytrigia geniculata* (Trin.) Nevski in Tr. Bot. inst. AN SSSR, ser. 1, 2 (1936) 82.—Ic.: Ledeb. Ic. pl. fl. ross. 3 (1831) Table 247; Fl. SSSR, 2, Plate 47, fig. 3.

Described from Altay (Charysh river). Type in Leningrad.

On rocky slopes, rocks and talus; in lower and middle mountain belts.

IA. Mongolia: *Mong. Alt.* (Ilyasty river valley, Sept. 9, 1931—Bar.).

General distribution: West. Sib. (Altay), East. Sib. (south), Nor. Mong. (Mong.-Daur.).

Note. The typical specimen of *Triticum bungeanum* Trin. (also from Charysh river and preserved in Leningrad) has a puberulent sheath of lower leaves, tending in this respect towards the closely related species *A. nevskii* Ivanova ex Grub.

7. **A. kanashiroi** Ohwi in J. Jap. Bot. 19 (1943) 167.—*Roegneria alashanica* Keng, Fl. ill. sin., Gram. (1959) 375, diagn. sin.—Ic.: Keng, l.c. fig. 303.

Described from Inner Mongolia. Type in Tokyo.

On rocky slopes, rocks and talus; up to midmountain belt.

IA. **Mongolia:** *East. Mong.* ("Fu-Sheng-Chuang between Chining and Kueisui, No. 3907, leg. Kanashiro"—Ohwi, l.c., (type!); *East. Gobi* (near Bailinmyao, 1960—Ivan.); *Alash. Gobi* (Alashan mountain range, nor.-east. slope of Khote-Gol gorge, June 19, 1908—Czet.; south. part of Alashan mountain range 50 km along Inchuan-Bayan Khoto road, 1800 m, June 10, 1958—Petr.).

General distribution: China (Dunbei, Nor.).

Note. We lacked the opportunity to study the typical specimen of *A. kanashiroi* and hence are not fully confident that the specimens listed above are related to it.

8. **A. nevskii** Ivanova ex Grub. in Bot. mater. Gerb. Bot. inst. AN SSSR, 17 (1955) 4; Grubov, consp. fl. MNR (1955) 76.

Described from Mongolian Altay. Type in Leningrad.

On rocky slopes, rocks and talus; in lower and middle mountain belts.

IA. **Mongolia:** *Khobd.* (south. bank of Uryuk-Nur lake, Sept. 15, 1931—Bar.; midcourse of Tsagan-Nurin-Gol, nor. slope of Obgor-Ul hill, Aug. 2, 1945—Yun.); *Mong. Alt.* (between Ikhes-Nur and Tonkhil'-Nur lakes, July 24, 1897—Klem.; Saksa-Gol river, Aug. 1, 1909—Sap.; between Khatu and Bukhu-Muren rivers, Aug. 7, 1909—Sap.; type!; Uinchi, on granites, Sept. 13, 1930—Bar.; Khasagtu-Khairkhan mountain range, east. slopes of Kukhengir hill on Dundu-Seren-Gol river, Sept. 18, 1930—Pob.; Tonkhil' somon, July 16; south. slope of Adzhi-Bogdo mountain range, Indertiin-Gol river valley, Aug. 6—1947, Yun.; Shargain-Gobi—Tonkhil'somon road, Sept. 6, 1948—Grub.).

IIA. **Junggar:** *Jung. Gobi* (east.: nor. slope of Baga-Khabtak-Nuru mountain range, Sept. 14, 1948—Grub.).

General distribution: endemic.

Note. *A. nevskii*, *A. kanashiroi* Ohwi and *A. geniculatum* (Trin.) C. Koch are very closely related and may be regarded as subspecies of *A. geniculatum* (Trin.) C. Koch s.l. S.A. Nevski placed these species in his section Holopyron (Holmb.) Nevski, not in section Pseudoroegneria Nevski, ignoring their very close relationship with awned species of group *A. aegilopoides* Drob. s.l. It is interesting that, along with awned species of the group *A. spicatum* (Pursh) Scribn. et J.G. Smith s.l., the very closely related unawned species *A. inerme* (Scribn. et J.G. Smith) Rydb., fully corresponding to the Asian group *A. geniculatum* s.l. is also quite extensively distributed within Nor. America.

Section 3. Hyalolepis
(Nevski) Nevski

9. **A. alatavicum** Drob. in Feddes repert. 21 (1925) 43; Nevski in Fl. SSSR, 2 (1934) 633; Fl. Kirgiz. 2 (1950) 201; Fl. Kazakhst. 1 (1956) 294. — *Elytrigia alatavica* (Drob.) Nevski in Tr. Sredneaz. gos. univ., ser. 8b, 17 (1934) 60.

Described from Junggar Alatau. Type in Tashkent (TAK).

On rocky slopes, rocks, talus and in pebble beds; in middle and upper mountain belts.

IB. Kashgar: *West.* (70 km nor.-west of Kashgar along road to Torugart, 2240 m, June 19; 10–12 km nor.-east of Baikurt village, June 20; 10 km south-east of Baikurt village, June 21—1959, Yun.; same site, 2300 m, No. 9769, June 21, 1959—Lee et al.); *South.* (Keriya oasis region, Nura river gorge, 3300 m, June 30, 1885—Przew.; 4 km south of Polur settlement, No. 65, May 11, 1959—Lee et al.).

IIA. Junggar: *Tien Shan* (upper course of Muzart river, Sazlik area, Sept. 8; same site, 1–2 km below Yangimallya settlement, Sept. 11—1958, Yun.).

IIIB. Tibet: *Chang Tang* (nor. slope of Russky mountain range, upper course of Aksu river, about 4000 m, June 3, 1890—Rob.; upper course of Tiznaf-Darya river near Kyude settlement, June 4, 1959—Yun.); *South.* ("Goring Valley, 90°25′, 30°12′, about 4500 m, leg. Littledale"—Hemsley, l.c.).

IIIC. Pamir (Pas-Rabat settlement, July 3; Kok-Muinak settlement at exit of Tagarma valley, July 27, 1909—Divn.; near western margin of Tashkurgan, June 13, 1959—Yun.).

General distribution: Jung.-Tarb., Nor. and Cent. Tien Shan; Middle Asia (Alay).

10. **A. batalinii** (Krasn.) Roshev. in Izv. Bot. sada Petra Vel. 14, suppl. 2 (1915) 96; Nevski in Fl. SSSR, 2 (1934) 632; Fl. Kirgiz. 2 (1950) 200; Fl. Kazakhst. 1 (1956) 294.—*A. alaicum* Drob. in Tr. Bot. muzeya AN, 16 (1916) 138; Nevski, l.c. 633; Fl. Kirgiz. 2 (1950) 205.—*Triticum batalinii* Krasn. in Scripta Bot. Univ. Petrop. 2, 1 (1889) 21.—*Elytrigia batalinii* (Krasn.) Nevski in Tr. Sredneaz. gos. univ., ser. 8b, 17 (1934) 61; Fl. Tadzh. 1 (1957) 313; Ikonnikov, Opred. rast. Pamira [Key to Plants of Pamir] (1963) 71.—*E. argentea* Nevski, l.c. 61.—*Roegneria carinata* Ovcz. et Sidor. in Fl. Tadzh. 1 (1957) 505.—? *R. longiglumis* Keng, Fl. ill. sin., Gram. (1959) 406.—? *Agropyron striatum* auct. non Nees ex Steud.: Hemsley, Fl. Tibet (1902) 205.—Ic.: Fl. Kazakhst. 1, Plate 22, fig. 18.

Described from Tien Shan (Sary-Yassy river). Type in Leningrad.

On rocky and clayey slopes, rocks, talus; in upper mountain belt.

IB. Kashgar: Nor. (nor.-west. slope of Sogdyn-Tau mountain belt 10–12 km south-east of Akchii settlement on Kokshaal-Darya river, Sept. 18—1958, Yun.); *West.* (10–12 km nor. of Baikurt village on Kashgar-Torugart road, 2600 m, June 20, 1959—Yun.; 50 km nor. of Kashgar toward Baikurt, No. 9722, June 20, 1959—Lee et al.); *South.* (Keriya river basin 4 km south of Polur settlement, May 11, 1959—Yun.).

IIA. Junggar: *Tien Shan* (Khomote district, near Bartu—timber factory road, 2160 m, No. 6991, Aug. 3, 1958—Lee and Chu; upper course of Muzart river, Lyangar tributary, Sept. 12, 1958—Yun.).

IIIC. Pamir.

General distribution: Nor. and Cent. Tien Shan, East. Pam.; Middle Asia (West. Tien Shan, Pamir-Alay), ? China (South-West.), ? Himalayas.

Note. Highly polymorphic species. Specimens tending towards its very close relative *A. alatavicum* Drob. with very short awns and wide-set spikelets were described as an independent species—*A. alaicum* Drob. It is highly probable that these specimens belong to relatively very low mountain populations and merit recognition as an independent subspecies but no distinct demarcation whatsoever has been detected between them and typical specimens of *A. batalinii*.

11. **A. grandiglume** (Keng) Tzvel. comb. nova.—*Roegneria grandiglumis* Keng, Fl. ill. sin., Gram. (1959) 405, diagn. sin.—Ic.: Keng, l.c. fig. 334 (type!). Planta perennis, 40–100 cm alta, laxe caespitosa, innovationes sübterraneas breviter repentes emittens; vaginae glabrae et laeves; ligulae ad 0.6 mm lg.; laminae 1.5–3 mm lt., laxe conduplicatae vel partim planae, glabrae, subtus laeves, supra scabrae. Spicae 4–9 cm lg., laxiusculae, leviter secundae; glumae (5) 6–8 (9) mm lg., late lanceolatae, 3–5-nerviae, glabrae vel subglabrae, apice acutatae vel in mucrone ad 1.5 mm lg. transeuntes; lemmata 7–9 mm lg., sat dense et longe pilosa, apice in mucrone vel arista 1–5 mm lg. transeuntia; antherae 1.5–2.2 mm lg.

Described from Qinghai. Type in Nanking.

In sand.

IIIA. Qinghai: *Nanshan* (near Kukunor lake, 3400 m, June 27, 1880—Przew.; nor. bank of Kukunor lake, dunes south of Dere-Nor lake, above 3300 m, April 13, 1886—Pot.; "Qinghai, without exact indication of site"—Keng, l.c., type!).
General distribution: endemic.

Note. This wholly isolated species, very close to *A. thoroldianum* Oliver, is evidently endemic in Kukunor. Its diagnosis in Latin has been given above to supplement that given in Chinese and the illustration of its type (Keng, l.c.).

12. **A. kengii** Tzvel. nom. novum.—*Roegneria hirsuta* Keng, Fl. ill. sin., Gram. (1959) 407, diagn. sin., non *Agropyron hirsutum* (Bertol.) Skalicky et Jiras., 1959.—Ic.: Keng, l.c. fig. 336 (type!).—Planta perennis, 40–100 cm alta, laxe caespitosa, innovationes subterraneas breviter repentes emittens; vaginae glabrae et laeves; laminae 1.5–4 mm lt., planae. Spicae 4–9 cm lg., laxiusculae; glumae 4–6 mm lg., late lanceolatae, acutae, 3–5-nerviae, glabrae; lemmata 7.5–10 mm lg., glumis subduplo longiora, dense sed breviter pilosa, apice in arista 5–8 mm lg. subrecta transeuntia.

Described from Qinghai. Type in Nanking.

Probably in sand and pebble beds.

IIIA. Qinghai: *Nanshan* ("Huan"yuan' district, No. 5257"—Keng, l.c., type!).
General distribution: endemic.

Note. Judging from the above cited illustration of the type of this species, it is probably a hybrid between species of sections Hyalolepis and Agropyron.

13. **A. kokonoricum** (Keng) Tzvel. comb. nova.—*Roegneria kokonorica* Keng, Fl. ill. sin., Gram. (1959) 408, diagn. sin.—Ic.: Keng, l.c. fig. 338 (type!).—Planta perennis, 40–100 cm alta, laxe caespitosa, innovationes subterraneas breviter repentes emittens; vaginae glabrae et laeves; laminae 1.5–3 mm lt., vulgo laxe conduplicatae. Spicae 4–7 cm lg., set densae, inferne interdum interruptae; glumae 3.5–5 mm lg., lanceolatae, 1–3-nerviae, sat dense et longiuscule pilosae, carinatae, apice in arista 3–4 mm lg. transeuntes; lemmata 4.5–6 mm lg., sat dense et longiuscule pilosa, apice in arista 4–6 mm lg. transeuntia.

Described from Qinghai. Type in Nanking.
In sand.

IIIA. Qinghai: *Nanshan* ("Khuan"yuan' district, No. 5364"—Keng, l.c., type!).
General distribution: endemic.

Note. Possibly, this species is of hybrid origin representing an intergeneric hybrid. It is interesting that the specimen preserved in the herbarium of the Botanical Institute, Academy of Sciences of the USSR, from coastal sand of lake Hubsugul (Hubsugul lake, Hilin, July 2, 1902, V. Komarov), representing an intergeneric hybrid with the participation of *Agropyron cristatum* (L.) Beauv., is very similar to the illustration of the type of this species.

14. A. melantherum Keng in Sunyatsenia, 6, 1 (1941) 62.—*Roegneria melanthera* (Keng) Keng, Claves Gen. et Sp. Gram. Sin. (1957) 187 and Fl. ill. sin., Gram. (1959) 401.—Ic.: Keng, l.c. (1959) fig. 329 and 330 (var. *tahopaica* Keng).
Described from Nor.-East. Tibet. Type in Nanking.
In sand and pebble beds.

IIIA. Qinghai: *Nanshan* (crossing 86 km west of Xining, Aug. 5, 1959—Petr.).
IIIB. Tibet: *Weitzan* (on south. bank of Oring-Nor lake on clayey-sandy soil, 4300 m, July 18, 1884—Przew.; "vicinity of Oring-Nor, northern part of Baian-Khara Mountains, No. 49302, 12 VII 1935, leg. C.W. Yao"—Keng, l.c. [1941], type!).
General distribution: endemic.

Note. This species is very closely related to A. *batalinii* (Krasn.) Roshev., differing from it mainly in very dense, subpectinate spikes approaching the spikes of A. *cristatum* (L.) Beauv. in habit. Variety A. *melantherum* var. *tahopaica* (Keng) Tzvel. comb. nova [=*Roegneria melanthera* var. *tahopaica* Keng, Fl. ill. sin., Gram. (1959) 401, diagn. sin; a varietate typica glumis pilosis et lemmatis subinermibus differt. Typus: Keng, l.c. fig. 330] described from Qinghai, differs from the type in pilose (not glabrous) glumes and near total absence of awn on lemma.

15. A. muticum (Keng) Tzvel. comb. nova.—*Roegneria mutica* Keng, Fl. ill. sin., Gram. (1959) 408, diagn. sin.—Ic.: Keng, l.c. fig. 337 (type!). — Planta perennis, 30–80 cm alta; vaginae glabrae; laminae 1.5–4 mm lt., planae vel laxe conduplicatae. Spicae 3–8 cm lg., laxiusculae; glumae 4–6.5 mm lg., late lanceolatae, 3–5-nerviae, sparse pilosae vel glabrae, apice acutae; lemmata 8–11 mm lg., sat dense et breviter pilosa, apice acuta; antherae 2–2.6 mm lg.
Described from Qinghai. Type in Nanking.
Probably on rocky slopes.

IIIA. Qinghai: *Amdo* ("Guide district, No. 4"—Keng, l.c., type!).
General distribution: endemic.

Note. The author of this species places it and A. *kengii* Tzvel. in distinct series, *Hirsutae* Keng (l.c. 406). Judging from the illustrations of these species mentioned above, they are in fact quite similar but their genetic relationship with other species of section Hyalolepis is not clear to us.

16. **A. stenachyrum** (Keng) Tzvel. comb. nova.—*Roegneria stenachyra* Keng, Fl. ill. sin., Gram. (1959) 404, diagn. sin.—Ic.: Keng, l.c. fig. 339 (type!).—Planta perennis, 30–100 cm alta. Spicae 6–10 cm lg., laxae et paulo flexuosae; glumae 3–5.5 mm lg., lanceolatae, trinerviae, secus nervos scabrae, apice acutae; lemmata 8–10 mm lg., breviter pilosa, apice in arista 7–11 mm lg. subrecta transeuntia.

Described from Mongolia (Khesi). Type in Nanking.

Conditions of habitation not clearly reported.

IA. **Mongolia:** *Khesi* ("Tszyutsyuan' district, No. 12443"—Keng, l.c., type!).
General distribution: endemic.

Note. The affinity of this species to section Hyalolepis and to genus *Agropyron* is rather doubtful. The author places it in a monotypical series, *Stenachyrae* Keng (l.c. 404), but, judging from the illustration of the type, it approaches quite closely the species of section Elymus, genus *Elymus* L. It is highly possible that it is of hybrid origin.

17. **A. thoroldianum** Oliver in Hook. Ic. pl. (1893) Table 2262; Hemsley in J. Linn. Soc. London (Bot.) 30 (1894) 123; Deasy, In Tibet and Chin. Turk. (1901) 405; Hemsley, Fl. Tibet (1902) 205; Pilger in Hedin, S. Tibet, 6, 3 (1922) 94; Melderis in Bor, Grasses Burma, Ceyl., Ind. and Pakist. (1960) 667.—*Roegneria thoroldiana* (Oliver) Keng, Claves Gen. et Sp. Gram. Sin (1957) 188 and Fl. ill. sin., Gram. (1959) 404.—Ic.: Oliver, l.c.; Keng, l.c. (1959) fig. 332.

Described from Tibet. Type in London (K).

In sand, talus and pebble beds.

IB. **Qaidam:** *hilly* (Ritter mountain range, sand on slope towards Syrtn, 3700 m, June 25, 1894—Rob.).
IIIB. **Tibet:** *Chang Tang* (south. slope of Muzlyk-Atas hills, 4600–4700 m, in sand, Aug. 17; nor. foothills of Przewalsky mountain range, 4600–5000 m, Aug. 19—1890, Rob.; "Tibet, 5500 m, No. 108, leg. Thorold"—Oliver, l.c., type!; "Northern Tibet, 4965 m, 7 IX 1896"—Pilger, l.c.; "Northern Tibet, 35° 48', 82° 19', 5200 m, 16 VIII 1898"— Deasy, l.c.; "Eastern Tibet, 33° 32', 80° 52', 5127 m, 18 VII 1901"—Pilger, l.c.); *Weitzan* (sandy sites in midcourse of Dzhagyn-Gol river, July 11, 1884—Przew.); *South.* ("Khambajong, No. 32, leg. Younghusband"—Melderis, l.c.).
General distribution: endemic.

Note. The variety *A. thoroldianum* var. *dasyphyllum* Roshev. var. nova in litt., distinguished within this species, is highly characteristic of Tibet—A varietate typica foliorum laminis et vaginis dense, and breviter pilosis differt. Typus: "Tibet borealis, in solo arenoso ad medium decursum fl. Dshagyn-Gol, No. 320, 11 VII 1884, N. Przewalskii" in Herb. Inst. Bot. Ac. Sci. URSS—LE conservatur. Possibly, it merits the rank of subspecies. Another variety—*A. thoroldianum* var. *laxiusculum* Meld. (l.c. 696), its type represented by the specimen from South. Tibet cited above, has very long (5– 7 mm) awns of lemma and very short pubescence of spikes. It is possible that it represents an independent South. Tibetan species more closely related to *A. melantherum* Keng than *A. thoroldianum*.

○ **Kengyilia gobicola** Yen et Yang, in Canad. Journ. Bot. 68 (1990) 1897.

IIIB. Tibet.

K. habahenensis Baum, Yen et Yang, in Pl. Syst. Evol. 174 (1990) 103.

IIA. Junggar: *Cis-Alt.*

o K. longiglumis (Keng, Keng f. et Chen) Yang, Yen et Baum, in Hereditas, 1992, 116 (1992) 27.—*Roegneria longiglumis* Keng, Keng f. et Chen, in Acta Nanking Univ. (Biol.) 3 (1963) 83.

IIA. Junggar. *Cis-Alt.*

o K. nana Yang, Yen et Baum, in Canad. Journ. Bot. 71 (1993) 341.

IIIB. Tibet.

o K. pamirica Yang et Yen, in Journ. Sichuan Agric. Univ. 10, 4 (1992) 567.

IIIB. Tibet.

o K. tahelacana Yang, Yen et Baum, in Canad. Journ. Bot. 71 (1993) 339.

IIA. Junggar. *Tien Shan.*

o K. tridentata Yen et Yang, in Novon, 4 (1994) 310.

IIIA. Qinghai. *Nanshan.*

o K. zhaosuensis Yang, Yen et Baum, in Canad. Journ. Bot. 71 (1993) 341.

IIA. Junggar. *Tien Shan.*

Section Agropyron

18. **A. cristatum** (L.) Beauv. Ess. Agrost. (1812) 146; Franch. Pl. David. 1 (1884) 340; Danguy in Bull. Mus. nat. hist. natur. 17 (1911) 558 and 20 (1914) 148; Krylov, Fl. Zap. Sib. 2 (1928) 357; p.p., Pavlov in Byull. Mosk. obshch. ispyt. prir. 38 (1929) 28; Nevski in Fl. SSSR, 2 (1934) 661; Persson in Bot. notiser (1938) 281; Hao in Engler's Bot. Jahrb. 68 (1938) 579; Kitag. Lin. Fl. Mansh. (1939) 58; Norlindh, Fl. mong. steppe, 1 (1949) 117; Fl. Kirgiz. 2 (1950) 207; Grubov, Consp. fl. MNR (1955) 75; Fl. Kazakhst. 1 (1956) 293; Keng, Fl. ill. sin., Gram. (1959) 416.—*Bromus cristatus* L. Sp. pl. (1753) 78.—*B. distinchus* Georgi, Bemerk. Reise Russ. Reich, 1 (1775) 197; Bobrov in Bot. mater. Gerb. Bot. inst. AN SSSR, 20 (1960) 7.—?*Agropyron barginensis* Skvortz. in Gordeev and Jernakov in Acta pedol. sinica, 2, 4 (1954) 278, nom. nud.—Ic.: Fl. SSSR, 2, Plate 47a, fig. 5; Keng, l.c. fig. 347.

Described from East. Siberia. Type in London (Linn.).

In steppes, arid meadows, sand, on rocky slopes, in pebble beds and among rocks; up to upper mountain belt.

IA. **Mongolia:** *Khobd., Mong. Alt., Cis-Hing., Cent. Khalkha, East. Mong., Bas. lakes, Val. lakes, Gobi-Alt., East. Gobi, West. Gobi* (Atas-Bogdo-Ula, 1943—Yun.); *Alash. Gobi* (Alashan mountain range, Tszosto gorge, May 15, 1908—Czet.; same site, south. part of mountain range 50 km along Inchuan—Bayan-Khoto road, 1800 m, June 10, 1958—Petr.); *Khesi* (Chzhan"e district, 55 km west of Yunchan town, Aug. 10, 1958—Petr.).

IB. Kashgar: *Nor., West.*

IIA. Junggar: *Alt. region, Tarb., Jung. Alt., Tien Shan, Jung. Gobi* (nor.: 30 km south of Burchum, 500 m, No. 3261, Sept. 22, 1956—Ching).

IIIA. Qinghai: *Nanshan* (near Kukunor lake, 3400 m, June 27, 1880—Przew.; valley of Sharagol'dzhin river, June 11, 1894—Rob.; sand bank of Kukunor lake, 4000 m, Aug. 28, 1908—Czet.; "Tsi-Gi-Gan-Ba, 3440 m, 24 VIII 1930"—Hao, l.c.; Altyntag mountain range 15 km south of Aksai settlement, 2800 m, Aug. 2, 1958; meadow on east. bank of Kukunor lake, 3210 m, Aug. 5; 25 km east of Gunhe, Aug. 7—1959, Petr.).

IIIB. Tibet: *Weitzan* (Burkhan-Budda mountain range, Khatu gorge, 3700–4000 m, July 18, 1901—Lad.).

General distribution: Jung.-Tarb., Nor. and Cent. Tien Shan; Middle Asia (Pamir-Alay), West. Sib. (south-east.), East. Sib., Far East (south), Nor. Mong., China (Altay, Dunbei, North.); introduced in many other countries of both hemispheres.

Note. Highly polymorphic species. A variety with dense and long hairy spikelets—*A. cristatum* var. *hirsutissimum* (Kryl.) Tzvel. comb. nova [= *A. cristatum* f. *hirsutissima* Kryl. Fl. Alt. i Tomsk. gub. 7 (1914) 1699], is quite extensively distributed in the hills of southern Siberia and northern Cent. Asia. Binomial *Bromus distinchus* Georgi (l.c.) also belongs to this variety. The variety with subglabrous or totally glabrous spikelets—*A. cristatum* var. *glabrispiculatum* Tzvel. var. nova, is considerably rare—Spiculae glabrae vel subglabrae. Typus: "Kaschgaria, in fauce Ulan-Tuz systematis fl. Charlym, 27 VI 1909, D. Divnogorska" in Herb. Inst. Bot. Ac. Sci. URSS—LE conservatur. Another variety (or subspecies ?) found sporadically—*A. cristatum* var. *erickssonii* Meld. (in Norlindh, l.c. 118), described from Inner Mongolia ("Naiman-Ul, 25 VI 1928, No. 451, leg. Eriksson"), has short but densely hairy leaf blades, sheath and stem (throughout length or only nodes). The herbarium of the Botanical Institute, Academy of Sciences of the USSR, has specimens of this variety, in particular from Mong. Altay (Gulin-Tala plain, Aug. 26, 1943; 10 to 12 km south of Bain-Ulegei, Aug. 4, 1945—Yun.).

19. A. desertorum (Link) schult. et Schult. f. Mant. 2 (1824) 412; Pavlov in Byull. Mosk. obshch. ispyt. Prir. 38 (1929) 28; Nevski in Fl. SSSR, 2 (1934) 657; Melderis in Norlindh, Fl. mong. steppe, 1 (1949) 120; Grubov, Consp. fl. MNR (1955) 75, p.p.; Fl. Kazakhst. 1 (1956) 290; Fl. Tadzh. 1 (1957) 311; Keng, Fl. ill. sin., Gram. (1959) 416.—*A. sibiricum* var. *desertorum* (Fisch. ex Link) Boiss. Fl. or. 5 (1884) 667; Krylov, Fl. Zap. Sib. 2 (1928) 356.—*Triticum desertorum* Fisch. ex Link, Enum. pl. Horti Berol. 1 (1821) 97.—Ic.: Fl. SSSR, 2, Plate 47a, fig. 4; Keng, l.c. fig. 346.

Described from Nor. Caucasus (Kuma river basin). Type in Berlin; isotype in Leningrad.

In clayey and rocky steppes; up to lower mountain belt.

IA. Mongolia: *Cent. Khalkha* (near Del'ger-Tsogtu somon, rocky knolls, June 15, 1950—Kal.); *East. Mong.* (Baishintin-Sume region, Urgo hill on way to Guntu collective, Aug. 18; Guntu area, Aug. 18; Tukhumyin-Bayan-Obo in Dariganga region, Aug. 25—1927, Zam.; "50 km ad mered. versus a Mandaltein-Sume, No. 950, 17 VIII 1935; 40 km ad bor.-occid. a. Khadain-Sume, No. 1161, 20 VIII 1936, leg. Eriksson"—

Melderis, l.c.); *East. Gobi* ("Ikhen-Gung, No. 5608, June 3, 1931, leg. Mühlenweg; Khujirtu-Gol, No. 1151, 19 VII 1927, leg. Hummel"—Melderis, l.c.; plantations 1 km south-east of Dalan-Dzadagad, July 28, 1951—Kal.; near Bailinmyao, 1959—Ivan.).

IIA. Junggar: *Alt. region* (25 km from Shara-Sume along road to Shipati, left bank of Kran river, July 7, 1959—Yun.); *Jung. Alt.* (45 km south-west of Toli, No. 1405, Aug. 8, 1957—Huang); *Zaisan* (Mai-Kapchagai, June 6, 1914—Schisch.).

General distribution: Aral-Casp., Balkh. region; Europe (south-east), Caucasus (Fore Caucasus), Middle Asia, West. Sib. (south).

Note. Possibly of hybrid origin from *A. cristatum* × *A. fragile* and occasionally hybridises with these assumed parent species. The result of hybridisation with *A. cristatum* is evidently the variety *A. desertorum* var. *pilosiusculum* Meld. (l.c. 121) with somewhat hairy lemma. Further, specimens of this variety cited by Melderis (l.c.) can even be direct hybrids from *A: cristatum* × *A. fragile*.

20. **A. fragile** (Roth) Nevski in Tr. Sredneaz. gos. univ., ser. 8b, 17 (1934) 53 and in Fl. SSSR, 2 (1934) 656; Fl. Kazakhst. 1 (1956) 289.—*A. sibiricum* (Willd.) Beauv. Ess. Agrost. (1812) 146; Krylov, Fl. Zap. Sib. 2 (1928) 356, excl. var.; Nevski in Fl. SSSR, 2 (1934) 657; Keng, Fl. ill. sin., Gram. (1959) 414.—*A. mongolicum* Keng in J. Wash. Ac. Sci. 28, 7 (1938) 305 and l.c. (1959) 414.—*Triticum fragile* Roth, Catalecta Bot. 2 (1800) 7.— *T. sibiricum* Willd. Enum. pl. Horti Berol. 1 (1809) 135.—*T. variegatum* Fisch. in Spreng. Pl. Pugill. 2 (1815) 24.—*T. angustifolium* Link, Enum. pl. Horti Berol. 1 (1821) 97.—*T. dasyphyllum* Schrenk in Bull. phys.-math. Ac. Sci. St.-Pétersb. 10 (1842) 356.—*Agropyron desertorum* auct. non Schult. et Schult. f.: Grubov, Consp. fl. MNR (1955) 75, p.p.—Ic.: Nevski, l.c. 2, Plate 47a, Fig. 3; Keng, l.c. (1938) fig. 4 and (1959) figs. 344–345.

Described probably from Kazakhstan. Site of type not known.

In sand, sand steppes; up to lower mountain belt.

IA. **Mongolia:** *Khobd.* (on Tolbo-Nur-Ulegei road, Aug. 4, 1945—Yun.); *Cent. Khalkha* (old Ulan-Bator—Dalan-Dzadagad road 15 km nor. of Erdeni-Dalai somon, July 13, 1948—Grub.); *East. Mong.* (Tukhumyin-Gobi in Dariganga region, July 23, 1927—Zam.; Moltsok-Elisu sand 3–4 km nor. of Moltsok-Khida, May 16, 1944—Yun.); *Val. lakes* (between Tuin-Gol and Tatsin-Gol rivers, Sept. 13, 1924—Pavl.; 5–7 km south of Torgatu somon along road to Guchin-Us somon, July 4, 1941—Yun.; Dzhinsetu somon, Bardzun-Khongor area, Aug. 17, 1949—Kal.); *East. Gobi, Alash. Gobi* (Tengeri sand near Bayan-Khoto, Aug. 4, 1958—Petr.); *Ordos* (on bank of Ulidu-Nor salt lake 60 km west of Ushin town, Aug. 2; 5 km nor.-west of Ushin town, Aug. 3; 18 km nor. of Dzhasak town; Aug. 15; 15 km east of Dzhasak town, Aug. 16— 1957, Petr.).

IB. **Kashgar:** *Nor.* (valley of Muzart river before its discharge into Bai basin, Sept. 12, 1958—Yun.).

IIA. **Junggar:** *Jung. Alt.* (45 km south-west of Toli, No. 1406, Aug. 8, 1957— Huang); *Jung. Gobi* (nor.: near Koktogoi, 950 m, No. 1034, June 10, 1959—Lee et al.; 47 km east of Burchum along road to Shara-Sume, July 5; 5–8 km nor. of Bulun-Tokhoi settlement in the lower courses of Urungu river, ridge sand, July 9—1959, Yun; 4 km nor.-west of Arak, No. 10624, July 12, 1959—Lee et al.); *Zaisan, Dzhark.*, (near Suidun settlement, July 1877—A. Reg.; right bank of Ili 7–8 km south-west of Suidun, Aug. 31, 1957—Yun.).

General distribution: Aral-Casp., Balkh. region; Europe (far south-east), Caucasus (Fore Caucasus), Middle Asia, West. Sib. (south).

Note. The type variety of this species with profusely hairy sheaths and blades of lower leaves is found only very sporadically like *A. fragile* var. *angustifolium* (Link) Tzvel. comb. nova =*Triticum angustifolium* Link, l.c.) with rather hairy lemma, probably of hybrid origin: *A. fragile* × *A. cristatum*. The most widely distributed, however, is the variety *A. fragile* var. *sibiricum* (Willd.) Tzvel. comb. nova (= *Triticum sibiricum* Willd, l.c.) with glabrous leaves and spikelets. From the latter, *A. mongolicum* Keng (l.c.), described from Inner Mongolia, whose isotype ("prov. Suiyuan, Payin-Obo about 90 li (90 × 633 yards) north-east of Peiling-Miao, Roerich Exped., No. 748, 9 VIII 1935, leg. Y.L. Keng") is preserved in the herbarium of the Botanical Institute, Academy of Sciences of the USSR, could not be clearly differentiated.

o *A. krylovianum* Schischk. in Animadv. Syst. Herb. Univ. Tomsk. 2 (1928) 2, Gubanov. Consp. Fl. Outer Mong. (1996) 16.

IA. Mongolia. *Khobd., Mong. Alt.*

21. *A. michnoi* Roshev. in Izv. Glavn. bot. sada SSSR, 28 (1929) 384; Nevski in Fl. SSSR, 2 (1934) 656; Grubov, Consp. Fl. MNR (1955) 75.

Described from Transbaikal. Type in Leningrad.

In sand in river and lake valleys, and sandy steppes.

IA. Mongolia; *Cent. Khalkha, East. Mong. Gobi-Alt.* (Artsatuin-Ama creek valley on south. slope of Ikhe-Bogdo mountain range, Sept. 7, 1943—Yun.); *Ordos* (50 km west of Dunshen town, Aug. 7, 1957—Petr.).

General distribution: East. Sib. (south), Nor. Mong. (Fore Hubs., Mong.-Daur.), China (Dunbei).

Note. Highly polymorphic species, possibly of hybrid origin (*A. cristatum* × ? *A. repens*), hybridising readily with closely related species, *A. cristatum* (L.) Beauv. and *A. fragile* (Roth) Nevski.

22. *A. pectinatum* (M.B.) Beauv. Ess. Agrost. (1812) 146.—*A. pectiniforme* Roem. et Schult. Syst. Veg. 2 (1817) 758; Nevski in Fl. SSSR, 2 (1934) 659; Fl. Kirgiz. 2 (1950) 206; Fl. Kazakhst. 1 (1956) 292; Fl. Tadzh. 1 (1957) 311; Melderis in Bor, Grass. Burma, Ceyl., Ind. and Pakist. (1960) 664.—*A. imbricatum* (M.B.) Roem. et Schult. l.c. 757; Nevski, l.c. 660.—*A. tarbagataicum* Plotn. in Tr. Omsk. sel'skokhoz. inst. 20 (1946) 44; Fl. Kazakhst. 1 (1956) 292.—*Triticum pectinatum* M.B. Fl. taur.-cauc. 1 (1808) 87, non R. Br. (1810).—*T. imbricatum* M.B. l.c. 88, non Lam. (1791). —Ic.: Fl. Kazakhst. 1, Plate 22, fig. 3.

Described from Crimea. Type in Leningrad.

In steppes, arid meadows, on rocky slopes, sand and rocks; up to upper mountain belt.

IA. Mongolia: *Cent. Khalkha* (Erdeni-Dalai somon, in vicinity of Amugulan-Ul hill on old Ulan Bator—Dalan-Dzadagad road, stipa steppe, July 13, 1948—Grub.).

IIA. Junggar: *Tien Shan* (Pilyuchi village nor. of Kul'dzha, May 19, 1877 — A. Reg.; upper course of Kunges river, about 3500 m, June 5, 1877—Przew.; between Ili and Dzhagastai, No. 3154, Aug. 7; 4 km east of Chzhaos, No. 812, Aug. 11; 11 km east of Aksu along road to Chzhaos, No. 1592, Aug. 15; 26 km from Chzhaos along road to Tekes, No. 990, Aug. 17; between Gunlyu and Shakhe, No. 3650, Aug. 18—1957, Huang).

General distribution: Aral-Casp., Balkh. region, Jung.-Tarb., Nor. and Cent. Tien Shan; Cent. and South. Europe, Mediterr., Balk.-Asia Minor, Near East, Caucasus, Middle Asia, West. Sib. (south), Nor. Mong. (Mong.-Daur.: mounds on left bank of Burgultai river 30 km from its discharge into Orkhon river, Aug. 1, 1950—Kal.); wild or introduced in many countries of both hemispheres.

Note. This species and *A. cristatum* (L.) Beauv. are probably more correctly treated as only subspecies of *A. cristatum* (L.) Beauv. s.l. Type *A. pectinatum* with glabrous spikelets as well as the rarer variety *A. pectinatum* var. *imbricatum* (Roem. et Schult.) Tzvel. comb. nova (= *A. imbricatum* Roem. et Schult. l.c.) with spikelets rather pilose but never so dense as in *A. cristatum*, are found in Cent. Asia. *A. tarbagataicum* Plotn. (l.c.) has been described from specimens with stems that are pilose under spikes. This characteristic, very often found in *A. pectinatum* from Middle Asia as well as Cent. Asia, is also known in specimens from the European part of the distribution range of this species and bears hardly any significant taxonomic importance. The two above-cited isolated reports of *A. pectinatum* in Mongolia are of interest from the phytogeographical view point if, indeed, this species was not introduced from plant crops.

o **A. pumilum** Candargy, in Arch. Biol. Veg. (Athenes) 1 (1901) 29, 49; Gubanov, Consp. Fl. Outer Mong. (1996) 16.

IA. Mongolia, *Khobd, Mong. Alt., East Mong.*
IIA. Junggar. *Cis-Alt.*

68. Eremopyrum (Ledeb.)
Jaub. et Spach

Ill. pl. or. 4 (1850–1853) 26.—*Triticum* sect.
Eremopyrum Ledeb. Fl. alt. 1 (1829) 12.

1. Spikes 8–18 mm long, 6–15 mm broad, breaking from base during fruiting; glumes glabrous but with very short (tuberculate) spinules along greatly thickened keel; lemma covered in lower part with very short hairs and, rather scabrous in upper part; palea with densely arranged but very short spinules along keels.
   .... 4. E. triticeum (Gaertn.) Nevski.

+ Spikes 20–45 mm long, 15–23 mm broad, with rachis breaking into segments during fruiting; glumes with fairly long spinules along keel; lemma glabrous or rather hairy; palea with fairly long interrupted spinules along keels .... 2.

2. Spikes during fruiting and on drying break very easily into segments; glumes as long as spikelet, densely pilose; lemma with dense and long hairs; tip of palea with two 0.7–2 mm long

terminal awns which almost reach base of awn of lemma; anthers
0.6–0.9 mm long .... 2. **E. distans** (C. Koch) Nevski.

+ Spikes with relatively poorly developed joints between rachial
segments which disintegrate less readily; glumes shorter than
spikelet; palea at tip with 2 lanceolate teeth up to 1 mm long, far
short of reaching base of awn of lemma; anthers 0.8–1.3 mm long.
.... 3.

3. Spikes glabrous, 2–4.5 cm long, 1.5–2.8 cm broad; glumes
relatively broad .... 1. **E. bonaepartis** (Spreng.) Nevski.

+ Spikes pilose (less dense than in *E. distans*), 1.5–3 cm long and
0.9–1.8 cm broad; glumes relatively narrower ....
.... 3. **E. orientale** (L.) Jaub. et Spach.

Section 1. Eremopyrum

1. **E. bonaepartis** (Spreng.) Nevski in Tr. Sredneaz. gos. univ., ser. 8b. 17
(1934) 52 and in Fl. SSSR, 2 (1934) 663; Fl. Kirgiz. 2 (1950) 208; Fl.
Kazakhst. 1 (1956) 310; Fl. Tadzh. 1 (1957) 319; Melderis in Bor, Grasses
Burma, Ceyl., Ind. and Pakist. (1960) 671.—*Triticum bonaepartis* Spreng. in
Nachr. Bot. Gart. Halle, 1 (1801) 40.—Ic.: Fl. Kazakhst. 1, Plate 23, fig. 4.

Described from Egypt. Type in Paris.

On rocky and melkozem slopes, in sand and pebble beds; up to lower
mountain belt.

IIA. Junggar: *Dzhark.* (Bayandai village near Kul'dzha, May 5, 1878—A. Reg.).

General distribution: Aral-Casp., Balkh. region, Nor. Tien Shan; Mediterr. (east.),
Balk.-Asia Minor, Near East, Caucasus (South. and East. Transcaucasus), Middle
Asia, Himalayas (west.).

2. **E. distans** (C. Koch) Nevski in Tr. Sredneaz. gos. univ., ser. 8b, 17
(1934) 52 and in Fl. SSSR, 2 (1934) 665; Fl. Tadzh. 1 (1957) 320; Melderis
in Bor, Grasses Burma, Ceyl., Ind. and Pakist. (1960) 672.—*Agropyron
distans* C. Koch in Linnaea, 21 (1848) 426.—*Eremopyrum orientale* auct. non
Jaub. et Spach: N. Kuznetzov in Fl. Kazakhst. 1 (1956) 310, p.p. —Ic.: Fl.
SSSR, 2, Plate 47a, fig. 6.

Described from South. Transcaucasus (Armenia). Type in Berlin.

On rocky and melkozem slopes, in sand and pebble beds; up to lower
mountain belt.

IB. Kashgar: *West.* (Kingtau mountain range trail 25 km south-west of Upal oasis,
June 9, 1959—Yun.; between Upal and Koshkulak, No. 586, June 8, 1959—Lee et al.).

IIA. Junggar: *Tien Shan* (Ili bank east of Kul'dzha, May 14, 1877; Taldy gorge,
1000–1200 m—A. Reg.); *Jung. Gobi* (between Urumchi and Khuanshan'-Kou, No.
526a, May 30; between Savan and Paotai, No. 769, June 10; between Paotai and
Syaed, No. 841, June 12—1957, Huang); *Dzhark.* (Ili bank south of Kul'dzha, June

30, 1877; Bayandai village near Kul'dzha, May 5; between Suidun and Ili river, May 7—1878, A. Reg.).

General distribution: Aral-Casp., Balkh. region, Nor. Tien Shan; Asia Minor, Near East, Caucasus (South. and East. Transcaucasus), Middle Asia, Himalayas (west.), Indo-Malay. (nor.-west.).

3. **E. orientale** (L.) Jaub. et Spach, Ill. pl. or. 4 (1850–1853) 26; Nevski in Fl. SSSR, 2 (1934) 664; Fl. Kirgiz. 2 (1950) 211; Fl. Kazakhst. 1 (1956) 310, p.p.; Keng, Fl. ill. sin., Gram. (1959) 418; Melderis in Bor, Grasses Burma, Ceyl., Ind. and Pakist. (1960) 672.—*Secale orientale* L. Sp. pl. (1753) 84.— *Agropyron orientale* (L.) Roem. et Schult. Syst. nat. 2 (1817) 757; Krylov, Fl. Zap. Sib. 2 (1928) 359. — Ic.: Jaub. et Spach, l.c. Table 319; Keng, l.c. fig. 348.

Described from Mediterranean. Type in London (Linn.).

In foothills and low-mountain steppes and semideserts, on sand, rocky slopes and pebble beds.

IIA. Junggar: *Alt. region* (2–3 km south-south-east of Shara-Sume settlement, left bank of Kran river, July 7, 1959—Yun.); *Jung. Gobi* (bank of Urmuchi river, No. 507, May 29; between Urumchi and Khuanshan'-Kou, No. 526, May 30–1957, Huang; Khobuk-Urungu interfluve 35 km from Shauzge state farm on Sulyugou road, July 11, 1959—Yun.); Dzhark. (Khoyur-Sumun village, May 27; Ili bank south of Kul'dzha, June 30—1877, A. Reg.).

General distribution: Aral-Casp., Balkh. region, Nor. Tien Shan; Europe (south-east.), Mediterr., Balk.-Asia Minor, Near East, Caucasus, Middle Asia, West. Siberia (south), Himalayas (west.), Afr. (nor.).

Section 2. M i c r o p y r o n Nevski

4. **E. triticeum** (Gaertn.) Nevski in Tr. Sredneaz. gos. univ., ser. 8b, 17 (1934) 52 and in Fl. SSSR, 2 (1934) 652; Fl. Kirgiz. 2 (1950) 208; Fl. Kazakhst. 1 (1956) 309; Fl. Tadzh. 1 (1957) 318; Keng, Fl. ill. sin., Gram. (1959) 419.—*Agropyron triticeum* Gaertn. in Novi Comm. Ac. Sci. Petrop. 14, 1 (1770) 540; Krylov, Fl. Zap. Sib. 2 (1928) 358.—Ic.: Keng, l.c. fig. 349.

Described from Kazakhstan (valley of Ural river). Type location not known.

In steppes and semideserts, on rocky and clayey slopes, in sand and solonetz; up to midmountain belt.

IIA. Junggar : *Tien Shan* (near Dzhin village, 1600–2000 m, June 5, 1879—A. Reg.; 10–12 km south of Urumchi along road to Davanchin-Daban, June 2, 1957—Yun.); *Jung. Gobi* (bank of Urumchi river, No. 505, May 9; near Shikhedza settlement, No. 683, June 7; westward of Syaedi village, No. 1113, June 14; near Syaedi village, June 16—1957, Huang; left bank of Manas river 20–23 km west of Syaedi state farm, June 13, 1957—Yun.); *Dzhark.* (near Kul'dzha town, April 28; Chimpansi village west of Kul'dzha, May 8; Ili bank south of Kul'dzha, June 30—1877, A. Reg.).

General distribution: Aral-Casp., Balkh. region, Nor. Tien Shan; Europe (south-east.), Balk.-Asia Minor, Near East, Caucasus, Middle Asia, West. Sib. (south).

69. Triticum L.
Sp. pl (1753) 85.

1.   Glumes with keel pectinate in upper part; not pectinate or almost so in lower part...1. T. aestivum L.
+   Glumes with highly prominent pectinate keel right up to base ... . T. durum Desf.

1. T. aestivum L. Sp. pl. (1753) 85; Nevski in Fl. SSSR, 2 (1934) 687; Hao in Engler's Bot. Jahrb. 68 (1938) 580; Norlindh, Fl. mong. steppe, 1 (1949) 135; Fl. Kirgiz. 2 (1950) 215; Grubov, Consp. fl. MNR (1955) 77; Fl. Kazakhst. 1 (1956) 317; Fl. Tadzh. 1 (1957) 327; Keng, Fl. ill. sin., Gram. (1959) 420; Bor, Grasses Burma, Ceyl., Ind. and Pakist. (1960) 679.—*T. sativum* Lam. Encycl. 2 (1786) 554; Krylov, Fl. Zap. Sib. 2 (1928) 361.—*T. vulgare* Vill. Hist. Pl. Dauph. 2 (1787) 153; Henderson and Hume, Lahore to Yarkand (1873) 341; Franch. Pl. David. 1 (1884) 340; Danguy in Bull. Mus. nat. hist. natur. 6 (1911) 7; Persson in Bot. notiser (1938) 275.—Ic.: Keng, l.c. fig. 351.

Described from Europe. Type in London (Linn.).

Extensively cultivated in oases and sometimes found as introduced or escaped along roadsides and around irrigation ditches, in wastelands; up to midmountain belt.

IA. Mongolia: *Mong. Alt., East. Mong., Bas. Lakes, Val. lakes* (Baidarik river valley, July 26, 1927—Simukov); *Gobi Alt., East. Gobi* (Mandal-Obo somon 25 km south of Bain-Dzak, July 27, 1951—Kal.); *Alash. Gobi* (Dynyuanin [Bayan-Khoto] oasis, June, 1908—Czet.); *Khesi* (near Gaotai town, June 8 and 13, 1886—Pot.).

IB. Kashgar: *West.* ("Jarkand"—Henderson and Hume, l.c., "Kentalek, 2700 m, 21 VII 1931 "—Persson, l.c.); *South.* (nor. foothills of Kunlun, 1600 m, July 13, 1889–Rob.).

IIA. Junggar: *Alt. region* (near Koktogoi, 1200 m, No. 2096, Aug. 18, 1956—Ching); *Tien Shan, Jung. Gobi, Dzhark.*

IIIA. Qinghai: *Nanshan, Amdo.*

IIIB. Tibet: *Weitzan* (Yangtze river basin, Chzherku temple, 3800 m, Aug. 11, 1900; Burkhan Budda mountain range, Khatu gorge, 3500 m, June 16, 1901—Lad.).

General distribution: Widely cultivated in many other countries but nature of Palaeo-Mediterranean.

Note. We prefer to place wheat *T. compactum* Host [Gram. Austr. 4 (1809) 4] with highly shortened spikelets, cultivated in some regions of Cent. Asia, among *T. aestivum* as a group of varieties (convarietas).

T. durum Desf. Fl. Atlant. 1 (1798) 114; Danguy in Bull. Mus. nat. hist. natur. 6 (1911) 7; Nevski in Fl. SSSR, 2 (1934) 685; Fl. Kirgiz. 2 (1950) 214; Fl. Kazakhst. 1 (1956) 318; Fl. Tadzh. 1 (1957) 334; Bor, Grasses Burma, Ceyl., Ind. and Pakist. (1960) 679.

Described from Nor. Africa. Type in Paris.

Occasionally cultivated in oases.

IB. **Kashgar:** *Nor.* ("Koutchar, VI 1907, leg. Vaillant"—Danguy, l.c.).
General distribution: cultivated in many countries but native of Palaeo-Mediterranean.

70. Secale L.
Sp. pl. (1753) 84.

1. **S. cereale** L. Sp. pl. (1753) 84; Krylov, Fl. Zap. Sib. 2 (1928) 360; Nevski in Fl. SSSR, 2 (1934) 667; Fl. Kirgiz. 2 (1950) 211; Grubov, Consp. fl. MNR (1955) 76; Fl. Kazakhst. 1 (1956) 313; Fl. Tadzh. 1 (1957) 323; Keng, Fl. ill. sin., Gram. (1959) 421; Melderis in Bor, Grasses Burma, Ceyl., Ind. and Pakist. (1960) 677.—Ic.: Fl. Kazakhst. 1, Plate 23, fig. 8; Keng, l.c. fig. 352. Described from Europe. Type in London (Linn.).

Cultivated in oases and sometimes found as introduced or escaped along roadsides and in various other crops.

IIA. **Junggar:** *Alt. region* (Altay [Shara-Sume], No. 2363, Aug. 15, 1956—Ching); *Jung. Gobi* (south.: between Paotai and Syaedi, in sand, No. 819, June 12; near Syaedi village, state farm of 9th regiment, No. 1514, June 16; left bank of Manas river, state farm of 23rd regiment. No. 331, July 3—1957, Huang).

General distribution: Aral-Casp., Balkh. region; Europe, Mediterr., Balk.-Asia Minor, Near East, Caucasus, Middle Asia, West Sib., East. Sib., Far East, Nor. Mong. (Hent.), China (Dunbei), Korea, Nor. Amer.; occasionally cultivated in many other countries also but the Palaeo-Mediterranean .

71. Hordeum L.
Sp. pl. (1753) 84.

1. Cultivated annual but sometimes wild with large and broad (10–20 mm broad) spikes; awns of lemma usually very long, rarely almost lacking ....2.
+   Wild perennial with relatively small and narrow (3–8 mm broad) spikes; awns of lemma up to 12 mm long ....4.
2. Spikes with two rows of spikelets, in groups of three each, only middle one bisexual and fertile; lateral barren (with rudimentary or staminate floret); lower lemma of middle spikelet with long awn ....6. **H. distinchon** L.
+   Spikes with six rows of spikelets, all three of each group fertile, with bisexual floret...3.
3. Tip of lemma with tripartite appendage, whose lateral robes are not as large and usually somewhat pendent while central lobe notably larger, spur-like and rather curved, sometimes with additional floret ....5. **H. aegiceras** Nees ex Royle.
+   Tip of lemma with long awn. ....7. **H. vulgare** L.

4. Awn of lemma in fertile (middle one in each group) spikelets (4) 5–10 (12) mm long; anthers in fertile spikelets 0.7–2 mm long ....
....5.

+ Awn of lemma in fertile spikelets up to 3 (4) mm long; anthers in fertile spikelets 2.8–4.3 mm long ....6.

5. Stems with puberulent nodes; lemma covered throughout outer surface with elongated spinules; anthers 1.2-2 mm long...............
.... 1. H. bogdanii Wilensky.

+ Stems with glabrous nodes; lemma with short spinules only in upper part; anthers 0.7–1.2 mm long ....
.... 3. H. roshevitzii Bowden.

6. Lemma with short spinules only in upper part, lower part smooth or almost so (with dispersed spinules barely discernible even under high magnification); nodes of stems usually glabrous, occasionally puberulent.......... 2. H. brevisubulatum (Trin.) Link.

+ Lemma covered throughout surface with rather elongated spinules, often transforming into short hairs; nodes of stems usually puberulent, occasionally glabrous ....
.... 4. H. turkestanicum Nevski.

Section 1. Stenostachys
Nevski

1. **H. bogdanii** Wilensky in Izv. Sarat. opytn. st. 1, 2 (1918) 13; Nevski in Fl. SSSR, 2 (1934) 724 and in Tr. Bot. inst. An SSSR, ser. 1, 5 (1941) 163 ; Melderis in Norlindh, Fl. mong. steppe, 1 (1949) 136; Fl. Kirgiz. 2 (1950) 230; Grubov, consp. fl. MNR (1955) 78; Fl. Kazakhst. 1 (1956) 331; Fl. Tadzh. 1 (1957) 281; Keng, Fl. ill. sin., Gram. (1959) 439; Krylov, Fl. Zap. Sib. 12, 1 (1961) 3136.   —Ic.: Nevski, l.c. (1941) fig. 30; Keng, l.c. fig. 373.
Described from Transvolga. Type in Leningrad.

In saline meadows and solonchak, sand and pebble beds of river and lake valleys; up to midmountain belt.

IA. **Mongolia**: *Mong. Alt., Bas. lakes, Gobi-Alt.* (south-east of Dalan-Dzadagad town, July 17, 1950—Kal.); West. Gobi (Bain-Gobi somon, Shar-Khulusun area, Aug. 9, 1943—Tsebigmid; Noyan somon, Bilgekhu-Bulak area 30–35 km east of Tsagan-Bogdo hill, Aug. 31, 1943—Yun.); *Alash-Gobi* (Chagan-Beli area, July 4 and 5; Khara-Sukhai area along left bank of Edzin-Gol—1866, Pot.; "prov. Ning-Hsia, Dash-Obo, No. 7016, 5 VII 1928, leg. Söderbom"—Melderis l.c.; border road south of Tostu-Nuru mountain range, near Khubdyin-Bulak spring, Aug. 15, 1948—Grub.; Tengeri sand near Bayan-Khoto, July 13, 1958—Petr.); *Khesi* (Yanchi village, June 17, 1886—Pot.; near Tszyutsyuan' town, June 25, 1890—Marten).

IB. **Kashgar**: *Nor.* (20 km south-west of Karashar along road to Kurl', June 28, 1958—Yun.); *Takla-Makan* (Cherchen oasis, June 3, 1890—Rob.).

IC. **Qaidam**: *plain* (lower belt of Burkhan Budda mountain range, Aug. 7, 1884—Przew.).

IIA. Junggar: *Tien Shan* (Nan'shan'-Kou picket, May 29, 1877—Pot.); *Jung. Gobi* (St. Kuitun settlement 3–4 km east of Kuitun state farm, June 30, 1957—Yun.); *Zaisan* (right bank of Ch. Irtysh below Burchum river estuary, Kiikpai collective, June 15, 1914 —Schisch.).

General distribution: Aral-Casp., Balkh. region, June.-Tarb., Nor. and Cent. Tien Shan; Europe (Transvolga), Middle Asia, West. Sib., Nor. Mong. (Hang.), Himalayas (west).

2. H. brevisubulatum (Trin.) Link in Linnaea, 17 (1843) 391; Nevski in Fl. SSSR, 2 (1934) 724 and in Tr. Bot. inst. AN SSSR, ser. 1, 5 (1941) 175, p.p.; Kitag. Lin. Fl. Mansh. (1939) 79; Melderis in Norlindh, Fl. mong. steppe, 1 (1949) 137; Fl. Kirgiz. 2 (1950) 229; Grubov, Consp. fl. MNR (1955) 78; Fl. Kazakhst. (1956) 331; Fl. Tadzh. 1 (1957) 282, p.p.; Keng, Fl. ill. sin., Gram. (1959) 439; Fl. Zap. Sib. 12, 1 (1961) 3137, p.p.; Ikonnikov, Opred. rast. Pamira [Key to Plants of Pamir] (1963) 74.—*H. secalinum* var. *brevisubulatum* Trin. Sp. Gram. Icon. et Descr. 1 (1828) Table 4.—? *H. macilentum* Steud. Synops. Pl. Glum. 1 (1954) 352; Nevski, l.c. (1941) 181.— *H. secalinum* subsp. *brevisubulatum* (Trin.) Kryl. Fl. Zap. Sib. 2 (1928) 364, p.p.—*H. pratense* auct. non Huds.: Franch. Pl. David. 1 (1884) 341.—*H. secalinum* auct. non Schreb.: Pavlov in Byull. Mosk. obshch. ispyt. prir. 38 (1929) 28.—Ic.: Trin. l.c.; Nevski, l.c. (1941) figs. 36 and 38; Keng, l.c. fig. 372.

Described from East. Siberia (Irkutsk town region). Type in Leningrad. In saline meadows and solonchak; up to upper mountain belt.

IA. Mongolia: *Khobd., Mong. Alt., Cent. Khalkha, East. Mong., Bas. lakes, Val. Lakes, Gobi-Alt.* (Dundu-Saikhan hills, July 2, 1909—Czet.; Dzun-Saikhan hills, Tsagan-Gol river in Yalo creek valley, Aug. 23, 1931—Ik.-Gal.; Khongor somon, Tergetu-Khuduk area, June 20, 1945—Yun.); *East. Gobi* ("Khonin-Chagan-Chŏlo-Nor, No. 6720, 22 VII 1927, leg. Söderbom; No. 1311, 1 VIII 1927, leg. Hummel"— Melderis, l.c.; Batu-Khalga river, Roerich Exped., Nos. 506 and 508, July 25, 1935— Keng; 1 km east of Dalan-Dzadagad, May–July, 1939—Surmazhab; near Bailinmyao, 1960—Ivan.).

IB. Kashgar: *West.* (70 km from Kashgar on road to Ulugchat, June 17, 1959— Yun.); *East.* (in gorge near Khami oasis, May 12, 1879—Przew.).

IIA. Junggar: *Alt. region* (west of Koktogoi, 1200 m, No. 1851, Aug. 13, 1956— Ching); *Jung. Alt.* (south. slope of Barlyk mountain range, Arba-Kezen' river valley, July 9, 1905—Obruchev); Tien Shan (Dzagastai-Gol, 3000 m, Sept. 5, 1879—A. Reg.); *Zaisan* (between Burchum and Kaba rivers, June 15, 1914—Schisch.); *Dzhark.* (Ili river bank west of Kul'dzha, May, 1877—A. Reg.).

General distribution: Aral-Casp., Balkh. region, Jung.-Tarb., Nor. and Cent. Tien Shan, East. Pam.; Europe (south-east.), Middle Asia, West. Sib. (south), East. Sib., Far East (south-west.), Nor. Mong., China (Dunbei, Nor.).

Note. Within the range of species adopted by us, S.A. Nevski [l.c. (1941)] distinguishes two ecogeographical races: a more western race with puberulent nodes of stems to which he assigned the name *H. brevisubulatum* Link and the more eastern race with glabrous nodes, named by him *H. macilentum* Steud. However, the above-cited variety of Trinius (type from Irkutsk town region) represents the basionym of *H. brevisubulatum* for which reason this name primarily belongs to the eastern race. The western race evidently nevertheless merits recognition at least as a variety—*H. brevisubulatum* var. *nevskianum* (Bowden) Tzvel. comb. nova [= *H. nevskianum* Bowden in Canad. J. Genet. and Cytol. 7 (1965) 396] whose eastern boundary runs through south-west. Altay and East. Tien Shan.

3. **H. roshevitzii** Bowden in Canad. J. Genet. and Cytol. 7 (1965) 395.— *H. secalinum* subsp. *typicum* Kryl. Fl. Zap. Sib. 2 (1928) 363.—*H. sibiricum* Roshev. in Izv. Glavn. bot. sada SSSR, 28 (1929) 385, non Schenk, 1908; Nevski in Fl. SSSR, 2 (1934)724 and in Tr. Bot. inst. AN SSSR, ser. 1, 5 (1941) 160; Melderis in Norlindh, Fl. mong. steppe, 1 (1949) 136; Grubov, Consp. fl. MNR (1955) 78; Fl. Kazakhst. 1 (1956) 331; Krylov, Fl. Zap. Sib. 12, 1 (1961) 3137.—Ic.: Nevski, l.c. (1941) fig. 28.

Described from East. Siberia. Lectotype (Transbaikal, Dzhida river basin, valley of Ichotoi river, No. 630, July 30, 1912—V. Smirnov) in Leningrad.

In saline meadows, solonchak, sand and pebble beds of river valleys; up to midmountain belt.

IA. **Mongolia:** *East. Mong.* ("Dabasun-Nor. No. 399a, 10 VIII 1919, leg. Andersson; Gungrik, No. 123, July 27, 1924, leg. Eriksson"—Melderis, l.c.; Baishintin-Sume, Orle area, Aug. 17, 1927—Zam.; Shavorte-Nur lake 50–60 km east of Erentsab, 19 VIII 1949—Yun.); *Bas. lakes* (south. bank of Khara-Us-Nur lake, Aug. 6, 1879—Pot.; Tuguryuk river valley, Aug. 14, 1930—Bar.); *Val. lakes* (Orok-Nur lake, Aug. 20, 1886—Pot.); *Gobi-Alt.* (valley of Legin-Gol river, Aug. 19, 1927—Simukova; Bain-Tukhum area, Aug. 4, 1931—Ik.-Gal.; Tsagan-Gol river near Bain-Gobi somon post, July 27, 1948—Grub.); *Ordos* (20 km east of Otok town near Ulantsaidenmyao lake, July 2; 20 km west of Dzhasak town, Aug. 16 — 1957, Petr.); *Khesi* (near Kheikho river close to Khuanchen village, June 15, 1886 — Pot.; Suanshantan valley, July 19, 1908—Czet.).

IIIA. **Qinghai:** *Nanshan* (lower forest belt along Tetung river, 2500 m, July 18, 1880 —Przw.).

General distribution: West. Sib. (south), East. Sib. (south and along Aldan river), Far East (south), Nor. Mong. (Mong.-Daur.), China (Dunbei).

4. **H. turkestanicum** Nevski in Tr. Sredneaz. gos. univ., ser. 8b, 17 (1934) 45, in Fl. SSSR, 2 (1934) 725 and in Tr. Bot. inst. AN SSSR, ser. 1, 5 (1941) 171; Persson in Bot. notiser (1938) 275; Fl. Kirgiz. 2 (1950) 230; Grubov, Consp. fl. MNR (1955) 78; Fl. Kazakhst. 1 (1956) 332; Fl. Tadzh. 1 (1957) 283; Melderis in Bor, Grasses Burma, Ceyl., Ind. and Pakist. (1960) 677; Ikonnikov, Opred. rast. Pamira [Key to Plants of Pamir] (1963) 74.—*H. secalinum* var. *brevisubulatum* f. *puberulum* Kryl. Fl. Alt. i Tomsk. gub. 7 (1914) 1703.—*H. secalinum* subsp. *brevisubulatum* var. *puberulum* (Kryl.) Kryl. Fl. Zap. Sib. 2 (1828) 364.—*H. brevisubulatum* var. *puberulum* (Kryl.) Meld. in Norlindh, Fl. mong. steppe, 1 (1949) 138.—*H. secalinum* auct. non Schreb.: Henderson and Hume, Lahore to Yarkand (1873) 341; Pilger in Hedin, S. Tibet, 6, 3 (1922) 95; Pampanini, Fl. Carac. (1930) 80.—*H. nodosum* auct. non L.: ? Hao in Engler's Bot. Jahrb. 68 (1938) 580; Ching in Contribs. U.S. Nat. Herb. 28 (1941) 597.—Ic.: Nevski, l.c. (1941) fig. 34.

Described from Middle Asia (Alay). Type in Leningrad.

In grasslands pebble beds, on rocky slopes, talus; up to upper mountain belt.

IA. **Mongolia:** *Mong. Alt.* (Urten-Gol river, June 20, 1877—Pot.; near Dain-Gol lake, July 5, 1906—Sap.); *Cent. Khalkha, East. Mong., Val. lakes* (Ongiin-Gol river, Khaileste ravine, July 21, 1920—Lis.; Ongiin-Gol river valley 10 km beyond Khoshu-Khid, July 8, 1941—Tsatsenkin); *East. Gobi, Khesi* (nor. Nanshan foothills, June 16, 1879—Przew.).

IB. **Kashgar:** *Nor.* (Uchturfan oasis, Yamansu village, June 9, 1908—Divn.); *West.*

IC. **Qaidam:** *hilly* (Ichegyn-Gol river, 3300 m, June 25, 1895—Rob.; vicinity of Kurlyk-Nor and Toso-Nor lakes, 2700 m, June 3, 1901—Lad.).

IIA. **Junggar:** *Tien Shan, Jung. Gobi* (nor.: between Barbagai and Burchum, No. 2896, Sept. 11, 1956—Ching; south.: St. Kuitun settlement 3–4 km east of Kuitun state farm, June 30, 1957—Yun.); *Dzhark.* (near Suidun town, July 18, 1877—A. Reg.).

IIIB. **Tibet:** *Chang Tang, Weitzan* (nor. slope of Burkhan Budda mountain range, Nomokhun-Gol gorge, 3300-3800 m, 1884—Przew.; same site, Khatu gorge, 3500 m, July 8, 1901 — Lad.).

IIIC. **Pamir** (Kashka-Su river gorge, June 28; Tagarma valley, July 23—1913, Knorring).

General distribution: Jung.-Tarb., Nor. and Cent. Tien Shan, East. Pam., Middle Asia (West. Tien Shan, Pamir-Alay), Himalayas (west., Kashmir).

Note. We recognise this species in a broader sense than S.A. Nevski and include in it all specimens genetically related to *H. brevisubulatum* (Trin.) Link s.l. with lemma densely covered with fairly long spinules or very short hairs. Other differences noted by S.A. Nevski between this species and *H. brevisubulatum* (s.s.) (extent of reduction of lateral spikelets, size of lemma, etc.) are not consistent and correlate poorly with each other. In general, it would be more appropriate to treat *H. turkestanicum* as one of the subspecies of *H. brevisubulatum* s.l. Like other species of the genus, *H. turkestanicum* hybridises quite readily with species of *Elymus* L. Among such hybrids, the most widespread and well known is hybrid *H. turkestanicum* × *Elymus nutans* = × *Elyhordeum schmidii* (Meld.) Meld. In Køie and Rech. f. Symb. Afghan. 6 (1965) 89 [= *Elymordeum schmidii* Meld. in Bor, Grasses Burma, Ceyl., Ind. and Pakist. (1960) 677]. The following specimens from Cent. Asia in particular belong to it: foothills of Chubaty hill in Sairam lake region, 2700–3000 m, Aug., 1878—A. Reg., nor. slope of Tien Shan, basin of Manas river, upper course of Ulan Usu river, July 24, 1957—Yun.

Section 2. Hordeum

5. **H. aegiceras** Nees ex Royle, Illustr. Bot. Himal. (1838) 418; Grubov, Consp. fl. MNR (1955) 78; Melderis in Bor, Grasses Burma, Ceyl., Ind. and Pakist. (1960) 675.—*H. coeleste* var. *trifurcatum* Schlecht. in Linnaea, 11 (1837) 543; Nevski in Tr. Bot. inst. AN SSSR, ser. 1, 5 (1941) 246.—*H. vulgare* var. *trifurcatum* (Schlecht.) Koern. in Koern. u. Wern. Handb. Getreideb. 1 (1885) 170; Keng, Fl. ill. sin., Gram. (1959) 442.—Ic.: Royle, l.c. Table 97, fig. 2; Keng, l.c. fig. 375 (4).

Described from the Himalayas. Type in London (K).

Cultivated in oases and sometimes found as introduced or escaped along roadsides, around irrigation ditches, and in various field crops; up to upper mountain belt.

IA. **Mongolia:** *Mong. Alt.* (along Bidzhiin-Gol river, Aug. 7, 1896—Klem); *Alash. Gobi* (Dyn'yuanin oasis [Bayan-Khoto], Sept. 6, 1908—Czet.).

IIIA. **Qinghai:** *Nanshan* (14 km south of Xining, Aug. 4, 1959—Petr.).

General distribution: China (Nor., Nor.-West., Cent., South-West.), Himalayas, Indo-Malay. (nor.-east); occasionally also cultivated in other countries of both hemispheres.

6. **H. distichon** L. Sp. pl. (1753) 85; Nevski in Fl. SSSR, 2 (1934) 728 and in Tr. Bot. inst. AN SSSR, ser. 1, 5 (1941) 241; Fl. Kirgiz. 2 (1950) 228; Fl. Kazakhst. 1 (1956) 334; Fl. Tadzh. 1 (1957) 290.—*H. vulgare* var. *distichon* (L.) Hack. in Engler u. Prantl, Pflanzenfam. 2, 2 (1886) 86; Krylov, Fl. Zap. Sib. 2 (1928) 362.—Ic.: Nevski, l.c. (1941) fig. 71.

Described from Europe. Type in London (Linn.).

Cultivated in oases and sometimes found as weed or escape among other crops, along roadsides and around irrigation ditches; up to upper mountain belt.

IA. **Mongolia:** *Gobi-Alt.* (submontane flat on south. slope of Dzun-Saikhan mountain range 45 km nor.-east of Bain-Dalai somon, Sept. 15, 1951—Kal.); *East. Gobi* (Mandal-Obo somon, spring 20 km west of Ulan-Bator—Dalan—Dzadagad old road and 25 km south of Bain-Dzak, July 27, 1951—Kal.).

General distribution: cultivated in many countries but native of the palaeo-Mediterranean.

7. **H. vulgare** L. Sp. pl. (1753) 84; Kanitz in Széchenyi, Wissensch. Ergebn. 2 (1898) 737; Danguy in Bull. Mus. nat. hist. natur. 6 (1911) 7; Nevski in Fl. SSSR 2 (1934) 728 and in Tr. Bot. inst. AN SSSR, ser. 1, 5 (1941) 243; Fl. Kirgiz. 2 (1950) 228; Grubov, Consp. fl. MNR (1955) 79; Fl. Kazakhst. 1 (1956) 333; Fl. Tadzh. 1 (1957) 290; Keng, Fl. ill. sin., Gram. (1959) 440; Melderis in Bor, Grasses Burma, Ceyl., Ind. and Pakist. (1960) 677.—*H. hexastichon* L. Sp. pl. (1753) 85; Franch. Pl. David. 1 (1884) 341; Diels in Filchner, Wissensch. Ergebn. (1908) 248; Danguy, l.c. 7.—*H. sativum* var. *vulgare* (L.) Hack. in Engler u. Prantl, Pflanzenfam. 2, 2 (1886) 87; Krylov, Fl. Zap. Sib. 2 (1928) 362.—*H. sp.* cult.: Henderson and Hume, Lahore to Yarkand (1873) 342.—Ic.: Keng, l.c. fig. 375.

Described from Europe. Type in London (Linn.).

Extensively cultivated in oases and sometimes found as a weed or escape along roadsides and around irrigation ditches, and in various crops; up to upper mountain belt.

IA. **Mongolia:** *Mong. Alt., Bas. lakes, Gobi-Alt.* (Leg lowland near Bain-Leg somon post, July 25, 1948—Grub.); *East. Gobi, Khesi.*

IB. **Kashgar:** *West.* ("Jarkand, to 3000 m"—Henderson and Hume, l.c.).

IIA. **Junggar:** *Tien Shan, Jung. Gobi* (near Bulun-Tokha town, Aug. 20, 1876—Pot.); *Dzhark.*

IIIA. Qinghai: *Nanshan.*

General distribution: widely cultivated in many countries but native of the Palaeo-Mediterranean.

Note. Apart from typical tetrastichous types of this species, a hexastichous type is cultivated in Cent. Asia. The latter is sometimes treated as a distinct subspecies, *H. vulgare* subsp. *hexastichon* (L.) Čelak. (= *H. hexastichon* L.) but, in our view, represents only a group of varieties (convarietas).

72. Psathyrostachys Nevski

in Fl. SSSR, 2 (1934) 712 and in Tr. Bot. inst. AN SSSR, ser. 1, 2 (1936) 57.

1.   Spikes extremely brittle down to base; rachial segments with 1–1.5 mm long hairs along broad-winged lateral ribs; glumes and lemma villous due to fairly long and dense hairs; stems scabrous or somewhat pilose under spikes ....
....4. **P. lanuginos** (Trin.) Nevski.

+   Spikes usually brittle in upper and middle parts; rachial segments with hairs up to 0.6 (0.8) mm long along narrow-winged lateral ribs; glumes and lemma scabrous or rather pilose ....2.

2.   Stems scabrous under spikes due to spinules or smooth. Glumes and lemma usually covered with elongated spinules, sometimes only partly transforming into short hairs ....
....2. **P. juncea** (Fisch.) Nevski.

+   Stems densely puberulent under spikes. Glumes and lemma pilose throughout surface or only in lower part ....3.

3.   Leaf blades densely covered with short spinules on both surfaces. Glumes and lemma usually pilose only in lower part; anthers 2.8–3.8 mm long................................1. **P. hyalantha** (Rupr.) Tzvel.

+   Leaf blades covered on both surfaces with scattered, very short spinules, often almost smooth in lower part . Glumes and lemma densely pilose; anthers 4–6.5 mm long ....
....3. **P. kronenburgii** (Hack.) Nevski.

1. **P. hyalantha** (Rupr.) Tzvel. comb. nova.—*Elymus hyalanthus* Rupr. in Osten-Sacken et Rupr. Sert. Tiansch. (1869) 36.—*E. kokczetavicus* Drob. in Tr. Bot. muzeya AN, 14 (1915) 131.

Described from Cent. Tien Shan (Dzhaman-Tau hills). Type in Leningrad.

On rubble and rocky slopes and rocks; up to midmountain belt.

IA. Mongolia: *Khesi* (Beidashan' hills nor. of Lan'chzhou, June 24; 67 km nor. of Lan'chzhou, June 29, 1957; Gulan, Kheishan' coal pits, June 22; promontory of Beidashan' hills 15 km nor. of Yunchan town, June 28; Nanshan foothills 15 km west of Yunchan town, July 10, 1958—Petr.).

IB. **Kashgar:** *Nor.* (near Uchturfan, June 6, 1908—Divn.; Bai district 15 km nor. of Davanchin village, 1800 m, No. 818, Sept. 4, 1958—Lee and Chu; near Khobsair, 1750–1800 m, No. 10477 and 10570, June 23 and 24, 1959—Lee et al.); *West.* (Bostan-Terek village, July 10, 1929—Pop.); *East.* (Bogdoshan' hills near Turfan, 1800 m, No. 5750, June 19, 1958—Lee and Chu).

IIA. **Junggar:** *Tien Shan* (Urumchi town region, No. 544, May 31, 1957—Huang).

General distribution: Balkh. region; Nor. and Cent. Tien Shan; Middle Asia (Alay), West. Sib. (far south).

2. **P. juncea** (Fisch.) Nevski in Fl. SSSR, 2 (1934) 714; Norlindh, Fl. mong. steppe, 1 (1949) 135; Fl. Kirgiz. 2 (1950) 223; Grubov, Consp. fl. MNR (1955) 78; Ikonnikov, Opred. rast. Pamira [Key to Plants of Pamir] (1963) 73.—? *P. perennis* Keng, Fl. ill. sin., Gram. (1959) 435, diagn. sin.— *Elymus junceus* Fisch. in Mém. Soc. natur. Moscou, 1 (1806) 45; ? Hemsley, Fl. Tibet (1902) 206; Krylov, Fl. Zap. Sib. 2 (1928) 371; Fl. Kazakhst. 1 (1956) 328.—*E. albertii* Regel in Acta Horti Petrop. 7 (1881) 561.—Ic.: Fisch. l.c. Table 4.

Described from Transvolga. Type in Leningrad.

In steppes, on rocky slopes, in sand, pebble beds, solonetz, sometimes as weed along roadsides and around irrigation ditches, in fields; up to midmountain belt.

IA. **Mongolia:** *Khobd., Mong. Alt., Bas. lakes, Gobi-Alt., East. Gobi* (Delger-Khangai crossing, on road to Mandal-Obo, Aug. 12, 1962—Lavr. et al.).

IIA. **Junggar:** *Alt. region, Tarb.* (east. fringe of Khobuk valley 30 km nor. of Kosh-Tologoi village along road to Altay [Shara-Sume], July 4, 1959—Yun.); *Jung. Alt., Tien Shan, Jung. Gobi.*

IIIB. **Tibet:** *Chang Tang* ("In 90° 45' and 35° 16', 5300 m, 6. VIII 1896, leg. Wellby and Malcolm"—Hemsley, l.c.; valley of Raskem-Darya river near Mazar settlement 244 km from Kargalyk on Tibet highway, June 4, 1959—Yun.).

General distribution: Aral-Casp., Balkh. region, Jung.-Tarb., Nor. and Cent. Tien Shan, East. Pam.; Europe (south-east), Near East (Iran), Middle Asia (Pamir-Alay), West. Sib. (south), East. Sib. (south-west.), Nor. Mong. (Hang.).

Note. *P. perennis* Keng (l.c.) described from around Urumchi town, judging from the illustration of its type (Keng, l.c. fig. 368), does not differ significantly from *P. juncea.*

3. **P. kronenburgii** (Hack.) Nevski in Fl. SSSR, 2 (1934) 713; Fl. Kirgiz. 2 (1950) 223; Fl. Tadzh. 1 (1957) 276; Keng, Fl. ill. sin., Gram. (1959) 435.— *Hordeum kronenburgii* Hack. in Allgem. Bot. Zeitschr. (1905) 133.—? *Elymus lanuginosus* auct. non Trin.: Hemsley, Fl. Tibet (1902) 206.—Ic.: Fl. Tadzh. 1, Plate 36, figs 5–7; Keng, l.c. fig. 369.

Described from Middle Asia (Alay mountain range, Taldyk pass). Type in Vienna; isotype in Leningrad.

On talus, rocks and rocky slopes; in middle and upper mountain belts.

IB. **Kashgar:** *Nor.* (nor.-west. part of Bai basin, Adyr mountain range 10 km nor.-west. of Bazar settlement, Sept. 4; same site, 15 km south of Kzyl-Bulak settlement along road to Vat, 1875 m, Sept. 13—1958, Yun.); *West.* (Kingtau mountain range 4–

5 km nor. of Koshkulak settlement, June 9, 1959—Yun.; in Upal settlement region, No. 584, June 9, 1959—Lee et al.); East. (nor.-west of Toksun settlement, No. 7363, June 20, 1958—Lee and Chu).

IIA. Junggar: *Tien Shan* (Iren-Khabirga mountain range, Naryn-Gol near Tsagan-Usu, 2000—2700 m, June 10; Mengute hill, 3300–3700 m, July 4—1879, A. Reg.; along Urumchi-Karashar road, 1150 m, No. 5954, July 21, 1958—Lee and Chu; Muzart river valley 10–12 km beyond its discharge into Bai basin, 2100 m, Sept. 7; Muzart river valley 1–2 km below Oiterek area, Sept. 11—1958, Yun.).

IIIB. Tibet: *Chang Tang* ("In 88° 20' and 35° 20', 5500 m, 29 VII 1896, leg. Wellby and Malcolm"—Hemsley, l.c.).

General distribution: Jung.-Tarb., Nor. and Cent. Tien Shan; Middle Asia (Pamir-Alay).

4. P. lanuginosa (Trin.) Nevski in Fl. SSSR, 2 (1934) 714.—*Elymus lanuginosus* Trin. in Ledeb. Fl. alt. 1 (1829) 121; Krylov, Fl. Zap. Sib. 2 (1928) 370; Fl. Kazakhst. 1 (1956) 328.—Ic.: Ledeb. Ic. pl. fl. ross. 3 (1831) Table 250.

Described from south. Altay. Type in Leningrad.

On rocky and rubble slopes, rocks and talus; up to midmountain belt.

IIA. Junggar: *Alt. region* (near Koktogoi, 1200 m, No. 10423, June 8, 1959—Lee et al.).

General distribution: Aral-Casp., Balkh. region; West. Sib. (far south and Altay).

73. L e y m u s Hochst.

in Flora, 7 (1848) 118, in adnot.—*Aneurolepidium* Nevski
in Fl. SSSR, 2 (1934) 697.—*Elymus* auct.,
non L.: Nevski in Fl. SSSR, 2 (1934) 694.

1. Spikes 10–35 cm long, 17–35 mm broad, with greatly thickened rachis; spikelets in lower and middle parts of spikes arranged in groups of 4–6 and, in upper, usually narrower part, in 3(2); palea glabrous and smooth along keels or only with few very short spinules in upper part of keels. Plant in sand has very thick stems and leaf blades 4–15 mm broad ....
....8. L. racemosus (Lam.) Tzvel.

+ Spikes smaller, usually up to 15, occasionally up to 25 cm long, and up to 20 mm broad; spikelets usually grouped 2–3, less often single; palea covered with fairly large spinules along keels ...... 2.

2. Lemma glabrous, smooth or covered with very short spinules; stalks of spikelets glabrous, smooth or somewhat covered with short spinules ....3.

+ Lemma largely covered with short hairs or elongated spinules; stalks of spikelets puberulent; spikelets grouped 2–3 ....7.

3. Glumes 12–20 mm long, linear-subulate, covered with short spinules; lemma 11–15 mm long, somewhat covered with very short spinules; anthers 5.5–7.5 mm long; spikelets grouped 2–3.

Tall (up to 120 cm) plant, forming dense mats with short creeping subsurface shoots .... 11. **L. tianschanicus** (Drob.) Tzvel.

+ Glumes up to 10 mm long; anthers 2.5–4.5 mm long .... 4.

4. Plant forming very dense mats, without creeping subsurface shoots; leaf blades 1.5–3 mm broad. Glumes 4–7 mm long, linear-subulate, somewhat covered with short spinules; lemma smooth or largely covered with very short spinules; spikelets grouped 2–3... 7. **L. petraeus** (Nevski) Tzvel.

+ Plants with long creeping subsurface shoots not forming mats. Lemma smooth or almost so, with a few short spinules only in upper part.. 5.

5. All spikelets grouped 2–3; lemma 4–7.5 mm long, perceptibly lustrous, with 1.5–4 mm long cusp or awn at tip; glumes linear-subulate. Stems and rootstock fairly stubby ....
.... **4. L. multicaulis** (Kar. et Kir.) Tzvel.

+ Spikelets single or grouped 2(3) in midpart of spikes; lemma 6–9 mm long, not lustrous, with up to 1.5 (2) mm long cusp at tip ....
.... 6.

6. Glumes lanceolate-subulate, enlarged in lower part; spikelets single or grouped 2(3). Stems and rootstock fairly stubby; surface shoots at base usually covered with cap of sheaths of dead leaves
.... 2. **L. chinensis** (Trin.) Tzvel.

+ Glumes linear-subulate, with almost no enlargement in lower part; spikelets invariably arranged singly. Stems and rootstock very slender; surface shoots at base with or without few sheaths of dead **leaves**....................................... 9. **L. ramosus** (Trin.). Tzvel.

7(2). Ligules of cauline leaves 2.5–4 mm long; plant 70–130 cm tall with 3–7 mm broad leaf blades. Lemma usually puberulent only laterally................................... 3. **L. ligulatus** (Keng) Tzvel.

+ Ligules up to 1 mm long .... 8.

8. Glumes lanceolate-subulate, their bases slightly overlying each other; spikes relatively narrow, 8–25 cm long and 4–10 mm broad; lemma 9–14 mm long, somewhat covered with short semipressed hairs or spinules .... 1. **L. angustus** (Trin.) Pilg.

+ Glumes linear-subulate, their bases not overlying each other; spikes broader on average.. 9.

9. Glumes linear-subulate, very narrow, with no trace of membranous margin; lemma 6–9 mm long, densely covered with short hairs, acute at tip or with up to 1.5 mm long cusp ....
.... 6. **L. paboanus** (Claus) Pilg.

+ Glumes subulate with linear-lanceolate, sometimes, base with membranous margin; lemma 7–12 mm long, usually covered with scattered short hairs or spinules, sometimes subglabrous, with 0.7–4 mm long cusp or awn at tip .... 10.

10. Spikes 4–10 cm long and 12–20 (25) mm broad, short and stubby, very dense, usually narrowing somewhat towards both ends .... 5. **L. ovatus** (Trin.) Tzvel.

+ Spikes 6–16 cm long and 7–20 mm broad, very long and slender, linear, usually less dense .... 10. **L. secalinus** (Georgi) Tzvel.

○**L. aemulans** (Nevski) Tzvel. in Notul. Syst. Herb. Inst. Bot. Acad. Sci. URSS, 20 (1960) 430.

IIA. **Junggar.** *Tien Shan.*

1. **L. angustus** (Trin.) Pilg. in Engler's Bot. Jahrb. 74 (1947) 6.—*Elymus angustus* Trin. in Ledeb. Fl. alt. 1 (1829) 119; Krylov, Fl. Zap. Sib. 2 (1928) 368; Pavlov in Byull. Mosk. obshch. ispyt. prir. 38 (1929) 29; Fl. Kazakhst. 1 (1956) 321.—*E. kirghisorum* Drob. in Tr. Bot. muzeya AN, 14 (1915) 135.—*E. turgaicus* Roshev. in Fl. Yugo-Vost. 2 (1928) 253, p. max. p.—*E. kugalensis* E. Nikit. in Fl. Kirgiz. 2 (1950) 218, diagn. ross.—*E. angustiformis* Pavl. in Vestn. AN Kaz. SSR, 5 (1952) 8o, non Drob. 1941.— *E. kuznetzovii* Pavl. in Fl. Kazakhst. 1 (1956) 322.—*Aneurolepidium angustum* (Trin.) Nevski in Fl. SSSR, 2 (1934) 700; Grubov, Consp. fl. MNR (1955) 77; Keng, Fl. ill. sin., Gram. (1959) 434.—Ic.: Ledeb. Ic. pl. fl. ross. 3 (1831) Plate 229; Keng, l.c. fig. 367.

Described from Altay (Chui river valley). Type in Leningrad.

In solonetz steppes, sand and pebble beds of river and lake valleys, in solonetz; up to midmountain belt.

IA. **Mongolis:** *Mong. Alt., Bas. lakes. Val. lakes* (Barun-Bayan-Ulan somon, Ologoi-Nurym-Khongor area, July 29, 1952—Davazhamts); *Gobi-Alt.* (on foothills of Dundu-Saikhan and Botugun-Shanda hills, Aug. 28, 1931—Ik.-Gal.); *Alash. Gobi* (Noyan somon, Khukhu-Tologoin-Bulak 21 km west of Obotu-Khural, Aug. 13, 1948—Crub.).

IB. **Kashgar:** *Nor.* (8 km nor. of Aksu, 2550 m, No. 8485, Sept. 26, 1958—Lee and Chu).

IIA. **Junggar:** *Alt. region, Tarb.* (nor. trails of Saur mountain range 30 km west-nor.-west of Kheisangou settlement on Burchum-Karamai road, July 10, 1959— Yun.); *Jung. Alt., Tien Shan, Jung. Gobi* (south.: near Kuitun village, No. 261, June 29; Savan district, near Nyutsyuan'tsz, No. 1659, July 19; 15 km nor.-west of Urumchi, No. 2247, Sept. 21—1957, Huang); *Zaisan* (near Burchum village, No. 3246, Sept. 20, 1956—Ching); *Dzhark.* (Khoyur-Sumun village south of Kul'dzha, 600 m, May 26, 1877—A. Reg.); *Balkh.-Alak.* (16 km nor. of Durbul'dzhin, 630 m, No. 1718, Aug. 16, 1957—Huang).

General distribution: Aral-Casp., Balkh. region, Jung.-Tarb., Nor. and Cent. Tien Shan; Europe (extreme south-east), Middle Asia (West. Tien Shan), West. Sib. (south), Nor. Mong. (Hang.).

Note. Highly polymorphic species which can be differentiated into several varieties. Thus, *L. angustus* var. *kirghisorum* (Drob.) Tzvel. comb. nova (= *Elymus kirghisorum*

Drob. l.c.) has very large (12–20 mm long) subglabrous lemma but is associated with the type variety through a long series of transitional forms and has no distinct distribution range.

o **L. bruneostachyus** Cui, in Clav. Pl. Xinjiang, 1 (1989) 184.

IIA. Junggar. *Tien Shan.*

2. **L. chinensis** (Trin.) Tzvel. comb. nova.—*L. pseudoagropyrum* (Trin. ex Griseb.) Tzvel. in Bot. mater. Gerb. Bot. inst. AN SSSR, 20 (1960) 430.—*L. regelii* (Roshev.) Tzvel. l.c.—*Triticum chinense* Trin. in Bunge, Enum. pl. China bor. (1832) 146.—*T. pseudoagropyrum* Trin. ex Griseb. in Ledeb. Fl. Ross. 4 (1852) 343.—*Elymus pseudoagropyrum* (Trin.) Turcz. in Bull. Soc. natur. Moscou, 29 (1856) 63; Pavlov in Byull. Mosk. obshch. ispyt. prir. 38 (1929) 28.—*E. regelii* Roshev. in Izv. Bot. sada AN SSSR, 30 (1932) 781; Fl. Kazakhst. 1 (1956) 326.—*E. chinensis* (Trin.) Keng in Sunyatsenia, 6, 1 (1941) 66; Melderis in Norlindh, Fl. mong. steppe, 1 (1949) 132.— *Agropyron pseudoagropyrum* (Trin.) Franch. Pl. David. 1 (1884) 340; Danguy in Bull. Mus. nat. hist. natur. 20 (1914) 148; Krylov, Fl. Zap. Sib. 12, 1 (1961) 3135.—*Aneurolepidium regelii* (Roshev.) Nevski in Fl. SSSR, 2 (1934) 709.—*A. pseudoagropyrum* (Trin.) Nevski, l.c. 710; Grubov, Consp. fl. MNR (1955) 78.—*A. chinense* (Trin.) Kitag. in Rept. Inst. Sci. Res. Manch. 2 (1938) 281; ej. Lin. Fl. Mansh. (1939) 61; Keng, Fl. ill. sin., Gram. (1959) 432.—*Agropyron repens* auct. non Beauv.: Forbes and Hemsley, Index Fl. Sin. 3 (1904) 432, p.p.—Ic.: Keng, l.c. fig. 364.

Described from Nor. China. Type in Leningrad.

In steppes, on solonetz, in saline meadows, sand and pebble beds of river valleys, sometimes as weed around irrigation ditches in various crops; up to lower mountain belt.

IA. Mongolia: *Cis-Hing., Cent. Khalkha, East. Mong., Bas. lakes* (Gumburde area, July 18, 1879—Pot.); *Val. lakes* (right bank of Ongiin-Gol river, July 27, 1893—Klem.); *East. Gobi* (15 km nor.-west of Delger somon, July 29, 1950—Kuznetsov).

IIA. Junggar: *Tien Shan* (Cent. Khorgos, 1000–1700 m, May 15, 1878—A. Reg.; Bogdoshan' hills south of Urumchi, No. 172, July 14, 1956—Ching); *Jung. Gobi* (south.); *Dzhark.* (Pilyuchi village near Kul'dzha, June 19; Khoyur-Sumun village south of Kul'dzha, May 27, 1877—A. Reg.); *Balkh.-Alak.* (Sary-Khulsyn south of Durbul'dzhin, July 22, 1947—Shum.).

General distribution: Balkh. region (east.), Jung.-Tarb.; West. Sib. (Altay), East. Sib. (south), Far East (south), Nor. Mong., China (Dunbei, Nor., Nor.-West.).

Note. We could detect no significant differences between types *Triticum chinense* Trin., *T. pseudoagropyrum* Trin. ex Griseb. and *Elymus regelii* Roshev. and hence consider all of these names synonyms for the same species. Among the varieties of this species, one may note *L. chinensis* var. *pumilus* (Meld.) Tzvel. comb. nova (= *Elymus chinensis* var. *pumilus* Meld. in Norlindh, l.c. 134) characterised, aside from small plant size, by sheaths and blades of lower leaves with short but dense hairs. It was described from a specimen from East. Gobi: "Bulung-Khuduk, dist. Dzun-Gung, prov. Suiyuan, No. 6609, 17 VI 1927, leg. G. Söderbom". *L. chinensis* hybridises readily with *L. secalinus*

(Georgi) Tzvel. to form clones that are rather intermediate in characteristics and, in the west, with *L. multicaulis* (Kar. et Kir.) Tzvel. as well.

3. **L. ligulatus (Keng) Tzvel. comb. nova.**—*Elymus dasystachys* var. *ligulatus* Keng in Sunyatsenia 6, 1 (1941) 65.

Described from Qinghai. Type in Nanking.

In grasslands, on rocky slopes and in pebble beds; in midmountain belt.

IIIA. **Qinghai:** *Nanshan* (nor. slopes of hills to south of Tetung river, July 18, 1872—Przew.; "Yao-Kai, Ping-Fan, Kansu prov., about 2000 m, No. 253, 4 VII 1923, leg. R.C. Ching"—Keng, l.c., type!).

General distribution: endemic.

4. **L. multicaulis (Kar. et Kir.) Tzvel.** in Bot. mater. Gerb. Bot. inst. AN SSSR, 20 (1960) 430.—*Elymus multicaulis* Kar. et Kir. in Bull. Soc. natur. Moscou, 14 (1841) 868; Fl. Kirgiz. 2 (1950) 222; Fl. Kazakhst. 1 (1956) 325; Fl. Tadzh. 1 (1957) 270; Krylov, Fl. Zap. Sib. 12, 1 (1961) 3139.—*E. aralensis* Regel in Bull. Soc. natur. Moscou, 41, 2 (1868) 285; Krylov, Fl. Zap. Sib. 2 (1928) 373.—*Aneurolepidium multicaule* (Kar. et Kir.) Nevski in Fl. SSSR, 2 (1934) 708.—**Ic.:** Fl. SSSR, 2, Plate 50, fig. 1.

Described from Junggar (Tarbagatai). Type in Leningrad.

In saline meadows, solonchaks, along banks of water reservoirs, often as weed around irrigation ditches, in farms and various crops; up to lower mountain belt.

IB. **Kashgar:** *Nor.* (around Uchturfan, June 18, 1908—Divn.); *West.* (around Yangigissar, May 25 and 26, 1909—Divn.; 36 km from Kargalyk on road to Tiznaf, No. 188, June 1; between Artush and Khalatsa, No. 9793, June 22—1959, Lee et al.).

IIA. **Junggar:** *Jung. Gobi, Dzhark.* (Chimpansi village west of Kul'dzha, May 8; Pilyuchi village near Kul'dzha, May 17—1877, A. Reg.).

General distribution: Aral-Casp., Balkh. region, Jung.-Tarb., Nor. Tien Shan; Middle Asia, West. Sib. (? Altay).

5. **L. ovatus (Trin.) Tzvel.** in Bot. mater. Gerb. Bot. inst. AN SSSR, 20 (1960) 430.—*Elymus ovatus* Trin. in Ledeb. Fl. alt. 1 (1829) 121; Krylov, Fl. Zap. Sib. 2 (1928) 371.—*Aneurolepidium ovatum* (Trin.) Nevski in Fl. SSSR, 2 (1934) 707; Grubov, Consp. fl. MNR (1955) 77.—**Ic.:** Ledeb. Ic. pl. fl. ross. 3 (1831) Table 251.

Described from Altay (Chulyshman river). Type in Leningrad.

In saline meadows, sand and pebble beds of river and lake valleys; up to upper mountain belt.

IA. **Mongolia:** *Mong. Alt.* (Tsitsirriin-Gol river valley, June 28, 1877—Pot.; along Katu river, tributary of Bukhu-Muren river, Aug. 5; Bukhu-Muren river, Aug. 8—1909, Sap.); *East. Mong.* (Dariganga region, Ongon-Elisu sand, Sept. 17, 1931—Pob.); *Bas. lakes, Gobi-Alt.* (south. slope of Ikhe-Bogdo mountain range, Khush' creek valley, Sept. 7, 1943—Yun.; east. extremity of Gichigine-Nuru mountain range 3 km south-west of Amani-Bulak spring., Aug. 27, 1948—Grub.).

IB. **Kashgar:** *Nor.* (Bai district 10 km south of Keinlak village, 2400 m, No. 8254, Sept. 8, 1958—Lee and Chu).

IIA. Junggar: *Alt. region* (Chingil' settlement, No. 771, Aug. 1, 1956—Ching); *Jung. Alt.* (upper course of Borotala river, Aug., 1878—A. Reg.); *Tien Shan* (Nilki gorge, June 30; Mengete hill, 3000 m, July 3; Algoi, 2000–2700 m, Sept. 12—1879, A. Reg.); *Zaisan* (left bank of Ch. Irtysh, Dzhelkaidar area, June 8, 1914—Schisch.).

IIIA. Qinghai: *Nanshan* (along lake Kukunor bank, Aug. 28, 1907—Czet.).

General distribution: West. Sib. (Altay), East. Sib. (south-west.), Nor. Mong. (Hang.).

Note. This species is found sporadically within the distribution range of very closely related *L. secalinus* (Georgi) Tzvel. with no significant difference from it assertainable even with respect to ecology. It is therefore highly probable that the short and very stubby spikes of *L. ovatus* represent only a consequence of their abnormal growth (induced by disease or selective growth of spikelets) within an entire clone or population; this abnormality prompted the establishment of an independent species. This assumption has been confirmed somewhat by a specimen from Gobi Altay (Bain-Tukhum lake, at the boundary of puff solonchaks, Aug. 6, 1931—Ik.-Gal.) with the habit of *L. ovatus* but with the distinctly manifest vivipary of spikelets.

6. **L.₋paboanus** (Claus) Pilg. in Engler's Bot. Jahrb. 74 (1947) 7.—*Elymus paboanus* Claus in Beitr. pflanzenk. Russ. Reich. 8 (1851) 170; Fl. Kirgiz. 1 (1956) 327; Fl. Kazakhst. 1 (1956) 327; Krylov, Fl. Zap. Sib. 12, 1 (1961) 3139.—*E. dasystachys* var. *salsuginosus* Griseb. in Ledeb. Fl. Ross. 4 (1852) 333. —*E. salsuginosus* (Griseb.) Turcz. ex Steud. Synops. Pl. Glum. 1 (1854) 350; Krylov, l.c. 2 (1928) 374; Pavlov in Byull. Mosk. obshch. ispyt. prir. 38 (1929) 30.—*Aneurolepidium paboanum* (Claus) Nevski in Fl. SSSR, 2 (1934) 707; Grubov, Consp. fl. MNR (1955) 78.—Ic.: Fl. Kazakhst. 1, Plate 24, fig. 5.

Described from Volga region (Kinel' river). Type in Leningrad.

In saline meadows and solonchak; up to midmountain belt.

IA. Mongolia: *Mong. Alt., Bas. lakes, Val. lakes, Gobi-Alt., East. Gobi* (nor. of Del'ger-Khangai, Aug. 1, 1931—Ik.-Gal.; 45 km nor.-east of Khan-Bogdo, Sept. 22, 1940—Yun.).

IB. Kashgar : *Nor.* (Muzart river valley in midcourse, Sazlik area, Sept. 9, 1958—Yun.).

IC. Qaidam : *hilly* (Orogyn-Gol river, 3200 m, June 2, 1895—Rob.).

IIA. Junggar : *Jung. Gobi* (Uiinchi somon, Borotsonchzhi area, Sept. 13, 1948—Grub.; valley of Darbata river near its intersection with Karamai-Altay road, June 20, 1957—Yun.; between Barkul' lake and Beisyan, No. 4911, Sept. 26, 1957—Huang); *Zaisan* (Inkhatai valley, Zimunai district, 650 m, No. 10574, June 25, 1959—Lee et al.).

General distribution: Aral-Gasp., Balkh. region, Jung.-Tarb., Nor. Tien Shan; Europe (far south-east), West. Sib. (south), East. Sib. (south-west.), Nor. Mong. (Hang.).

Note. Compared to closely related species *L. secalinus* (Georgi) Tzvel., this species inhabits far more saline soils.

7. **L. petraeus** (Nevski) Tzvel. in Bot. mater. Gerb. Bot. inst. AN SSSR, 20 (1960) 429.—*Aneurolepidium petraeum* Nevski in Fl. SSSR, 2 (1934) 705 and in Tr. Bot. inst. AN SSSR, ser. 1, 2 (1936) 70.—*Elymus petraeus* (Nevski) Pavl. in Fl. Kazakhst. 1 (1956) 325.

Described from Kazakhstan (Zaisan lake). Type in Leningrad.

On rocks, talus and rocky slopes; in lower and middle mountain belts.

IIA. Junggar: *Tien Shan* (10 km nor. of Shuvutin-Daba settlement along Urtaksar—Sairam-Nur lake road, limestone rocks, Aug. 18; Ketmen' mountain range 1 km nor. of Sarbushin settlement along Ili—Kzyl-Kura road, Aug. 21—1957, Yun.; Dzinkho village 10 km south of Sairam-Nur lake, Aug. 31, 1959—Petr.).

General distribution: Balkh. region (east.), Jung.-Tarb.

Note. This species belongs to a small group of species of the genus *Leymus* Hochst., closest to the genus *Psathyrostachys* Nevski and, through it, to *Hordeum* L. as well.

8. **L. racemosus** (Lam.) Tzvel. in Bot. mater, Gerb. Bot. inst. AN SSSR, 20 (1960) 429.—*L. giganteus* (Vahl) Pilg. in Engler's Bot. Jahrb. 74 (1947) 7.— *Elymus racemosus* Lam. Tabl. Encycl. méth. 1 (1792) 207.—*E. giganteus* Vahl, Symb. bot. 3 (1794) 10; Simpson in J. Linn. Soc. London (Bot.) 41 (1913) 454; Danguy in Bull. Mus. nat. hist. natur. 20 (1914) 148; Pavlov in Byull. Mosk. obshch. ispyt. prir. 38 (1929) 29; Nevski in Fl. SSSR, 2 (1934) 696; Fl. Kirgiz. 2 (1950) 218; Grubov, Consp. fl. MNR (1955) 77; Fl. Kazakhst. 1 (1956) 321.—*E. sabulosus* M.B. Fl. taur.-cauc. 1 (1809) 81.—*E. arenarius* var. *giganteus* (Vahl) Schmalh. Fl. Sredn. i Yuzhn. Rossii, 2 (1897) 667; Krylov, Fl. Zap. Sib. 2 (1928) 369.—*E. arenarius* var. *sabulosus* (M.B.) Schmalh. l.c. 667; Krylov, l.c. 370.—Ic.: Fl. SSSR, 2, Plate 49, fig. 11.

Described from South. Siberia (and possibly Kazakhstan ?). Type in Paris.

In sand and sandy steppes.

IA. Mongolia: *Cent. Khalkha* (Borokhchin area, June 23, 1895—Klem.; Borokhchin lake, July 6, 1924—Pavl.; Tola river valley in Butszinkhe area, Aug. 9, 1925—Gus.; near Borokhchin river on way to Zain-Gegen from Urga, July 15, 1926—Pavl.; right bank of Borokhchin river, Sept. 3, 1926—Proz.; between Lyun somon and Khadasyn along Ulan Bator-Tsetserleg road 25–30 km east of somon, Sept. 4, 1942; same site, Khadasyn region, July 6, 1947—Yun.; south. fringe of Tsagan-Nur lake lowland along Ulan Bator-Tsetserleg road, June 25, 1948—Grub.; 125 km from Ulan Bator facing Navan-Khure near Tola river meander scroll, July 1, 1949—Yun.); *Bas. lakes* (Borig-Del' sand 4–5 km south-east of Baga Nur lake, July 25, 1945; 40 km south of Ulyasutai along road to Tsagan-Ol, Shurygin-Gol area, July 17, 1947—Yun.).

IIA. Junggar: *Alt. region* (near Altay [Shara-Sume] settlement, No. 2278, Aug. 20, 1956—Ching); *Jung. Gobi* (nor.: 5–8 km nor. of Bulun-Tokha settlement in lower courses of Urunga river, July 9; left bank of Urunga river 60–65 km beyond Din'syan along road to Ertai, July 13—1959, Yun.); *Zaisan* (Blandy-Kul' sand, July 6, 1900—Reznichenko; "Bords de l'Irtich, 30 VIII 1895, leg. Chaffanjon"—Danguy, l.c.; "sandy plains of the upper Irtysh basin, leg. Price"—Simpson, l.c.).

General distribution: Aral-Casp., Balkh. region, Nor. Tien Shan (Issyk-Kul' lake); Europe (south-east.), Caucasus, West. Sib. (south), East. Sib. (south), Nor. Mong. (Hang.).

Note. Variety *L. racemosus* var. *sabulosus* (M.B.) Tzvel. comb. nova (=*Elymus sabulosus* M.B. l.c.) with stems entirely smooth under spikes, reported just recently from Cent. Asia, might possibly merit the rank of a subspecies.

9. **L. ramosus** (Trin.) Tzvel. in Bot. mater. Gerb. Bot. inst. AN SSSR, 20 (1960) 430.—*Triticum ramosum* Trin. in Ledeb. Fl. alt. 1 (1829) 114.— *Agropyron ramosum* (Trin.) Richt. Pl. Eur. 1 (1890) 126; Krylov, Fl. Zap. Sib. 2 (1928) 355; Fl. Kazakhst. 1 (1956) 300.—Ic.: Ledeb. Ic. pl. fl. ross. 3 (1831) Table 245.

Described from South. Altay. Type in Leningrad.

In steppes, on solonetz, in sand and pebble beds of river valleys, often as weed along roadsides, around irrigation ditches and in farms; up to lower mountain belt.

IIA. **Junggar:** *Balkh.-Alak.* (Chuguchak basin 2 km west of Durbul'dzhin on road to Chuguchak, Aug. 7, 1957—Yun.).

General distribution: Aral-Casp., Balkh. region; Europe (south-east.), Caucasus (Fore Caucasus), West. Sib. (south), East. Sib. (south-west.).

10. **L. secalinus** (Georgi) Tzvel. comb. nova.—*L. dasystachys* (Trin.) Pilg. in Engler's Bot. Jahrb. 74 (1947) 6; Ikonnikov, Opred. rast. Pamira [Key to Plants of Pamir] (1963) 73.—*Triticum secalinum* Georgi, Bemerk, einer Reise, 1 (1775) 198.—*T. littorale* Pall. Reise, 3 (1776) 287.—*Elymus dasystachys* Trin. in Ledeb. Fl. alt. 1 (1829) 120, p.p.; Henderson and Hume, Lahore to Yarkand (1873) 341; Franch. Pl. David. 1 (1884) 341; Hemsley in J. Linn. Soc. London (Bot.) 30 (1894) 120; Deasy, In Tibet and Chin. Turk. (1901) 405; Hemsley, Fl. Tibet (1902) 205; Forbes and Hemsley, Index Fl. Sin. 3 (1904) 433; Keissler in Ann. Naturhist. Hofmuseums Wien, 22 (1907) 32; Danguy in Bull. Mus. nat. hist. natur. 17 (1911) 7; Pilger in Hedin, S. Tibet, 6, 3 (1922) 95; Krylov, Fl. Zap. Sib. 2 (1928) 373; Pavlov in Byull. Mosk. obshch. ispyt. prir. 38 (1929) 29; Pampanini, Fl. Carac. (1930) 80; Persson in Bot. notiser (1938) 275; Ching in Contribs, U.S. Nat. Herb. 28 (1941) 597; Melderis in Norlindh, Fl. mong. steppe, 1 (1949) 124; Fl. Kirgiz. 2 (1950) 221; Fl. Kazakhst. 1 (1956) 327; Fl. Tadzh. 1 (1957) 269; Melderis in Bor, Grasses Burma, Ceyl., Ind. and Pakist. (1960) 669.—*E. litoralis* Turcz. ex Steud. Synops. Pl. Glum. 1 (1854) 350.—*E. glaucus* Regel in Acta Horti Petrop. 7 (1870) 585.—*Aneurolepidium dasystachys* (Trin.) Nevski in Fl. SSSR, 2 (1934) 706; Kitag. Lin. Fl. Mansh. (1939) 61; Grubov, Consp. fl. MNR (1955) 77; Keng, Fl. ill. sin., Gram. (1959) 432.—*Agropyron chino-rossicum* Ohwi in Acta phytotax. et geobot. 10, 2 (1941) 100.—Ic.: Gmel. Fl. sib. 1 (1747) Table 251; Ledeb. Ic. pl. fl. ross. 3 (1831) Table 249; Melderis, l.c. (1949) fig. 12bc, fig. 14 (var. *parviflorus*), fig.15 (var. *mongolicus*), fig. 16 (f. *nudus*), fig. 17 (var. *pubiculmis*); Fl. Tadzh. 1, Plate 36, figs. 1–4; Keng, l.c. fig. 365.

Described from East. Siberia (Baikal lake). The above-cited illustration of Gmelin (l.c.) represents the type. Map 10.

In saline meadows, coastal pebble beds and sand, on rocky slopes and talus; up to upper mountain belt.

IA. Mongolia: *Khobd., Mong. Alt., Cent. Khalkha, East. Mong., Bas. lakes, Val. lakes, Gobi-Alt., East. Gobi, West. Gobi* (Bilgekhu-Bulak area 30–35 km east of Tsagan-Bogdo, July 31, 1943—Yun.); *Alash. Gobi, Ordos, Khesi.*

IB. Kashgar: *Nor. West., South., East.* (Khami oasis, May 20, 1879—Przew.; Karashar basin 15 km south of Ushan-Tuz village, July 26, 1958—Yun.); *Takla Makan* (Tarim river valley near its confluence with Khotan-Darya, Shochaakty area, Sept. 23, 1957—Yun.).

IC. Qaidam: *plains, hilly* (Orogyn-Gol river, 3500 m, July 2, 1895—Rob.; vicinity of Kurlyk Nor lake, 2600 m, July 3, 1901—Lad.).

IIA. Junggar: *Alt. region* (27 km south of Shara-Sume, 800 m, No. 2693, Aug. 6, 1956—Ching); *Jung. Alt., Tien Shan, Jung. Gobi, Zaisan* (left bank of Ch. Irtysh facing Cherektas hill, June 11; right bank of Ch. Irtysh below Burchum river between Sary-Dzhasak area and Kiikpai collective, June 15—1914, Schisch.).

IIIA. Qinghai: *Nanshan, Amdo* ("Jao-Chieh"—Ching, l.c.; 20 km east of Gunhe town, 7 VIII 1959"—Petr.).

IIIB. Tibet: *Chang Tang, Weitzan, South.* ("Shekar, No. 193, leg. Kingston"—Melderis in Bor, l.c.).

IIIC. Pamir.

General distribution: Balkh. region (east.), Jung.-Tarb., Nor. and Cent. Tien Shan, East. Pam.; Near East (Afghanistan), Middle Asia (Pamir-Alay), West. Sib. (Altay), East. Sib. (south and Cent. Yakutia), Nor. Mong. (Fore Hubs., Hang., Mong.-Daur.), China (west. Dunbei, Nor., Nor.-West.), Himalayas.

Note. *L. secalinus* is a highly polymorphic species with a few varieties which perhaps merit the rank of subspecies. One of them, *L. secalinus* var. *pubescens* (O. Fedtsch.) Tzvel. comb. nova [= *Elymus dasystachys* var. *pubescens* O. Fedtsch. Fl. Pam. (1903) 203; = *E. dasystachys* f. *tomentella* Hemsley, l.c. (1894) 124; = *E. dasystachys* var. *pubiculmis* Meld. in Norlindh, l.c. 131] has sheaths, leaf blades on both sides and often stems that are very short but profusely pilose. Almost all specimens of *L. secalinus* from Pamir, Tibet and Qaidam as well as many specimens from Inner Mongolia belong to this variety. Another variety, *L. secalinus* var. *mongolicus* (Meld.) Tzvel. comb. nova (= *Elymus dasystachys* var. *mongolicus* Meld. in Norlindh, l.c. 128), has very broad glumes occasionally somewhat overlying each other in the lower part; in this respect, it approaches *L. angustus* (Trin.) Pilg. and is distributed mainly in Inner Mongolia and Qinghai (its type "20 km ad occid. versus a Shara-Muren, No. 788, 19 VIII 1934,. leg. Eriksson" is preserved in Stockholm). Within this variety, its author distinguishes two more forms: *Elymus dasystachys* var. *mongolicus* f. *angustiformis* Meld. (l.c. 130), closest to *L. angustus* in the form of glumes, and *E. dasystachys* var. *mongolicus* f. *nudus* Meld. (l.c. 129) with glabrous lemma. The latter form has evidently been described from hybrids *Leymus chinensis* × *L. secalinus* with lemma glabrous or weakly pilose only on sides and is quite extensively distributed in all regions of common occurrence of these two species in southern Siberia as well as in Cent. Asia. Another variety, *Elymus dasystachys* var. *parviflorus* Meld., (l.c. 126), also established from a specimen from Inner Mongolia ("Khonin-Chagan-Cholo-Gol, No. 1282, 1 VIII 1927, leg. Hummel") with relatively small, profusely pilose lemma, is very similar to type *Elymus dasystachys* Trin., differing from it only in sheaths of lower leaves with short pubescence (possibly the result of hybridisation with var. *pubescens*).

11. L. tianschanicus (Drob.) Tzvel. in Bot. mater. Gerb. Bot. inst. AN SSSR, 20 (1960) 469.—*Elymus tianschanicus* Drob. in Feddes repert. 21 (1925) 45; Fl. Kazakhst. 1 (1956) 322.—*Aneurolepidium tianschanicum*

(Drob.) Nevski in Fl. SSSR, 2 (1934) 703; Keng, Fl. ill. sin., Gram. (1959) 432.—Ic.: Keng, l.c. fig. 366.

Described from West. Tien Shan (Brich-Mulla village). Type in Tashkent (TAK).

On rocky slopes, talus and rocks; in lower and middle mountain belts.

IIA. Junggar: *Tien Shan* (nor. slope of Ketmen' mountain range 1 km nor. of Sarbushin settlement along Ili—Kzyl-Kura road Aug. 21, 1957—Yun.).

General distribution: Nor. Tien Shan; Middle Asia (West. Tien Shan).

74. Elymus L.

sp. pl. (1753) 83, p.p.—*Roegneria* C. Koch in Linnaea, 21 (1848) 413.

1. Lemma without awn or with erect up to 6 mm long awn; glumes not more than 2/3 as long as contiguous lemma; spikes erect or slightly pendent .... 2.
+ Lemma with 7–30 (40) mm long awn .... 7.
2. Spikelets grouped 2 (3) all along spikes; lemma scabrous, with 2–5 mm long awn at tip; anthers 1.5–2.3 mm long...............................
.... 4. E. dahuricus Turcz. ex Griseb. (var. **brevisetus** Ohwi).
+ Spikelets invariably single along spikes .... 3.
3. Lemma glabrous and smooth throughout most of surface, with few short spinules only in upper part along ribs; anthers 1–2 (2.3) mm long .... 4.
+ Lemma covered with spinules throughout most of surface; spinules sometimes elongated and transformed into short hairs.
.... 5.
4. Rachilla of spikelet densely covered with very short setaceous spinules; spikes 4–12 cm long, fairly dense and erect. Leaf blade usually pilose, less often in upper part ....
.... 8. E. kronokensis (Kom.) Tzvel.
+ Rachilla of spikelet short but densely hairy; spikes 8–20 cm long, quite sparse and slightly pendent. Leaf blade scabrous in upper part............................... E. trachycaulus (Link) Gould et Shinners.
5. Anthers 2–2.7 mm long; glumes 7–12 mm long, lanceolate, gradually acuminate, forming up to 3 mm long cusp or awn; spikes dense, usually somewhat secund. Leaf blade usually with scattered hair in upper part, less often only scabrous ....
.... 10. E. mutabilis (Drob.) Tzvel.
+ Anthers 1–2 mm long; glumes 6.5–9 mm long, lanceolate-elliptical, quite abruptly transforming into up to 1.5 (2) mm long cusp at tip. Leaf blade scabrous in upper part.............................. 6.
6. Spikes not secund, relatively sparse; lemma scabrous due to short spinules .... 12. E. praecaespitosus (Nevski) Tzvel.

+ Spikes somewhat secund, denser; lemma covered with elongated spinules, transforming into short hairs ....
.... 15. **E. transbaicalensis** (Nevski) Tzvel.

7(1) Lemma puberulent along margin and in upper part, elsewhere (excluding callus) glabrous and smooth, with flexuose 13–23 mm long awn at tip; glumes 6–9 mm long, with 5 ribs; anthers 1.6–2.2 mm long; spikes 10–22 cm long, pendent, with very sparse, solitary spikelets. Forest plant 50–120 cm tall with flat and broad (4–10 mm broad) leaf blade ....
.... 11. **E. pendulinus** (Nevksi) Tzvel.

+ Lemma glabrous, smooth or rather scabrous due to spinules, less often [among *E. czilikensis* (Drob.) Tzvel.] with capilliform elongated spinules uniformly covering surface of lemma .... 8.

8. Spikes lax or fairly dense but with rather flexuose rachis, somewhat secund and pendent; glumes more than 2/3 as long as contiguous lemma, often slightly displaced towards one side of spikelets; lemma rather scabrous, with laterally recurved, more than 10 mm long awn at tip .... 9.

+ Spikes erect or only slightly pendent, bilateral or somewhat secund; glumes less than 2/3 as long as contiguous lemma, not displaced towards one side of spikelet; lemma smooth or rather scabrous with awn erect or recurved laterally .... 18.

9. Both sides of leaf blade with very short but dense hairs. Spikes quite dense, secund, with unevenly disposed single or sometimes paired spikelets .... 25. **E. pamiricus** Tzvel.

+ Leaf blade glabrous, smooth or somewhat scabrous due to short spinules on lower side .... 10.

10. Spikes arculately pendent, somewhat secund, with fairly sparse but evenly spaced (only lowest spikelets more widely spaced), few (usually up to 10), invariably single spikelets; glumes identical in all spikelets, lanceolate-elliptical, 1/2–2/3 as long as contiguous lemma; anthers 2–3.2 mm long. Plant 25–70 cm tall, forming very dense mat .... 11.

+ Spikes with rather flexuose rachis, pendent, secund, with usually more numerous, unevenly arranged, single or sometimes paired spikelets; glumes differ greatly in form and size even within a single spike, lanceolate, usually half or more as long as contiguous lemma; anthers 1–2.3 mm long. Plant 30–120 cm tall, usually lax-caespitose .... 13.

11. Awn of lemma 30–60 mm long; rachilla of spikelets greatly flattened on back; very brittle; anthers 2.2 to 3.2 mm long ....
.... 21. **E. canaliculatus** (Nevski) Tzvel.

+ Awn of lemma 13–28 mm long; rachilla of spikelets less flattened on back and less brittle; anthers 2–2.6 mm long ....12.

12. Rachilla of spikelet covered with capilliform elongated spinules; lemma smooth on back; palea smooth; or almost so, between keels .... 20. E. **burchan-buddae** (Nevski) Tzvel.

+ Rachilla of spikelet covered with very short but slender spinules; lemma with scattered spinules on back; palea with scattered spinules between keels. ....
.... 23. E. **czimganicus** (Drob.) Tzvel.

13(10). Almost all spikes with paired spikelets in midportion (although not many) ....14.

+ All spikes exclusively with single spikelets...............................16.

14. Leaf blade 1.5–2.5 mm broad, usually convoluted longitudinally, smooth in lower part, somewhat scabrous in upper. Spikes quite sparse; anthers 1.5–2 mm long...
.... 19. E. **atratus** (Nevski) Hand.-Mazz.

+ Leaf blade (2.5) 3–8 (12) mm broad, flat or somewhat longitudinally convoluted, rather scabrous in lower part, scabrous in upper, occasionally with additional scattered hairs.
....15.

15. Spikes 5–12 cm long, with spikelets densely set, stalks of spikelets 0.5–1.2 mm long; anthers 1.5–2.3 mm long ....
.... 24. E. **nutans** Griseb.

+ Spikes 7–20 cm long, with spikelets quite wide-set; stalks of spikelets 0.4–0.5 mm long; anthers 1–1.7 mm long ....
.... 27. E. **sibiricus** L.

16(13). Spikes 5–12 cm long, with spikelets densely set; stalks of spikelet 0.4–1 mm long, highly variable in length on same spike; anthers 1.5–2 mm long...
.... 26. E. **schrenkianus** (Fisch. et Mey.) Tzvel.

+ Spikes 7–20 cm long, with spikelets quite wide-set; stalks of spikelets 0.4–0.6 mm long ....17.

17. Lemma 1.5–2 mm long, with 1(3) ribs, palea 3–5 mm long, with 3 ribs; anthers 1.2–1.7 mm long. Radical leaves numerous, 1–2.5 mm broad, convoluted longitudinally like bristles ....
.... 18. E. **antiquus** (Nevski) Tzvel.

+ Lemma 2.5–5 mm long, with (1) 3 ribs, palea 4–7.5 mm long with 3–5 ribs; anthers 2–2.7 mm long. Radical leaves usually numerous, 3–10 mm broad, flat ....
.... 22. E. **confusus** (Roshev.) Tzvel.

18(8). Spikes with spikelets grouped 2(3) throughout length or only in midportion ....19.

+ Spikes only with single spikelets throughout length ....21.

19. Leaf blade 8–18 mm broad, with scattered hairs or less often glabrous in upper part or on both sides. Lemma smooth, or almost so, on back, rather scabrous along sides and in upper part; anthers 2.5–3.6 mm long ....5. **E. excelsus** Turcz. ex Griseb.

+ Leaf blade 3–9 mm broad, usually glabrous, less often with scattered hairs in upper part. Lemma scabrous throughout surface; anthers 1.5–2.3 mm long .... 20.

20. Stems and sheath glabrous and smooth; very rarely sheaths of lower leaves puberulent. Awn 10–20 mm long, usually rather laterally recurved .... 4. **E. dahuricus** Turcz. ex Griseb.

+ Stems under panicle, at and under nodes scabrous due to very short but dense setavous spinules; sheaths usually slightly scabrous. Awn 7–14 mm long, suberect ....
.... 13. **E. tangutorum** (Nevski) Hand.-Mazz.

21(18). Lemma only in upper part and close to margin with few spinules; glabrous and smooth elsewhere on surface (except callus) ...... 22.

+ Lemma covered with spinules throughout surface; spinules sometimes transformed into short hairs .... 24.

22. Anthers 2.3–3 mm long; awn of lemma 12–20 mm long, somewhat flexuose. Forest plant with usually flat 3–11 mm broad leaf blade
.... 2. **E. caninus** (L.) L.

+ Anthers 1.2–2 mm long. Plant of rocky slopes and pebble beds ..
.... 23.

23. Awn of lemma distinctly laterally recurved. Leaf blade 1.5–3 mm broad, usually longitudinally convoluted, glabrous and smooth in lower part, puberulent in upper ....
.... 16. **E. varius** (Keng) Tzvel.

+ Awn of lemma not laterally recurved, somewhat flexuose. Leaf blade 2.5–8 mm broad, flat or faintly convoluted longitudinally, rather scabrous on both sides due to short spinules ....
.... 17. **E. vernicosus** (Nevski ex Grub.) Tzvel.

24(21). Lemma with highly thickened, laterally recurved 20–40 mm long awn .... 25.

+ Lemma with very slender, erect or flexuose, less often [among *E. abolinii* (Drob.) Tzvel.] with slightly laterally recurved 7–15 (18) mm long awn .... 26.

25. Glumes broadly lanceolate, 9–16 mm long, usually almost as long as contiguous lemma; anthers 2.8–4 mm long ....
.... 9. **E. macrolepis** (Drob.) Tzvel.

+ Glumes lanceolate, 7–11 mm long, usually almost 2/3 as long as contiguous lemma; anthers 2–2.7 mm long ....
.... 6. **E. gmelinii** (Ledeb.) Tzvel.

26. Anthers 2.5–3.5 mm long; palea with scattered elongated spinules between keels .... 27.

+ Anthers 1.5–2.2 mm long; palea glabrous and almost smooth between keels .... 28.

27. Spikes not secund; their rachis not only along ribs, but also on back with scattered spinules; awn of lemma usually somewhat laterally recurved; glumes with 2–4 mm long awn at tip ....
.... 1. E. abolinii (Drob.) Tzvel.

+ Spikes somewhat secund; their rachis scabrous usually only along ribs; awn of lemma erect or somewhat flexuose; glumes with up to 2 mm long awn at tip . 3. E. czilikensis (Drob.) Tzvel.

28. Glumes 6–10 mm long, with 5–7 ribs, with 2–7 mm long awn at tip. Leaf blade 3–8 mm broad, usually flat, often diffusely hairy in upper part. .... 7. E. komarovii (Nevski) Tzvel.

+ Glumes 8–12 mm long, with 3–5 ribs, with up to 2 mm long cusp at tip. Leaf blade 1.5–3.5 mm broad, usually longitudinally convoluted, glabrous .... 13. E. scabridulus (Ohwi) Tzvel.

Section 1. Turczaninovia
(Nevski) Tzvel[7]

1. **E. abolinii** (Drob.) Tzvel. comb. nova.—*Agropyron abolinii* Drob. in Feddes repert. 21 (1925) 42; Fl. Kazakhst. 1 (1956) 303.—*Roegneria abolinii* (Drob.) Nevski in Fl. SSSR, 2 (1934) 611; Fl. Kirgiz. 2 (1950) 192.—**Ic.**: Fl. Kazakhst. 1, Plate 22, fig. 13.

Described from Nor. Tien Shan (Alma Ata region). Lectotype in Tashkent; isotype in Leningrad.

In grasslands among shrubs, on rocky slopes and in pebble beds; in middle and upper mountain belts.

IIA. Junggar: *Tien Shan* (Maralty village on Kunges river, 2700 m, Aug., 1879.—A. Reg.; east. part of Ketmen' mountain range, Sarbushin crossing on road to Kzyl-Kura from Ili, Aug. 23, 1957—Yun.).

General distribution: Jung.-Tarb., Nor. and Cent. Tien Shan.

Note. Only the variety with long awns, *E. abolinii* var. *divaricans* (Nevski) Tzvel. comb. nova [=*Roegneria abolinii* f. *divaricans* Nevski in Fl. SSSR, 2 (1934) 612], has been reported thus far from Cent. Asia. Its specimens possibly represent hybrids of the typical variety with short awn and closely related species *E. macrolepis* (Drob.) Tzvel.

2. **E. caninus** (L.) L. Fl. Suec., ed. 2 (1755) 39.—*Triticum caninum* L. Sp. pl. (1753) 86.—*Agropyron caninum* (L.) Beauv. Ess. Agrost. (1812) 146; ? Forbes and Hemsley, Index Fl. Sin. 3 (1904) 431, p.p.; Krylov, Fl. Zap. Sib.

---

[7]*Elymus* sect. Turczaninovia (Nevski) Tzvel. comb. nova.—*Clinelymus* sect. Turczaninovia nevski in Izv. Bot. sada AN SSSR, 30 (1932) 645.

2 (1928) 348; Pavlov in Byull. Mosk. obshch. ispyt. prir. 38 (1929) 27; Fl. Kazakhst. 1 (1956) 303; Melderis in Bor, Grasses Burma, Ceyl., Ind. and Pakist. (1960) 660.—*Roegneria canina* (L.) Nevski in Fl. SSSR, 2 (1934) 617; Fl. Kirgiz. 2 (1950) 195; Fl. Tadzh. 1 (1957) 309.

Described from Europe. Type in London (Linn.).

In forests, forest glades and among shrubs; in midmountain belt.

IIA. **Junggar:** *Tien Shan* (Talkibakh, July 1877; Borgata brook on nor. flank of Kash river valley, 1300–2000 m, July 5, 1879—A. Reg.; nor. flank of water divide between Tsanma and Kunges rivers toward Kunges valley, spruce thicket, Aug. 7, 1958—Yun.).

General distribution: Jung.-Tarb., Nor. Tien Shan; Europe, Mediterr., Balk.-Asia Minor, Near East, Caucasus, Middle Asia (West. Tien Shan, West. Pam.), West. Sib., East. Sib. (south.-west.), Nor. Mong. (? Hang.), ? Himalayas.

3. **E. czilikensis** (Drob.) Tzvel. comb. nova.—*Agropyron czilikense* Drob. in Feddes repert. 21 (1925) 43.—*A. tianschanicum* Drob. l.c. 42, p.p.; Fl. Kazakhst. 1 (1956) 305; krylov, Fl. Zap. Sib. 12, 1 (1961) 3127.—*Roegneria tianschanica* (Drob.) Nevski in Fl. SSSR, 2 (1934) 615; Fl. Kirgiz. 2 (1950) 195; Fl. Tadzh. 1 (1957) 307.—? *R. platyphylla* Keng, Fl. ill. sin., Gram. (1959) 370, diagn. sin.—Ic.: Fl. Kazakhst. 1, Plate 22, fig. 15; Keng, l.c. fig. 299.

Described from Tien Shan. Type in Tashkent; isotype in Leningrad.

In meadows, forest glades, on rocky slopes, in riverine pebble beds; in middle and upper mountain belts.

IIA. **Junggar:** *Tien Shan* (Kyzymchek hills north of Kul'dzha, 2700–3000 m, July 29, 1878—A. Reg.).

General distribution: Jung.-Tarb., Nor. and Cent. Tien Shan; Middle Asia (hilly).

4. **E. dahuricus** Turcz. ex Griseb. in Ledeb. Fl. Ross. 4 (1852) 331; Franch. Pl. David. 1 (1884) 342; Forbes and Hemsley, Index Fl. Sin. 3(1904) 433; Krylov, Fl. Zap. Sib. 2 (1928) 368; Pavlov in Byull. Mosk. Obshch. ispyt. prir. 38 (1929) 29; Pampanini, Fl. Carac. (1930) 80; Kitag. Lin. Fl. Mansh. (1939) 74; Fl. Kazakhst. 1 (1956) 320; Melderis in Bor, Grasses Burma, Ceyl., Ind. and Pakist. (1960) 669.—*E. dahuricus* var. *cylindricus* Franch. l.c.— *E. cylindricus* (Franch.) Honda in J. Fac. Sci. Univ. Tokyo, sect. sect. III, 3, 1 (1930) 17, non Pohl, 1810; Kitag. l.c.—*Clinelymus dahuricus* (Turcz.) Nevski in Izv. Bot. sada AN SSSR, 30 (1932) 645 and in Fl. SSSR, 2 (1934) 691; ? Ching in Contribs. U.S. Nat. Herb. 28 (1941) 597; Melderis in Norlindh, Fl. mong. steppe, 1 (1949) 123; Fl. Kirgiz. 2(1950) 216; Grubov, Consp. fl. MNR (1955) 77; Fl. Tadzh. 1 (1957) 274; Keng, Fl. ill. sin., Gram. (1959) 427.—Ic.: Fl. SSSR, 2, Plate 49, fig. 3; Keng, l.c. figs. 360 and 361.

Described from Transbaikal. Type in Leningrad.

In meadows and pebble beds, among shrubs, in forest glades, sometimes as weed around irrigation ditches and in various crops, along roadsides; up to midmountain belt.

IA. **Mongolia:** *Khobd.* (Bukhu-Muren river flood plain 4–5 km nor. of Bukhu-Muren somon, July 31, 1945—Yun.); *Mong. Alt., Cent. Khalkha* (Kerulen river valley near Bain-Erkhet; same site, near Dalai-Beis—1899, Pal.); *East. Mong.* (Ulan-Morin river, Aug. 11; Tai-Tukhai area, Aug. 17—1884, Pot.; near Hailar town, No. 2977, 1954—Wang; near Shilin-Khoto and near Hailar town—1959, Ivan.); *Bas. lakes* (Khobdo river valley on Khobdo-Ulangom road, Aug. 23, 1944; stream of Kharkira river 3–4 km south of Ulangom, July 28, 1945—Yun.); *Gobi-Alt.* (Dundu-Saikhan hills, Ulan-Khundei creek valley, Aug. 18, 1931—Ik.-Gal.; Dundu- and Dzun-Saikhan hills, July-Aug., 1933—Simukova; nor. slope of Ikhe-Bogdo, Batyuten-Ama creek valley, Sept. 12, 1943—Yun.); *East. Gobi* ('Khonin-Chaghan-Chölö-Gol, No. 1283, 1 VIII, 1927. leg. Hummel"—Norlindh, l.c.; Temur-Khada, Roerich Exped., No. 303, Aug. 25; Khar-Sair, Roerich Exped.; No. 493, Aug. 23—1935, Keng); *Ordos* (10 km south-west of Ushin town, Aug. 4, 1957—Petr.); *Khesi* (55 km west of Yunchan town, Aug. 16, 1958—Petr.).

IB. **Kashgar:** *Nor.* (Yakka-Aryk village, Aug. 13, 1929—Pop.; near Karashar town, No. 6853, July 28, 1958—Lee and Chu; valley of Khaiduk-Gol river 4–8 km west of 5th Regiment State Farm, July 29; upper course of Kyzyl river nor.-west of Kucha town, Sept. 2—1958, Yun.).

IIA. **Junggar:** *Alt. region* (near Altay settlement [Shara-Sume], No. 2362, Aug. 25; Altay district, 1400 m, No, 2486, Aug. 27—1956, Ching); *Jung. Alt., Tien Shan. Jung.-Gobi* (south.: near Savan village, 500 m, No. 3720, Oct. 4, 1956—Ching; Savan district, between Katszyvan and Tsian'tsyuan' villages, No. 1694, July 21, 1957—Huang).

IIIA. **Qinghai:** *Nanshan* ("Ni-Ma-Lang-Kou"—Ching, l.c.; val. of Tetung river near stud farm, 2800 m, Aug. 20, 1958—Dolgushin).

IIIC. **Pamir** (Pas-Robat village, Aug. 3, 1909—Divn.).

General distribution: Jung.-Tarb., Nor. and Cent. Tien Shan; Middle Asia (Pamir-Alay), West. Sib. (Altay), East. Sib. (south), Far East (south), Nor. Mong. (Hang., Mong.-Daur.), China (Dunbei, Nor., Nor.-West., Cent., South-West.), Korea, Japan.

Note. It would perhaps be more correct to treat this species and the closely related *E. excelsus* Turcz. ex Griseb. and *E. tangutorum* (Nevski) Hand.-Mazz. as subspecies of the same species—*E. dahuricus* Turcz. ex Griseb. s.l. Further, variety *E. dahuricus* var. *brevisetus* Ohwi [in J. Jap. Bot. (1943) 168] with very short (2–5 mm long) awns of lemma, described from Nor. China (Shanxi province), possibly represents an independent subspecies. Specimens from Khesi and Qinghai (Tetung river valley) cited above as well as the specimen from Tien Shan (in the region of Tyan'chi lake, No. 4255, Sept. 19, 1957—Huang) belong to the latter variety.

5. E. excelsus Turcz. ex Griseb. in Ledeb. Fl. Ross. 4 (1852) 331; Forbes and Hemsley, Index Fl. Sin. 3 (1904) 434; Kitag. Lin. Fl. Mansh. (1939) 74. —*E. dahuricus* var. *excelsus* (Turcz.) Roshev. in Bot. mater. Gerb. Glavn. bot. sada RSFSR, 4 (1923) 138.—*Clinelymus excelsus* (Turcz.) Nevski in Izv. Bot. sada AN SSSR, 30 (1932) 646 and in Fl. SSSR, 2 (1934) 692; Keng, Fl. ill. sin. Gram. (1959) 426. —Ic.: Fl. SSSR, 2, Plate 49, fig. 4; Keng, l.c. fig. 358.

Described from Transbaikal. Type in Leningrad.

In meadows and forest glades, among shrubs; up to lower mountain belt.

IA. **Mongolia:** *East. Mong.* (Ourato, hautes montagnes du centré, July 1866—David; near Trekhrech'e, 1951—Li et al.).

General distribution: East. Sib. (south-east.), Far East (south), China (Dunbei, Nor.), Korea, Japan.

6. **E. gmelinii** (Ledeb.) Tzvel. comb. nova.—*Triticum caninum* var. *gmelinii* Ledeb. Fl. alt. 1 (1829) 118.—*Agropyron gmelinii* (Ledeb.) Scribn. et J.G. Smith in Bull. U.S. Departm. Agric. Div. Agrost. 4 (1897) 30, quoad nom.; Krylov, Fl. Alt. 7 (1914) 1695; Pavlov in Byull. Mosk. obshch. ispyt. prir. 38 (1929) 27.—*A. turczaninovii* Drob. in Tr. Bot. muzeya AN, 12 (1914) 47; Krylov, Fl. Zap. Sib. 2 (1928) 347; Grubov, Consp. fl. MNR (1955) 76; Fl. Kazakhst. 1 (1956) 302.—*Roegneria turczaninovii* (Drob.) Nevski in Fl. SSSR, 2 (1934) 607; Melderis in Norlindh, Fl. mong. steppe, 1 (1949) 122; Fl. Kirgiz. 2 (1950) 191; Keng, Fl. ill. sin., Gram. (1959) 393.—*Agropyron strigosum* auct. non Boiss.: Franch. Pl. David. 1 (1884) 340.—Ic.: l.c. fig. 320.

Described from Altay. Type in Leningrad.

In meadows and forest glades, sparse larch forests and riverine pebble beds; up to midmountain belt.

IA. **Mongolia:** *Cis-Hing.* (Khuntu somon 35 km south-east of Bain-Tsagan, Aug. 6; same site, 5 km west of Toge-Gol river, Aug. 7–1949, Yun.); *Cent. Khalkha, East. Mong.* (Erdaotszin'tszy village south of Manchuria station, 700 m, No. 948, June 26, 1951—Li et al.; near Shilin-Khoto, 1959—Ivan.); *Gobi-Alt.* (Dzun-Saikhan hills, Aug. 24, 1931—Ik.-Gal.).

IIA. **Junggar:** *Alt. region* (Altay district, 1350 m, No. 3223, Sept. 16, 1956—Ching); *Jung. Alt.* (1 km south of Toli settlement, No. 2042, Aug. 22, 1957—Huang); *Tien Shan* (6 km south-west of Chzhaos, 1760 m, No. 827, Aug. 11; 15 km nor. of Ulastai, No. 3935, Aug. 29—1957, Huang; Ketmen' mountain range, Sarbushin crossing on Ili-Kzyl-Kura road, Aug. 23, 1957—Yun.).

General distribution: Jung.-Tarb.,? Nor. Tien Shan; West. Sib. (Altay), East. Sib., Far East, Nor. Mong. (Hent., Hang., Mong.-Daur.), China (Dunbei, Nor.), Korea, Japan.

7. **E. komarovii** (Nevski) Tzvel. comb. nova.—*Agropyron komarovii* Nevski in Izv. Bot. sada AN SSSR, 30 (1932) 620; Grubov, Consp. fl. MNR (1955) 75; Fl. Kazakhst. 1 (1956) 305; Krylov., Fl. Zap. Sib. 12, 1 (1961) 3127.—*Roegneria komarovii* (Nevski) Nevski in Fl. SSSR, 2 (1934) 615; Melderis in Norlindh, Fl. mong. steppe, 1 (1949) 122.—Ic.: Fl. SSSR, 2, Plate 45, fig. 4.

Described from East. Sayan (valley of Oka river). Type in Leningrad.

In meadows, larch forests and on rocky slopes; in middle and upper mountain belts.

IA. **Mongolia:** *Mong. Daur.* (near Tsenkher-Mandal somon facing Tsenkheriin-Duganga, July 23, 1949—Yun.); *Gobi-Alt.* (Dzun-Saikhan hills, Yalo upper creek valley, Aug. 25 and 26, 1931—Ik. Gal.).

IIA. **Junggar:** *Jung. Alt.* (between Syaeda and Ven'tsyuan, 2200–2900 m, No. 1406, Aug. 13, 1957—Huang); Tien Shan (Krisu village, 1800 m, No. 1842, July 17, 1957—Huang).

General distribution: Jung.-Tarb.; West. Sib. (Altay), East. Sib. (south-west.), Nor. Mong. (Hent., Hang.).

8. E. kronokensis (Kom.) Tzvel. comb. nova.—*Agropyron kronokense* Kom. in Feddes repert. 13 (1915) 87.—*A. boreale* (Turcz.) Drob. in Tr. Bot. muzeya AN, 16 (1916) 84; Krylov, Fl. Zap. Sib. 12, 1 (1961) 3130.—*Triticum boreale* Turcz. in Bull. Soc. natur. Moscou, 29, 1 (1856) 58, non *Elymus borealis* Scribn. 1900.—*Roegneria borealis* (Turcz.) Nevski in Fl. SSSR, 2 (1934) 624.—*R. scandica* Nevski l.c.—*R. leiantha* Keng, Fl. ill. sin., Gram. (1959) 373, diagn. sin.—*R. kronokensis* (Kom.) Tzvel. in Arkt. fl. SSSR, 2 (1964) 246.—Ic.: Fl. SSSR, 2, Plate 45, fig. 1; Keng, l.c. fig. 301.

Described from Kamchatka. Type in Leningrad.

In grasslands and on rocky slopes; in upper (bald peak) mountain belt.

IA. **Mongolia**: *Mong. Alt.* (Tumun somon, nor. slope of Taishiri-Ul mountain range, July 12, 1945—Yun.).

IIA. **Junggar**: *Jung. Alt.* (upper course of Borotala river, 2800 m, Aug. 1878—A. Reg.).

IIIA. **Qinghai**: *Nanshan* ("Da-Tung [Tetung] district"—Keng, l.c.).

General distribution: Arct., Europe (Scandinavian hills), West. Sib. (Altay), East. Sib., Far East (excluding south.), Nor. Mong. (Fore Hubs.), Nor. Amer. (Nor. Cordillera).

Note. *E. kronokensis* var. *borealis* (Turcz.) Tzvel. comb. nova. (=*Triticum boreale* Turcz. l.c.) with leaf blade glabrous on lower side and puberulent on upper side is the most extensively distributed variety of this species, which is highly characteristic of east Asian bald peaks. The type variety—*E. kronokensis* var. *kronokensis* with leaf blade having short but dense hairs on both sides has been reported thus far only from Kamchatka, Chukchi-Anadyr region and Mong. Altay (specimen cited above). Another variety, *E. kronokensis* var. *scandica* (Nevski) Tzvel. comb. nova (=*Roegneria scandica* Nevski l.c.), with glabrous leaf blade has not been reported so far in Cent. Asia. Species described from Qinghai, *Roegneria leiantha* Keng (l.c.), which we could not study, is provisionally placed among *E. kronokensis* (Kom.) Tzvel. s.l.

9. E. macrolepis (Drob.) Tzvel. comb. nova.—*Agropyron macrolepis* Drob. in Feddes repert. 21 (1925) 41, s.s.; Fl. Kazakhst. 1 (1956) 302; Melderis in Bor, Grasses Burma, Ceyl., Ind. and Pakist. (1960) 663; Krylov, Fl. Zap. Sib. 12, 1 (1961) 3126.—*A. curvatum* Nevski in Izv. Bot. sada AN SSSR, 30 (1932) 629.—*Roegneria curvata* (Nevski) Nevski in Fl. SSSR, 2 (1934) 608; Fl. Kirgiz. 2 (1950) 191; Fl. Tadzh. 1 (1957) 300.—Ic.: Fl. SSSR, 2, Plate 47, fig. 5.

Described from Nor. Tien Shan (Ketmen' mountain range). Type in Leningrad.

In grasslands, on rocky slopes and in riverine pebble beds; in middle and upper mountain belts.

IA. **Mongolia**: *Mong. Alt.* (south. flank of Indertiin-Gol valley near summer camp in Bulugun somon, July 24, 1947—Yun.).

IIA. **Junggar**: *Alt. region* (near Qinhe, 1800 m, No. 903, Aug. 2; between Qinhe and Chzhunkhaitsz, 2000 m, No. 1461, Aug. 7—1956, Ching; 20 km nor.-west of Shara-Suma, July 7; right bank of Kairta river 20 km nor.-west of Koktogoi, Kuidun river

valley, July 15—1959, Yun.); *Tarb.* (south. slope of Saur mountain range, Karagaitu river valley, Bain-Tsagan creek valley, June 23, 1957—Yun.); *Tien Shan.*
General distribution: Jung.-Tarb., Nor. and Cent. Tien Shan; Middle Asia (hilly), West. Sib. (Altay), Himalayas (West.).

10. **E. mutabilis** (Drob.) Tzvel. comb. nova.—*Agropyron mutabile* Drob. in Tr. Bot. muzeya AN, 16 (1916) 88, s.s.; Krylov, Fl. Zap. Sib. 1 (1928) 350; Pavlov in Byull. Mosk. obshch. ispyt., prir. 38 (1929) 28; ? Melderis in Bor, Grasses Burma, Ceyl., Ind. and Pakist. (1960) 663.—*A. angustiglume* Nevski in Izv. Bot. sada AN SSSR, 30 (1932) 615; Grubov, Consp. fl. MNR (1955) 75; Fl. Kazakhst. 1 (1956) 306.—*Roegneria angustiglumis* (Nevski) Nevski in Fl. SSSR, 2 (1934) 618; Fl. Kirgiz. 2 (1950) 196.—*R. mutabilis* (Drob.) Hyl. in Uppsala Univ. Årsskr. 7 (1945) 36.

Described from Yakutia; lectotype (Chona river bank near Dushenka village, No. 317, 1914, V. Drobov) in Leningrad.

In meadows, forest glades, larch forests and riverine pebble beds; in middle and upper mountain belts.

IIA. Junggar: *Alt. region* (near Burchum settlement, 1400 m, No. 3023, Sept. 18, 1956—Ching); *Tien Shan* (Nizhnii Arystyn, 2300–2700 m, July 20, 1879—A. Reg.; along Dana river, 3200 m, No. 2205, July 23; 5 km south-west of Dzhagastai, 2180 m, No. 744, Aug. 8; hills in Barkul' lake region, 2350 m, No. 2221, Sept. 27—1957, Huang).
General distribution: Jung.-Tarb., Nor. Tien Shan; Europe (nor.-east.), West. Sib., East. Sib., Far East (Kamchatka), Nor. Mong. (Fore Hubs., Hent., Hang.), ? Himalayas (west.).

11. **E. pendulinus** (Nevski) Tzvel. comb. nova.—*Roegneria pendulina* Nevski in Fl. SSSR, 2 (1934) 616 and in Tr. Bot. inst. AN SSSR, ser. 1, 2 (1936) 50; Kitag. Lin. Fl. Mansh. (1939) 91; Keng. Fl. ill. sin., Gram. (1959) 362.—*Triticum caninum* var. *amurense* Korsh. in Acta Horti Petrop., 12 (1892) 414.—Ic.: Fl. SSSR, 2, Plate 45, fig. 6; Keng, l.c. fig. 291.

Described from Far East (valley of Amur river), Type in Leningrad.

In forests, forest glades and meadows; up to lower mountain belt.

IA. Mongolia: Cis-Hing. (near Hailar railway station, in forest, June 10, 1902—Litw.).
General distribution: Far East (south), China (Dunbei), Korea, ? Japan.

12. **E. praecaespitosus** (Nevski) Tzvel. comb. nova.—*Agropyron praecaespitosum* Nevski in Izv. Glavn. bot. sada SSSR, 29 (1930) 541; Fl. Kazakhst. 1 (1956) 308.—*Roegneria praecaespitosa* (Nevski) Nevski in Tr. Sredneaz. gos. univ., ser. 8b, 17 (1934) 70.—*R. oschensis* Nevski in Fl. SSSR, 2 (1934) 619; Fl. Tadzh. 1 (1957) 308.

Described from East. Tien Shan. Type in Leningrad.

In grasslands, on rocky slopes, rocks and in riverine pebble beds; in middle and upper mountain belts.

IIA. Junggar: *Alt. region* (near Qinhe settlement, 1800 m, No. 1236, Aug. 2, 1956—Ching; right bank of Kairta river 20 km nor. of Koktogoi, Kuidun river valley, July 15,

1959—Yun.); *Jung. Alt.* (Toli village region, No. 1132, Aug. 7; Ven'tsyuan' town region, 2400 m, No. 1949, Aug. 14; 30 km west of Ven'tsyuan', 2400 m, No. 2025, Aug. 25—1957, Huang); *Tien Shan* (among others; Yulty-Arystyn, July 7, 1879—A. Reg., type!).

General distribution: Jung.-Tarb., Nor. and Cent. Tien Shan; Middle Asia (West. Tien Shan, Pamir-Alay), Nor. Mong. (Fore Hubs., Hang).

13. **E. scabridulus** (Ohwi) Tzvel. comb. nova. —*Agropyron scabridulum* Ohwi in J. Jap. Bot. 19 (1943) 166.—? *Roegneria sinica Keng*, Fl. ill. sin., Gram. (1959) 367, diagn. sin. —Ic.: Keng, l.c. fig. 297.

Described from Inner Mongolia. Type in Tokyo.

In meadows, on rocky slopes and rocks.

IA. Mongolia: *East. Mong.* ("Fu-Sheng-Chuang between Chining and Kueisui, No. 3841, leg. T. Kunashiro"—Ohwi, l.c., type!).

IIIA. Qinghai: cited without precise locality (Keng, l.c.).

General distribution: China (Dunbei, Nor., Nor.-West., Cent.).

Note. Without reliable material on *Agropyron scabridulum* Ohwi and *Roegneria sinica* Keng, we are not entirely confident of these species being synonyms.

14. **E. tangutorum** (Nevski) Hand.-Mazz. Symb. Sin. 7, 5 (1936) 1292. —*Clinelymus tangutorum* Nevski in Izv. Bot. sada AN SSSR, 30 (1932) 647; Keng, Fl. ill. sin., Gram. (1959) 427.—Ic.: Keng, l.c. fig. 359.

Described from Qinghai. Type in Leningrad.

In forests, forest glades and among shrubs; in middle and upper mountain belts.

IIIA. Qinghai: *Nanshan* (forest belt of Yuzhno-Tetungsk mountain range, July 16, 1872—Przew., type!).

IIIB. Tibet: *Weitzan* (nor. slope of Burkhan-Budda mountain range, Khatu gorge, 3500 m, July 8 1901—Lad.).

General distribution: endemic.

○ **E. tenuispicus** (Yang et Zhou) Tzvel. comb. nova.—*Roegneria tenuispica* Yang et Zhou, Novon, 4 (1994) 307.

IIIB. Tibet.

E. trachycaulus (Link) Gould et Shinners in Rhodora, 56 (1954) 28.—*E. pauciflorus* (Schwein.) Gould in Madrono, 9 (1947) 126, non Lam. 1791.—*Triticum pauciflorum* Schwein. in Keating, Narr. Exped. St.-Peter's River, 2 (1824) 383.—*T. trachycaulon* Link, Hort. bot. Berol. 2 (1833) 189.—*Agropyron tenerum* Vasey in Bot. Gaz. 10 (1885) 258; Krylov, Fl. Zap. Sib. 12, 1 (1961) 3130.—*A. pauciflorum* (Schwein.) Hitchc. in Amer. J. Bot. 21 (1934) 132, non Schur, 1859.—*A. atbassaricum* Golosk. in Bot. mater. Gerb. Bot. inst. AN SSSR, 14 (1951) 73; Fl. Kazakhst. 1 (1956) 307.—*Roegneria trachycaulon* (Link) Nevski in Fl. SSSR, 2 (1934) 599.—*R. pauciflora* (Schwein.) Hyl. in Uppsala Univ. Årsskr. 7 (1945) 89; Keng, Fl. ill. sin., Gram. (1959) 374.

Described from cultivated specimens originating from Nor. America. Type in Berlin.

Cultivated as food plant and sometimes reported as introduced or escape along roadsides, around irrigation ditches and along farm boundaries; up to lower mountain belt.

IA. **Mongolia:** *East. Gobi* (brook 1 km south-east of Dalan-Dzadagad town, July 28; test plantations near Dalan-Dzadagad, Sept. 12—1951, Kal.).

General distribution: Nor. Amer.; wild in many countries of Europe and in more southern regions of the USSR.

15. **E. transbaicalensis** (Nevski) Tzvel. comb. nova.—*Agropyron transbaicalense* Nevski in Izv. Bot. sada AN SSSR, 30 (1932) 618; Grubov, Consp. fl. MNR (1955) 76; Krylov, Fl. Zap. Sib. 12, 1 (1961) 3129.—*A. sajanense* (Nevski) Grub. l.c. 76, quoad pl.—*Roegneria transbaicalensis* (Nevski) Nevski in Fl. SSSR, 2 (1934) 619.

Described from Fore Baikal (Angara river basin). Type in Leningrad.

In meadows, pebble beds and sand in river valleys; up to midmountain belt.

IA. **Mongolia:** *Gobi-Alt.* (Dzun-Saikhan hills, Yalo upper creek valley, Aug. 24 and 26, 1931—Ik.-Gal.)

General distribution: West. Sib. (Altay), East. Sib. (south), Nor. Mong. (Fore Hubs., Hang.—Khan-Khukhei mountain range, Mong.-Daur.).

16. **E. varius** (Keng) Tzvel. comb. nova.—*Roegneria varia* Keng, Fl. ill. sin., Gram. (1959) 397, diagn. sin.—Ic.: Keng, l.c. fig. 346 (type!).—Planta perennis 30–100 cm alta laxe caespitosa; culmi et vaginae vulgo glabrae et laeves; ligulae ad 0.6 mm lg.; laminae 1.5–3 mm lt., vulgo laxe convolutae, subtus glabrae et laeves, supra brevissime pilosae. Spicae 4–12 cm lg., erectae et sat densae; glumae 7–8.5 mm lg., 5-nerviae, scabrae, apice acutae vel in mucrone ad 0.7 mm lg. transeuntes; lemmata 7.5–10 mm lg., glabra et laevia, solum prope apicem scabriuscula, in arista paulo divergente 10–18 mm lg.; transeuntia; paleae lemmatis subaequilongae carinis scabris; antherae 1.2–1.8 mm lg.

Described from China (Gansu province). Type in Nanking.

On rocky and rubble slopes, in pebble beds and among shrubs; in midmountain belt.

IIIA. **Qinghai:** *Nanshan* (east. extremity of Nanshan hills 25 km south of Gulan, 2450 m, Aug. 12, 1958—Petr.).

General distribution: China (Nor.-West., Cent.).

17. **E. vernicosus** (Nevski ex Grub.) Tzvel. comb. nova.—*Agropyron vernicosum* Nevski ex Grub. in Bot. mater. Gerb. Bot. inst. AN SSSR, 17 (1955) 6; Grubov, Consp. fl. MNR (1955) 76.—*R. foliosa* Keng, Fl. ill. sin., Gram. (1959) 366, diagn. sin.—Ic.: Keng, l.c. 296.

Described from Mongolia. Type in Leningrad.

On rocky slopes, rocks, in pebble beds and among shrubs; up to midmountain belt.

IA. Mongolia: *Mong. Alt.* (Khara-Dzarga hills, Boro-Gol river valley, Aug. 24, 1930—Pob.); *East. Mong.* (near Hailar town, July 6, 1901—Lipsky), *Gobi-Alt.* (Bain-Tşagan mountain range; Tsubulyur creek valley, No. 3892, Aug. 5, 1931—Ik.-Gal., type!; Bain-Tsagan mountain range, Aug. 11; Dundu-Saikhan hills, Ulan-Khundei creek valley, Aug. 18; Dzun-Saikhan hills, Aug. 22, 24, 25 and 26; Barun-Saikhan hills, near Tegetu-Gol river, Sept. 20—1931, Ik.-Gal.; Bain-Tsagan mountain range, July-Aug., 1933—Simukova); *East. Gobi* (south of Tumur Hada, Roerich Exped., No. 620, Aug. 2; Madenii-Amok, Roerich Exped., No. 770, Aug. 10—1935, Keng).

General distribution: endemic.

Note. One of the above-cited East. Gobi specimens (Roerich Expedition, No. 770) is an isotype of *Roegneria foliosa* Keng.

Section 2. Elymus

18. **E. antiquus** (Nevski) Tzvel. comb. nova.—*Agropyron antiquum* Nevski in Izv. Bot. sada AN SSSR, 30 (1932) 515.—? *Roegneria brevipes* Keng, Fl. ill. sin., Gram. (1959) 378, diagn. sin.—Ic.: Keng, l.c. fig. 307.

Described from Tibet (Kam). Type in Leningrad.

In grasslands, forest glades and sparse forests; in upper mountain belt.

IIIA. Qinghai: *Nanshan* ("near Khuan"yuan', No. 5246, I.L. Gen and B. Ts. Gen"—Keng, l.c.).

General distribution: China (Cent., South-West.).

Note. This species, as well as *E. durus* (Keng) Tzvel., are closely related to *E. confusus* (Roshev.) Tzvel. The relations helps among these species are not clearly understood, however, due to inadequate study of the flora of Qinghai and East. Tibet to date.

19. **E. atratus** (Nevski) Hand.-Mazz. Symb. Sin. 7 (1936) 1292.—*Clinelymus atratus* Nevski in Izv. Bot. sada AN SSSR, 30 (1932) 644; Keng, Fl. ill. sin., Gram. (1959) 425.—? *Hystrix kunlunensis* Hao in Engler's Bot. Jahrb. 68 (1938) 580.—Ic.: Keng, l.c. fig. 357.

Described from Qinghai. Type in Leningrad.

In grasslands, riverine pebble beds, on rocky slopes; in upper mountain belt.

IIIA. Qinghai: *Nanshan* (Gulan town region, near Ushilin crossing, July 14, 1875—Pias., type!; east coast of Kukunor lake, July 5, 1959—Petr.); *Amdo* ("Kokonor, auf dem Plateau Da-Ho-Ba, 4000 m, No. 1058, 28 VIII 1934"—Hao, l.c.).

IIIB. Tibet: *Weitzan* ("Amne Matchin, auf dem Gebirge, 4500 m, No. 1131, 2 IX 1934"—Hao, l.c.).

General distribution: endemic.

20. **E. burchan-buddae** (Nevski) Tzvel. comb. nova.—*Agropyron burchan-buddae* Nevski in Izv. Bot. sada AN SSSR, 30 (1932) 514.—? *Roegneria breviglumis* Keng, Fl. ill. sin., Gram. (1959) 377, diagn. sin. —Ic.: Keng, l.c. fig. 306.

Described from Tibet. Type in Leningrad.

On rocky slopes, talus and in pebble beds; in upper mountain belt.

IIIA. Qinghai : cited without locality (Keng, l.c.).

IIIB. Tibet: *Weitzan* (nor. slope of Burkhan-Budda mountain range, Khatu gorge, sandy-rocky bed, 3300–4300 m, No. 234, July 12, 1901—Lad., type!).

General distribution : China (Cent., South-West.).

Note. This species is very close to *E. czimganicus* (Drob.) Tzvel. and can be treated as its subspecies.

21. **E. canaliculatus (Nevski) Tzvel. comb. nova.**—*Agropyron canaliculatum* Nevski in Izv. Bot. sada AN SSSR, 30 (1932) 509; Melderis in Bor, Grasses Burma, Ceyl., Ind. and Pakist. (1960) 659. —*A. flexuosissimum* Nevski, l.c. 510.—*Anthosachne jacquemontii* (Hook. f.) Nevski in Tr. Sredneaz. gos. univ., ser. 8b. 17 (1934) 65 and in Fl. SSSR, 2 (1934) 598, quoad pl.—*Roegneria jacquemontii* (Hook. f.) Ovcz. et Sidor. in Fl. Tadzh. 1 (1957) 295, quoad pl.; Ikonnikov, Opred. rast. Pamira [Key to Plants of Pamir] (1963) 70.—*Agropyron longearistatum* auct. non Boiss.; Hemsley, Fl. Tibet (1902) 205; Pilger in Hedin, S. Tibet, 6, 3 (1922) 94.

Described from Middle Asia (Peter the First mountain range). Type in Leningrad.

On rocky slopes, talus, rocks and in pebble beds; in upper mountain belt.

IIIB. Tibet: *Chang Tang* ("Northern Tibet, Japkaklil, Chimen-Tagh, 3998 m, 22, VII 1900"—Hedin, l.c.; along Tibetan highway, No. 10037, July 16, 1959—Lee et al.); *South.* ("between Gunda-Yaukti and Tazang, 5100–5400 m, leg. Strachey and Witterbottom"—Hemsley, l.c.).

IIIC. Pamir ("Jam-Bulak-Bashi glacier on Mus-tagh-ata, 4439 m, 29 VII, 1894"—Hedin, l.c.).

General distribution: East. Pam.; Middle Asia (Gissar-Darv., Pamir-Alay), Himalayas (west., Kashmir).

Note. According to Melderis (l.c.), this species differs from the closely related *E. jacquemontii* (Hook. f.) Tzvel. comb. nova [=*Agropyron jacquemontii* Hook. f. Fl. Brit. Ind. (1897) 369] in longer (9–12, not 7–8 mm long) smooth lemma as well as very long (2.2–3.2, not 1.5–2 mm long) anthers.

22. **E. confusus (Roshev.) Tzvel. comb. nova.**—*Agropyron confusum* Roshev. in Bot. mater. Gerb. Glavn. bot. sada RSFSR, 5 (1924) 150; Pavlov in Byull. Mosk. obshch. ispyt. prir. 38 (1929) 27; Grubov, Consp. fl. MNR (1955) 75.— *Roegneria confusa* (Roshev.) Nevski in Fl. SSSR, 2 (1934) 605; Keng, Fl. ill. sin., Gram. (1959) 381, excl. var.

Described from East. Siberia. Type in Leningrad.

In meadows, forest glades and larch forests; up to midmountain belt.

IA. Mongolia: *Mong. Alt.* (Tumun somon, nor. slope of Taishiri-Ul mountain range, July 12, 1945; Bulugun floodplain near discharge of Ulyaste-Gol river into it, July 20, 1947—Yun., meander scroll of Bulugun-Gol near Bulugun somon, Sept. 22, 1948—Grub.); *Cent. Khalkha* (Dzhargalante river basin, source of

Kharukhe river on nor. slope of Uste hill, Aug. 12, 1925—Krasch. and Zam.;
Sharakhain-Khundei lowland 10 km south of Gangin-Daba crossing, July 11, 1948—
Grub.; Tsinkir-Mandal somon, valley of Tsinkiriin-Gol river near Tsinkiriin-Duganga,
July 22, 1949—Yun.); *East. Mong.* (near Trekhrech, e, 700 m, No. 1880, July 10, 1951—
Li et al.); *Gobi-alt.* (Ikhe-Bogdo mountain range, Bityuten-Ama creek valley, Aug. 12,
1927—Simukova).

General distribution : East. Sib., Far East, Nor. Mong. (Fore Hubs., Hent., Hang.),
China (Dunbei).

Note. This species is not always clearly distinguishable from the closely related *E.
sibiricus* L., although the two were placed in different genera until recently. Variety
Roegneria confusa var. *breviaristata* Keng (l.c. 381, fig. 309) described from Sinkiang
evidently belong to *E. abolinii* (Drob.) Tzvel.

23. **E. czimganicus (Drob.) Tzvel. comb. nova.**—*Agropyron czimganicus*
Drob. in Feddes repert. 21 (1925) 40; Fl. Kazakhst. 1 (1956) 301.—*Roegneria
czimganica* (Drob.) Nevski in Fl. SSSR, 2 (1934) 604; Fl. Kirgiz. 2 (1950) 188;
Fl. Tadzh. 1 (1957) 296; Ikonnikov, Opred. rast. Pamira [Key to Plants of
Pamir] (1963) 71.—Ic.: Fl. SSSR, 2, Plate 46, fig. 3; Fl. Tadzh. 1 (1957) Plate
37, figs. 1–3.

Described from West. Tien Shan. Type in Tashkent (TAK).

On rocky slopes, talus and rocks; in middle and upper mountain belts.

IB. **Kashgar:** *West.* (10–12 km nor. of Baikurt settlement on Kashgar-Torugart
road, June 20, 1959—Yun.).

IIA. **Junggar:** *Jung. Alt.* (Dzhair mountain range, gorge 4–5 km south of Yamata
picket on Chuguchak road, Aug. 4, 1957—Yun.; 20 km south of Ven'tsyuan', 2810 m,
No. 1469, July 14, 1957—Huang); *Tien Shan.*

IIIC. **Pamir** (Muztag-Ata foothills, July 20, 1909—Divn.; between Muztag-Ata and
Sarykol mountain range along road to Tashkurgan from Kashgar, Ulug-Rabat
crossing, June 12; same site, 10–12 km south of Ulug-Rabat crossing, 4100 m, June
14—1959, Yun.).

General distribution: Jung.-Tarb., Nor. and Cent. Tien Shan, East. Pam.; Middle
Asia (hilly).

24. **E. nutans Griseb.** in Nachr. Gesellsch.Wissensch. u. Univ. Goett. 3
(1868) 72; Hand.-Mazz. Symb. Sin. 7, 5 (1936) 1292; Persson in Bot. notiser
(1938) 275; Fl. Kazakhst. 1 (1956) 321; Melderis in Bor, Grasses Burma,
Ceyl., Ind. and Pakist. (1960) 670; Ikonnikov, Opred. rast. Pamira [Key to
Plants of Pamir] (1963) 73.—*Clinelymus nutans* (Griseb.) Nevski in Izv. Bot.
sada AN SSSR, 30 (1932) 644 and in Fl. SSSR, 2 (1934) 691; Fl. Kirgiz. 2
(1950) 216; Grubov, Consp. fl. MNR (1955) 77; Fl. Tadzh. 1 (1957) 274;
Keng, Fl. ill. sin., Gram. (1959) 424. —Ic.: Fl. SSSR, 2, Plate 49, fig. 2.

Described from Himalayas. Type in London (K).

In grasslands, riverine pebble beds and sand; in middle and upper
mountain belts.

IA. **Mongolia:** *Mong.-Daur.* (hill ridge intersecting road to Dalan-Dzadagad 72 km
south of Ulan Bator, Aug. 7, 1951—Kal.); *Gobi-Alt.* (Dundu-Saikhan, hills, July 7,
1909—Czet.; pass between Dundu- and Dzun-Saikhan, July 22, 1943—Yun.).

IB. **Kashgar:** *West.* (Bostan-Terek village, July 10, 1929—Pop.; "Jerzil, 2800 m, 1, VII, 1930; Bostan-Terek, ca 2400 m, 4 VIII 1934;—Persson, l.c.)" *South.* (nor. foothills of Kunlun, 2500 m, June 16, 1889–Rob.).

IC. **Qaidam:** *hilly* (in wet sand near Ichegyn-Gol river, June 21, 1895–Rob.).

IIA. **Junggar:** *Jung. Alt.* (20 km south of Ven'tsyuan', 2810 m, No. 1469, Aug. 14, 1957—Huang); *Tien Shan.*

IIIA. **Qinghai:** *Nanshan*

IIIB. **Tibet.** *Chang Tang, Weitzan* (left bank of Dychu river, 4300–5000 m, June 19; south. bank of Russky lake, 4500–4700 m, July 6—1884, Przew.; nor.-west. bank of Russky lake, 4500 m, June 27, 1900; nor. slope of Burkhan-Budda mountain range, Khatu gorge, 3500 m, June 24, 1901—Lad.); *South.* ("Shekar, No. 364, leg. Kingston"—Melderis, l.c.).

IIIC. **Pamir** (Ulug-Tuz gorge in Charlym river basin, June 29, 1909—Divn.; Tagarma valley, July 23; Egin village, Aug. 11—1913, Knorring).

General distribution: Jung.-Tarb., Nor. and Cent. Tien Shan, East. Pam.; Middle Asia (West. Tien Shan, Pamir-Alay), East. Sib. (Sayans), Nor. Mong. (Hang.), China (Cent., South.-West.).

Note. Morphological differences between this species and the very closely related *E. schrenkianus* (Fisch. et Mey.) Tzvel. are not very sharp since some spikes of *E. nutans* occasionally have only single spikelets. However, there are obvious ecological differences between the two species: *E. nutans* usually occurs in meadows and pebble beds of river valley and gorge floors while *E. schrenkianus* is found on rocky slopes and talus.

25. **E. pamiricus** Tzvel. in Bot. mater. Gerb. Bot. inst. AN SSSR, 20 (1960) 425.—*Roegneria schrenkiana* auct. non Nevski: Ovczinnikov and Sidorenko in Fl. Tadzh. 1 (1957) 299, p.p.; Ikonnikov, Opred. rast. Pamira [Key to Plants of Pamir] (1963) 70.

Described from East. Pamir (basin of West. Pshart river). Type in Leningrad.

On rocky slopes, talus and rocks; in high mountain belt.

IA. **Mongolia:** *Mong. Alt.* (Adzhi-Bogdo mountain range, upper course of Indertiin-Gol river, 3100 m, Aug. 6; same site, Mainigtu-Ama creek valley, Aug. 7—1947, Yun.).

IIIB. **Tibet:** *Chang Tang* (Tibetan highway 7 km west of Seryk crossing, No. 507, June 4, 1959—Lee et al.).

General distribution: East. Pam.; Middle Asia (Pamir-Alay).

26. **E. schrenkianus** (Fisch. et Mey.) Tzvel. in Bot. mater. Gerb. Bot. inst. AN SSSR, 20 (1960) 428.—*Triticum schrenkianum* Fisch. et Mey. in Bull. phys.-math. Ac. Sci. St.-Pétersb. 3 (1845) 305.—*Agropyron schrenkianum* (Fisch. et Mey.) Drob. in Tr. Bot. muzeya AN, 16 (1916) 136; Grubov, Consp. fl. MNR (1955) 76; Fl. Kazakhst. 1 (1956) 301; Krylov Fl. Zap. Sib. 12, 1 (1961) 3125.—*Roegneria schrenkiana* (Fisch. et Mey.) Nevski in Fl. SSSR, 2 (1934) 605; Fl. Kirgiz. 2 (1950) 188; Fl. Tadzh. 1 (1957) 299, p.p. — Ic.: Fl. Kazakhst. 1, Plate 22, fig. 12; Fl. Tadzh. 1, Plate 37, figs. 4–7.

Described from East. Kazakhstan (Tarbagatai mountain range). Type in Leningrad.

On rocky slopes, in grasslands, talus and pebble beds; in middle and upper mountain belts.

IA. **Mongolia:** *Mong. Alt.* (val. of Kholbo-Nur lake, Aug. 4; upper course of Bor-Burgasa river, 1900 m, Aug. 10—1926, Bar.; Bulugun river basin, upper course of Ketsu-Sairin-Gol river, July 26, 1947—Yun.); *Gobi-Alt.* (Ikhe-Bogdo, Bityuten-Ama creek valley, Aug. 12, 1927—Simukova; same site, nor. slope of Shishkhid creek valley, June 30, 1945—Yun.; same site, south-east. slope of Narin-Khurimt gorge, 2900 m, Aug. 28, 1948—Grub.; Dundu-Saikhan hills, Aug. 19; Dzun-Saikhan hills, Aug. 26—1931, Ik.-Gal.; south. slope of Dundu-Saikhan hills, July 20, 1950—Kal.).

IIA. **Junggar:** *Alt. region* (between Kul'tai and Chzhunkhaitsz, 1500 m, No. 1148, Aug. 5, 1956—Ching); *Jung. Alt.* (south. slope of Jung. Alatau below Koketau, July 21, 1909—Lipsky; west of Ven'tsyuan', No. 2040, Aug. 25, 1957—Huang); *Tien Shan.*

General distribution: Jung.-Tarb., Nor. and Cent. Tien Shan; Middle Asia (Pamir-Alay), West. Sib. (Atlay), East. Sib. (Sayans).

27. **E. sibiricus** L. Sp. pl. (1753) 83; Franch. Pl. David. 1 (1884) 341; ? Hemsley in J. Linn. Soc. London (Bot.) 30 (1894) 120; ? Deasy, in Tibet and Chin. Turk. (1901) 405; ? Hemsley, Fl. Tibet (1902) 206; Forbes and Hemsley, Index Fl. Sin. 3 (1904) 434; Krylov, Fl. Zap. Sib. 2 (1928) 366; Pavlov in Byull. Mosk. obshch. ispyt. prir. 38 (1929) 29; Rehder in J. Arnold Arb. 14 (1933) 4; Hao in Engler's Bot. Jahrb. 68 (1938) 580; Kitag. Lin. Fl. Mansh. (1939) 75; ? Ching in Contribs. U.S. Nat. Herb. 28 (1941) 597; Grubov, Consp. fl. MNR (1955) 77; Fl. Kazakhst. 1 (1956) 320; Melderis in Bor, Grasses Burma, Ceyl., Ind. and Pakist. (1960) 671.—*E. krascheninnikovii* Roshev. in Izv. Bot. sada AN SSSR, 30 (1932) 780.—*Clinelymus sibiricus* (L.) Nevski in Izv. Bot. sada AN SSSR, 30 (1932) 641 and in Fl. SSSR, 2 (1934) 690; Fl. Kirgiz. 2 (1950) 216; Fl. Tadzh. 1 (1957) 272; Keng, Fl. ill. sin., Gram. (1959) 423.—*Ic.:* Nevski, l.c. (1932) fig. 1; Keng, l.c. fig. 353.

Described from Siberia. Type in London (Linn.).

In meadows, pebble beds and sand in river valleys, among shrubs, in forest glades; up to upper mountain belt.

IA. **Mongolia:** *Khobd., Mong. Alt., Cis-Hing., Cent. Khalkha, East. Mong.* (Argun' district, near Genkhetsyao bridge, No. 1174, July 8; near Trekhrech'e, 700 m, No. 1385, July 14—1951, Li et al.; Kerulen floodplain 30 km east of Engershand, Aug. 9, 1956—Dashnyam; near Hailar town, 1960—Ivan.); *Bas. lakes* (10 versts [35,000 ft] from Ubsu-Nur lake, July 3, 1892—Krylov; 20 km west of Ulangom, Aug. 11, 1931—Bar.); *Gobi-Alt.*

IIA. **Junggar:** *Jung. Alt.* (near Ven'tsyuan', No. 4597, Aug. 25, 1957—Huang); *Tien Shan, Zaisan* (between Kaba river and Besh-Kuduk coll., June 17, 1914—Schisch.).

IIIA. **Qinghai:** *Nanshan* (forest belt of Yuzhno-Tetungsk mountain range, July 24, 1872—Przew.; "am Ufer des Sees Kokonor, 3800 m, No. 1245 u. 1263, 12 IX 1934"—Hao, l.c.); *Amdo* ("Schalakutu, 3400 m, No. 874, 18 VIII; auf dem Plateau Da-Ho-Ba, 4000 m, No. 1058, 28 VIII—1934", Hao, l.c.).

IIIB. **Tibet:** *Weitzan* (nor. slope of Burkhan-Budda mountain range along Nomokhun-Gol river, 3300–3800 m, July 22, 1884—Przew.); ? *South.* ("Tisum, 5000 m, 1848, leg. Strachey and Witterbottom"—Hemsley, l.c.).

General distribution: Jung.-Tarb., Nor. and Cent. Tien Shan; Europe (extreme east), Middle Asia (Zeravshan mountain range), West. Sib., East. Sib., Far East. Nor. Mong., China (Dunbei, Nor., Nor.-West, Cent., East., South-West.), Himalayas, Korea, Japan, Nor. Amer. (Alaska).

Plate I.
1—Psammochloa villosa (Trin.) Bor; *2—Timouria saposhnikovii* Roshev.

Plate II.
1—Ptilagrostis pelliotii (Danguy) Grub.; *2—Ptilagrostis dichotoma* Keng; *3—Stipa mongolorum* Tzvel.; *4—Stipa glareosa* P. Smirn.; *5—S. gobica* Roshev.

276

Plate III.
1—*Calamagrostis altaica* Tzvel.; 2—*Calamagrostis przevalskyi* Tzvel.;
3—*Deschampsia ivanovae* Tzvel.

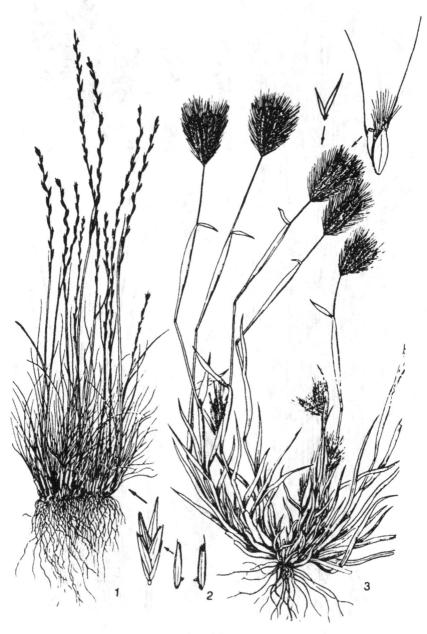

Plate IV.
1—*Tripogon chinensis* (Franch.) Hack.; 2—*Tripogon purpurascens* Duthie;
3—*Chloris virgata* Sw.

Plate V.
1—*Enneapogon borealis* (Griseb.) Honda; 2—*Aeluropus micrantherus* Tzvel.;
3—*Cleistogenes songorica* (Roshev.) Ohwi.

Plate VI.
1—*Cleistogenes kitagawai* Honda; 2—*Cleistogenes squarrosa* (Trin.) Keng.

Plate VII.

1—Melica tangutorum Tzvel.; *2—Melica kozlovii* Tzvel.; *3—Littledalea przevalsky*
Tzvel.

Plate VIII.
1—Puccinellia altaica Tzvel.; *2—Puccinellia roborovskyi* Tzvel.

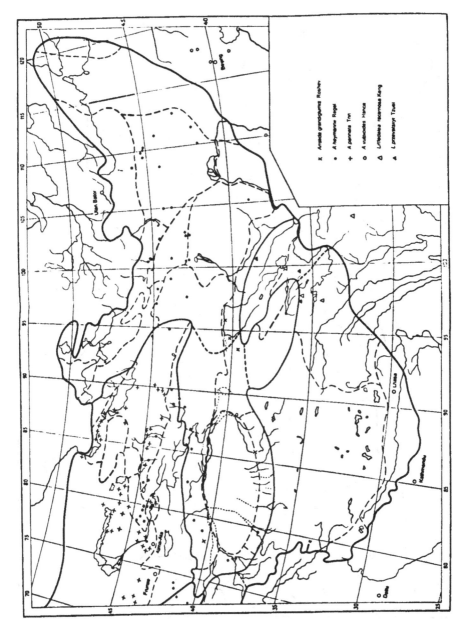

Map 1.

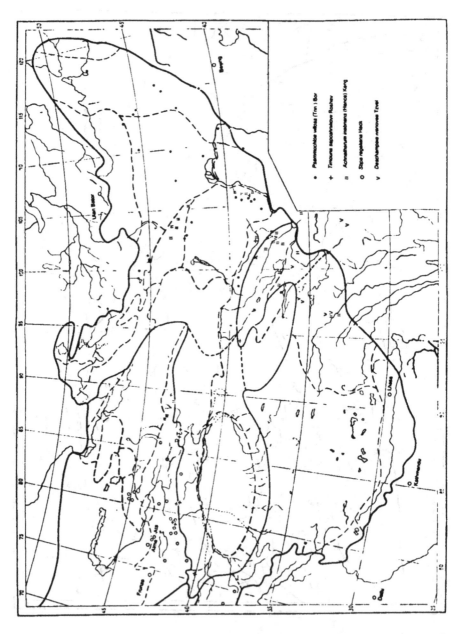

Map 2.

284

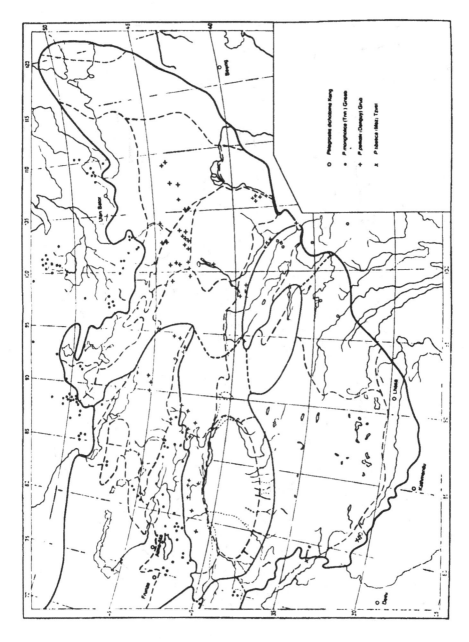

Map 3.

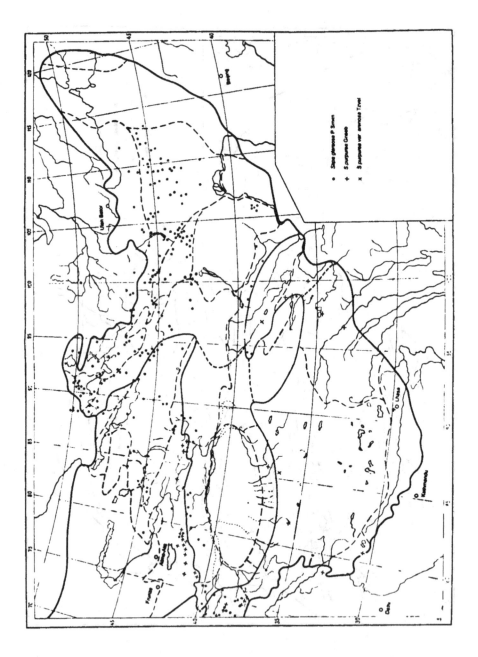

Map 4.

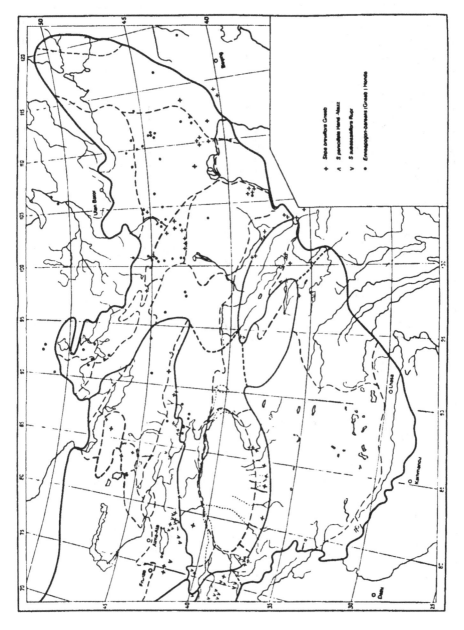

Map 5.

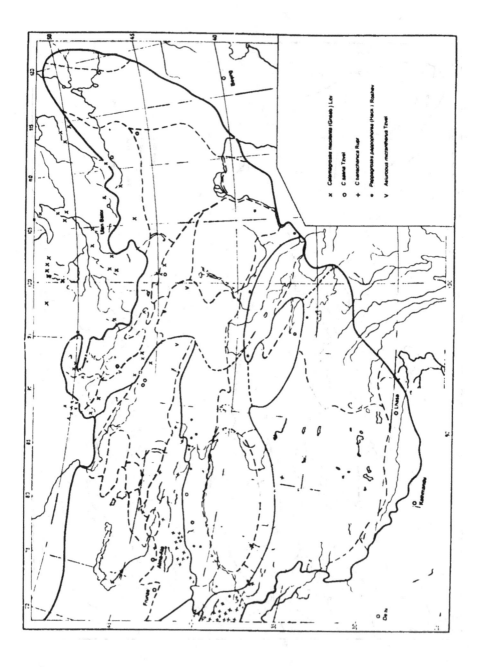

Map 6.

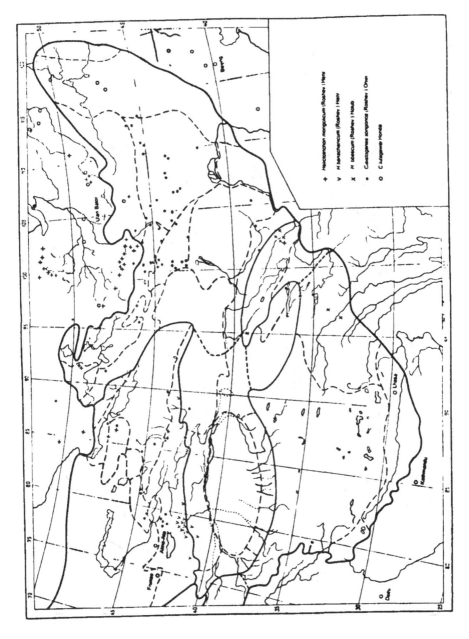

Map 7.

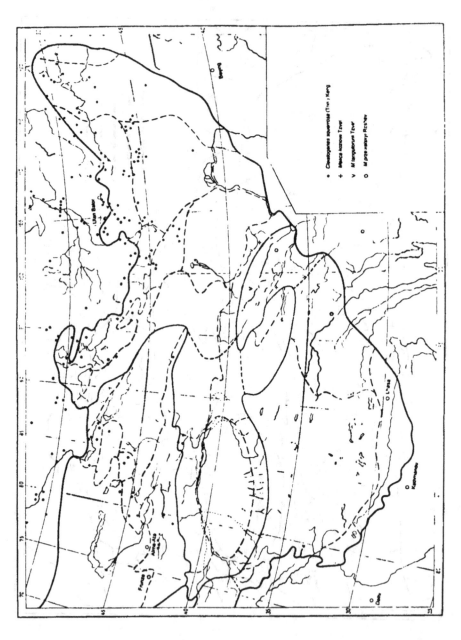

Map 8.

290

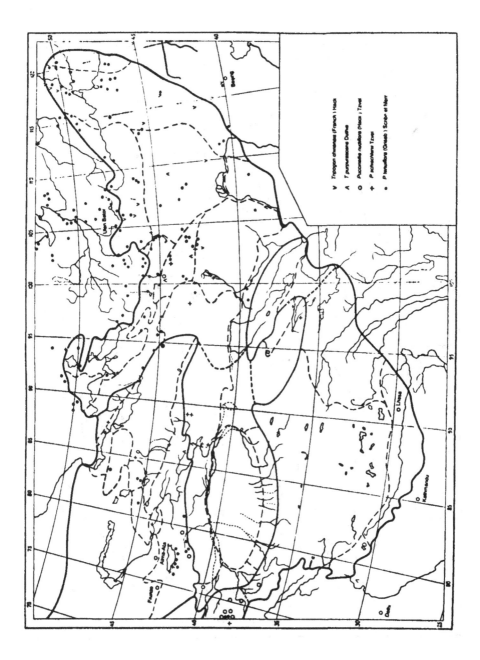

Map 9.

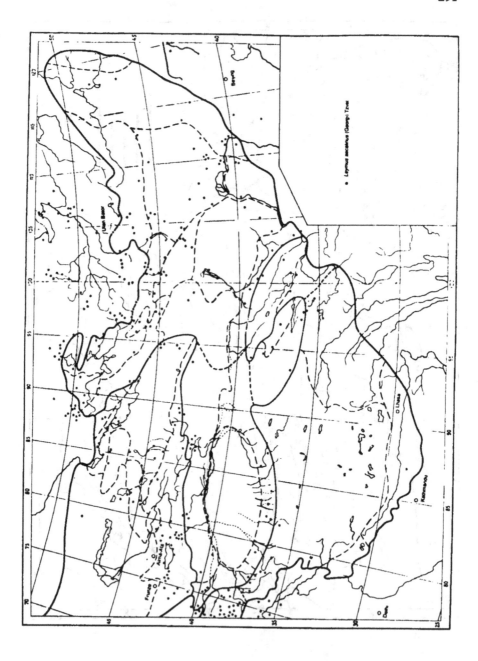

Map 10.

ADDENDUM[1]

Changes in Synonymy

p. 11. *Setaria glauca* auct. non (L.) Beauv. = **S. pumila** (Poir.) Schult. in Schult. et Schult. f., Mantissa, 2 (1824) 274.

p. 29. *Digitaria adscendens* (Kunth) Henr. = D. ciliaris (Retz.) Koel. Descr. Gram. Gallia, Germ. (1802) 27, quoad nom.

p. 36. *Typhoides arundinacea* (L.) Moench = **Phalaroides arundinacea** (L.) Rauschert, Feddes Repert. 79, 6 (1969) 409.

p. 39. *Hierochloe odorata* auct. non (L.) Beauv. = H. arctica C. Presl, in Reliq. Haenk. 1 (1830) 252.

p. 44. *Timouria saposhnikovii* Roshev. = **Achnatherum saposhnikovii** (Roshev.) Nevski, in Acta Inst. Bot. Acad. Sci. URSS, ser. 1, 4 (1937) 224.

p. 69. *Stipa nakaii* Honda, in Rep. First Sci. Exped. Manch., sect. 4, 4 (1936) 104 = **Achnatherum nakaii** (Honda) Tateoka ex Keng, in Fl. ill. Sin. Gram. (1959) 591.

p. 74. *Stipa sibirica* (L.) Lam. = **Achnatherum sibiricum** (L.) Keng ex Tzvel. in Probl. Ecol., Geobot., Bot. Geogr. Flor. (1977) 140.

p. 83. *Alopecurus alpinus* auct. non Smith = **A. turczaninovii** Nikif. in Bot. Zhurn. 73, 11 (1988) 1601; Gubanov, Consp. Fl. Outer Mong. (1996) 16.

p. 86. *Polypogon* Desf. sect. *Nowodworskya* (C. Presl) Tzvel. = **Polypogon** sect. **Polypogonagrostis** Aschers. et Graebn. Syn. Mitteleur. Fl. 2 (1899) 163.

p. 92. *Pentatherum dshungricum* Tzvel. in Pl. Asiae Centr. 4 (1968) 77. = **Agrostis dshungarica** (Tzvel.) Tzvel. comb. nova.

p. 133. *Phragmites communis* Trin. = **P. australis** (Cav.) Trin. ex Steud. Nomencl. Bot., ed. 2, 2 (1841) 324.

p. 133. *P. communis* var. *pseudodonax* Rabenh. = **P. altissimus** (Benth.) Nabille, in Rech. Pl. Corse, 2 (1869) 39.

p. 165. *P. pruinosa* auct. non Korotky = **P. tianschanica** (Regel) Hack. ex O. Fedtsch. in Acta Horti Petropol. 21 (1903) 441.

p. 180. *Glyceria plicata* (Fries) Fries = **G. notata** Cheval. Fl. Gener. Envir. Paris 2, 1 (1827) 174.

G. debilior auct. non (Fr. Schmidt) Kudo = **G. triflora** (Korsh.) Kom. in Fl. SSSR, 2 (1934) 459, 798.

---

[1] See comments of the series editor-in-chief on p. 7.

p. 205. *Festuca pseudosulcata* Drob. var. *litvinovii* Tzvel. = **F. litvinovii** (Tzvel.) E. Alexeev, in Novit. Syst. Pl. Vasc. 13 (1976) 31.

p. 208. *F. gigantea* (L.) Vill. = **Schedonorus giganteus** (L.) Soreng et Terrell, Phytologia, 83, 2 (1997) 86; Tzvelev, in Bot. Zhurn. 84, 7 (1999) 114.

F. orientalis (Hack.) Krecz. et Bobr. = **Schedonorus phoenix** (Scop.) Holub, Preslia, 70, 1 (1998) 113; Tzvelev, in Bot. Zhurn. 84, 7 (1999) 114. [Syn.: *S. arundinaceus* (Schreb.) Dumort., non Roem. et Schult.; *S. littoreus* (Retz.) Tzvel.; *Festuca arundinacea Schreb.*].

p. 211. *Zerna* Panz. nom. illeg. = **Bromopsis** Fourr. in Ann. Soc. Linn. Lyon, N.S. 17 (1869) 187.

p. 211. *Z. inermis* (Leyss.) Lindm. = **Bromopsis inermis** (Leyss.) Holub, in Folia Geobot. Phytotax. (Praha) 8 (1973) 167.

p. 212 *Z. korotkyi* (Drob.) Nevski = **Bromopsis korotkyi** (Drob.) Holub, in Folia Geobot. Phytotax. (Praha) 8 (1973) 168.

p. 212. *Z. plurinodis* Keng ex Tzvel. in Pl. Asiae Centr. 4 (1968) 176. = **Bromopsis plurinodis** (Keng ex Tzvel.) Tzvel. comb. nova.

Z. pumpelliana auct. non (Scribn.) Tzvel. = **Bromopsis sibirica** (Drob.) Peschk. In Novit. Syst. Pl. Vasc. 23 (1986) 26.

p. 222. *Agropyron elongatiforme* Drob. = *Elytrigia elongatiformis* (Drob.) Nevski, in Acta Univ. Asiae Med., ser. 8b, 17 (1934) 61.

p. 223. *A. repens* (L.) Beauv. = **Elytrigia repens** (L.) Nevski, Acta Inst. Bot. Acad. Sci. URSS, ser. 1, 1 (1933) 14.

p. 224. *A. aegilopoides* Drob. = **Elytrigia gmelinii** (Trin.) Nevski, in Acta Inst. Bot. Acad. Sci. URSS, ser. 1, 2 (1936) 78.

p. 225. *A. dshungaricum* (Nevski) Nevski = **Elytrigia dshungarica** Nevski, in Acta Inst. Bot. Acad. Sci. URSS, ser. 1, 2 (1934) 61.

p. 225. *A. ferganense* Drob. = **Elytrigia ferganensis** (Drob.) Nevski, in Acta Inst. Bot. Acad. Sci. URSS, ser. 1, 2 (1934) 61.

A. geniculatum (Trin.) C. Koch = **Elytrigia geniculata** (Trin.) Nevski, in Acta Inst. Bot. Acad. Sci. URSS, ser. 1,2 (1936) 82.

p. 226. *A. kanashiroi* Ohwi, in Journ. Jap. Bot. 19 (1943) 167. = **Elytrigia kanashiroi** (Ohwi) Tzvel. comb. nova.

A. nevskii Ivanova ex Grub. = **Elytrigia nevskii** (Ivanova ex Grub.) Ulzij, in Lat.-Mong.-Rus. Dictionary Pl. Vasc. Mong. (1983) 53.

p. 226. *Agropyron* sect. *Hyalolepis* (Nevski) Nevski = **Kengyilia** Yen et Yang, Canad. Journ.Bot. 68 (1990) 1897.

A. alatavicum Drob. = **Kengyilia alatavica** (Drob.) Yang, Yen et Baum, Canad. Journ. Bot. 71 (1993) 343.

p. 227. *A. batalinii* (Krassn.) Roshev. = **Kengyilia batalinii** (Krassn.) Yang, Yen et Baum, Canad. Journ. Bot. 71 (1993) 343.

p. 228. *A. grandiglume* (Keng, Keng f. et Chen) Tzvel. = **Kengyilia grandiglumis** (Keng, Keng f. et Chen) Yang, Yen et Baum, in Hereditas, 116 (1992) 28. [Syn.: *Roegneria grandiglumis* Keng, Keng f. et Chen, in Acta Nanking Univ. (Biol.) 3 (1963) 82].

A. kengii Tzvel.= **Kengyilia hirsuta** (Keng, Keng f. et Chen) Yang, Yen et Baum, in Hereditas, 116 (1992) 28. [Syn.: *Roegneria hirsuta* Keng, Keng f. et Chen, Acta Nanking Univ. (Biol.) 3 (1963) 87].

A. kokonoricum (Keng, Keng f. et Chen) Tzvel. = **Kengyilia kokonorica** (Keng, Keng f. et Chen) Yang, Yen et Baum, in Hereditas, 116 (1992) 27. [Syn.: *Roegneria kokonorica* Keng, Keng f. et Chen, in Acta Nanking Univ. (Biol.) 3 (1963) 88].

p. 229. *A. melantherum* Keng = **Kengyilia melanthera** (Keng) Yang, Yen et Baum, Hereditas, 116 (1992) 28. *A. muticum* (Keng, Keng f. et Chen) Tzvel. = **Kengyilia mutica** (Keng, Keng f. et Chen) Yang, Yen et Baum, Hereditas, 116 (1992) 28. [Syn.: *Roegneria mutica* Keng, Keng f. et Chen, 3 (1963) 87].

p. 230. *A. stenachyrum* (Keng, Keng f. et Chen) Tzvel. = **Kengyilia stenachyra** (Keng, Keng f. et Chen) Yang, Yen et Baum, in Hereditas, 116 (1992) 27. [Syn.: *Roegneria stenachyra* Keng, Keng f. et Chen, in Acta Nanking Univ. (Biol.) (1963) 79).

A. thoroldianum Oliver = **Kengyilia thoroldiana** (Oliver) Yang, Yen et Baum, in, Hereditas, 116 (1992) 27.

A. thoroldianum var. *dasyphyllum* Roshev. ex Tzvel. Pl. Asiae Centr. 4 (1968) 190. = **Kengyilia dasyphylla** (Roshev. ex Tzvel.) Tzvel. comb. nova.

A. thoroldianum var. *laxiusculum* Meld. in Bor, Grasses Burma, Ceyl., Ind. a. Pakist. (1960) 696. = **Kengyilia laxiuscula** (Meld.) Tzvel. comb. nova.

p. 232. *Agropyron cristatum* var. *ericksonii* Meld. = **A. ericksonii** (Meld.) Peschk. in Fl. Sibir. 2 (1990) 38.

p. 233. *A. fragile* (Roth) Nevski = **A. fragile** (Roth) Candargy, in Arch. Biol. Veg. (Athenes) 1 (1901) 58.

p. 253. *Leymus racemosus* (Lam.) Tzvel. var. *sabulosus* auct. non (Bieb.) Tzvel. = **L. racemosus** subsp. **crassinervius** (Kar. et Kir.) Tzvel. in Novit. Syst. Pl. Vasc. 8 (1971) 65; Peschkova, Fl. Sibir. 2 (1990) 50.

p. 254. *L. secalinus* auct. non (Georgi) Tzvel. = **L. littoralis** (Griseb.) Peschk. in Novit. Syst. Pl. Vasc. 24 (1988) 23.

p. 261. *Elymus czilikensis* auct. non (Drob.) Tzvel. = **E. uralensis** (Nevski) Tzvel. subsp. **tianschanicus** (Drob.) Tzvel. in Novit. Syst. Pl. Vasc. 10 (1973) 22.

p. 264. *E. macrolepis* auct. non (Drob.) Tzvel. = **E. fedtschenkoi** Tzvel. in Novit. Syst. Pl. Vasc. 10 (1973) 21.

p. 267. *E. vernicosus* (Nevski ex Grub.) Tzvel. = **E. brachypodioides** (Nevski) Peschk. in Fl. Stepp. Sib. Bajcal. (1972) 45; Gubanov, Consp. Fl. Outer Mong. (1996) 18.

ALPHABETICAL INDEX OF PLANT NAMES

INDEX OF PLANT DISTRIBUTION RANGES

INDEX OF PLANT DRAWINGS

Printed in the United States
by Baker & Taylor Publisher Services